Sustainable Catalysis in Ionic Liquids

Sustainable Catalysis in Ionic Liquids

Edited by
Pedro Lozano

CRC Press
Taylor & Francis Group
Boca Raton London New York

CRC Press is an imprint of the
Taylor & Francis Group, an **informa** business

The cover image shows a schematic illustration of an enzyme immersed into an structured ionic liquid medium, based on organized cations and anions. Courtesy of Dr. Pedro Lozano.

CRC Press
Taylor & Francis Group
6000 Broken Sound Parkway NW, Suite 300
Boca Raton, FL 33487-2742

First issued in paperback 2020

ISBN-13: 978-1-138-55370-5 (hbk)
ISBN-13: 978-0-367-73337-7 (pbk)

Visit the Taylor & Francis Web site at
http://www.taylorandfrancis.com

and the CRC Press Web site at
http://www.crcpress.com

Contents

SECTION I   Ionic Liquids in Organic Catalysis

SECTION II   Ionic Liquids in Biocatalysis and Biomass Processing

Foreword

One may wonder why so many chemists are that fascinated by ionic liquids. Indeed, this special class of molten salts is known for more than 50 years and, nowadays, represents one of the most popular areas in "ionic" solution chemistry. Certainly, one reason for this might be related to the non-trivial physical-chemical properties of these fluids at the molecular, nano, and bulk structural organization. Indeed, when applying a new ionic liquid, one is often surprised by a non-predictable result. A second reason for the popularity of these complex fluids is their anomalously high electrochemical and thermodynamic stabilities, as well as the negligible vapor pressure and, hence, the facility to work with them under non-conventional conditions of pressure and temperature. A number of chemists (this number, however, is rapidly decreasing) still believe that ionic liquids are simple "solvents" or too expensive. A third and more relevant reason is related with many important applications of these compounds that have been explored in various fields of chemistry, physics, material sciences, and biology. We believe that several new applications of ionic liquids and their mixtures will be discovered in the near future, thus, further demonstrating the great potentials of ionic liquids.

Several review articles covering all major aspects of the synthesis, physical-chemical properties, and applications of ionic liquids have been published, some of them recently. However, due to obvious space limitations, these monographs could not cover in depth all the aspects of synthesis, properties, and applications. Professor Pedro Lozano has ingeniously assembled a team of experts to review one of the most exciting applications of ionic liquids: sustainable catalysis. In particular, this book is concentrated on two domains that have scarcely been explored in recent reviews and books: organic catalysis, as well as biocatalysis and biomass processing.

I am confident that this book will provide valuable information for scientists currently working either permanently or for a limited amount of their time in an area that requires the use of ionic liquids in sustainable organic and biocatalysis.

Jairton Dupont
Porto Alegre

Preface

Chemistry is probably the science that has most contributed to the continuous improvements to our quality of life. However, the continuous growth of our knowledge of chemistry and its industrial applications, which has aided so much in satisfying many of the critical needs of our society (foods, medicines, materials, etc.), has been accompanied by an important environmental impact.

It has been said that all new developments in the chemical industry of this century should be sustainable, and that such sustainability in the chemical processes will follow two key axes: catalysis and the use of clean solvents. The selectivity in chemical transformations provided by efficient catalysts is directly related with several of the principles of green chemistry, for example, waste prevention, atomic economy, less hazardous synthesis, the reduction of derivatives, etc. The formation of wastes/contaminants and undesired by-products, inherent in classical synthetic processes, is minimized by using catalytic steps. The development of efficient catalytic processes also leads to significant savings in production costs for industry, as well as reduced environmental impacts. In this context, nature has always been a source of inspiration for chemists, and the transfer of the exquisite catalytic efficiency shown by enzymes in nature to chemical processes may constitute the most powerful toolbox yet for developing a clean and sustainable chemical industry. Furthermore, it is well accepted that the technological application of biocatalysts is greatly enhanced if they can be used in non-aqueous environments, because of the resulting expansion in the repertoire of enzyme-catalyzed transformations, as a result of *catalytic promiscuity*.

Solvents are auxiliary materials used in chemical processes, where they act as media for mass-transport, reaction, and product separation steps. However, volatile organic solvents (VOSs), which are responsible for a major part of the performance of processes in the chemical industry, strongly impact on environment, health, safety, and costs. It is estimated that several million tons of VOSs are discharged into the atmosphere each year because of all industrial processing operations. The search for new environmentally benign non-aqueous solvents, which can be easily recovered/recycled, as well as efficient catalysts, is a priority for the development of green/sustainable chemical processes. Nowadays, ionic liquids (ILs) are one of the non-aqueous green solvents that receive most attention worldwide. What are ILs? They are simply liquids composed entirely of ions at a temperature lower than 100°C. Using sodium chloride as an example, molten sodium chloride (up 800°C) is an ionic liquid, while an aqueous solution of this salt is an ionic solution. Typical room temperature ionic liquids (RTILs) are based on organic cations (e.g., 1,3-dialkylimidazolium, etc.), paired with a variety of anions that have a strongly delocalized negative charge (e.g., PF_6^-, etc.). These RTILs are colorless, low viscosity, and easily manipulable materials with very interesting solvent properties, such as negligible vapor pressure (they do not, therefore, evaporate), excellent thermal stability (up to 300°C in many cases), the ability to dissolve a wide range of organic and inorganic compounds, including gases (e.g., H_2, CO_2), a non-flammable nature, high conductivity, and a large electrochemical window. Consequently, the risk of exposure and the possibility that they will damage the atmosphere are practically non-existent. Moreover, their polarity, hydrophilicity/hydrophobicity, and solvent miscibility can be tuned by selecting the appropriate cation and anion and even by introducing functional groups (e.g., hydroxyl, amine, etc.) in the ion structure (also named task-specific ionic liquids), making them as useful tools for (bio)catalytic and/or product recovery processes. The key criteria concerning sustainability is that they can easily be recovered and reused, reducing the costs of most process in which they are applied. All of these properties point to ILs as ideal liquids solvents for developing sustainable chemical processes.

The figure shows the evolution of papers published in the field *ionic liquid** and *catal** over the last 20 years, as a result of searching in Web of Science—Clarivate Analytics.

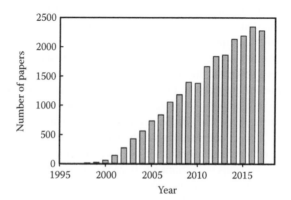

Because of the plethora of useful properties shown by ILs for developing chemical processes, their interest to academia and industry lies in the behavior of these new solvents, and their possible applications for catalytic process, as manifested by an increase in the number of references related to ILs year by year. Nowadays, many organic synthetic procedures are performed in ionic liquids (e.g., hydroformylation, hydrogenation, epoxidation, Friedel-Crafts alkylations, Diels-Alder cycloadditions, etc.), where it has been observed how ILs are able to improve reaction rates, selectivity and/or enantioselectivity of the processes, even playing a direct role as catalyst (e.g., task-specific ionic liquids). In the case of biocatalysis, ILs have been shown to be ideal non-aqueous solvents, because of the activity and stability of the enzymes, even under extremely harsh conditions, for example, 150°C, while maintaining all the enantioselective properties usually exhibited by them in synthetic reactions.

This book contains a collection of carefully chosen chapters, all written by recognized researchers in their respective fields, to present a global vision of the different catalytic applications of ionic liquids. The work is divided into two sections: Section I, Ionic liquids in organic catalysis, and Section II, Ionic liquids in biocatalysis and biomass processing. The book will serve to introduce the reader to the wide field of catalysis in ionic liquids and provide an insight into the current status of the area.

I would like to acknowledge all the contributors, as well as the efforts of those who have made the publication of this book possible. I particularly thank Hilary Lafoe and the other people at Taylor & Francis who have supported me during the whole adventure.

Pedro Lozano
Murcia

About the Editor

Pedro Lozano is Professor at the University of Murcia (Spain) and has served as Dean of the Faculty of Chemistry since 2014. Born in Ceutí, Spain in 1961, Professor Lozano received his PhD in Sciences (Chemistry) at the University of Murcia in 1988. Between 1990 and 1991, he spent two years at the Centre de Bioingénierie Gilbert Durand, Toulouse (France) as a post-doctoral fellow. In 1993, he returned to the University of Murcia (Spain) as lecturer in biochemistry and molecular biology, being finally promoted to full professor in 2004. His research activity has always been related with enzyme technology for sustainable processes in ionic liquids and supercritical fluids. In 2002, he published the first continuous green biocatalytic reactor, based on the combination of enzymes with ILs and scCO$_2$ able to directly provide pure products. More recently, Professor Lozano coined the term sponge-like ionic liquids, because of a new property discovered for these neoteric solvents, which permits the development of green biocatalytic processes with easy and clean recovery of pure products. As dean, he has built a permanent periodic table on the facade of the faculty of chemistry at the University of Murcia, 150 m^2 of surface, which could be the largest in the world.

Contributors

Elena Álvarez-González
Departamento de Bioquímica y Biología
  Molecular "B" e Inmunología
Facultad de Química
Universidad de Murcia
Murcia, Spain

Claudio Araya-López
Laboratory of Membrane Separation
  Processes (LabProSeM)
Department of Chemical Engineering
University of Santiago de Chile
Santiago, Chile

Juana M. Bernal
Departamento de Bioquímica y Biología
  Molecular "B" e Inmunología
Facultad de Química
Universidad de Murcia
Murcia, Spain

René Cabezas
Laboratory of Membrane Separation
  Processes (LabProSeM)
Department of Chemical Engineering
University of Santiago de Chile
Santiago, Chile

Antonio Donaire
Departamento de Química Inorgánica
Facultad de Química
Universidad de Murcia
Murcia, Spain

Caroline Emilie Paul
Laboratory of Organic Chemistry
Wageningen University & Research
Wageningen, the Netherlands

Cecilia García-Oliva
Department of Organic and
  Pharmaceutical Chemistry
Faculty of Pharmacy
Complutense University of Madrid
Madrid, Spain

Vincent Gauchot
Département de Chimie
Université de Montréal
Montréal, Québec, Canada

Vicente Gotor-Fernández
Department of Organic and Inorganic
  Chemistry
University of Oviedo
Oviedo, Spain

Jason P. Hallett
Department of Chemical Engineering
Imperial College London
London, UK

Allan D. Headley
Department of Chemistry
Texas A&M University-Commerce
Commerce, Texas

María J. Hernáiz
Department of Organic and
  Pharmaceutical Chemistry
Faculty of Pharmacy
Complutense University of Madrid
Madrid, Spain

Pilar Hoyos
Department of Organic and
  Pharmaceutical Chemistry
Faculty of Pharmacy
Complutense University of Madrid
Madrid, Spain

Ahmed Ali Hullio
Dr. M.A Kazi Institute of Chemistry
University of Sindh
Sindh, Pakistan

Toshiyuki Itoh
Department of Chemistry and
   Biotechnology
Graduate School of Engineering
Center for Research on Green
   Sustainable Chemistry
Tottori University, Japan

Dong Wook Kim
Department of Chemistry
Inha University
Incheon, South Korea

Sungyul Lee
Department of Applied Chemistry
Kyung Hee University
Seoul, South Korea

Pedro Lozano
Departamento de Bioquímica y Biología
   Molecular "B" e Inmunología
Facultad de Química
Universidad de Murcia
Murcia, Spain

Gastón Merlet
Laboratory of Membrane Separation
   Processes (LabProSeM)
Department of Chemical Engineering
University of Santiago de Chile
Santiago, Chile

Susana Nieto-Cerón
Departamento de Bioquímica y Biología
   Molecular "B" e Inmunología
Facultad de Química
Universidad de Murcia
Murcia, Spain

Toshiki Nokami
Department of Chemistry and
   Biotechnology
Graduate School of Engineering
Center for Research on Green
   Sustainable Chemistry
Tottori University, Japan

Julio Romero
Laboratory of Membrane Separation
   Processes (LabProSeM)
Department of Chemical Engineering
University of Santiago de Chile
Santiago, Chile

Andrea R. Schmitzer
Département de Chimie
Université de Montréal
Montréal, Québec, Canada

Wei-Chien Tu
Department of Chemical Engineering
Imperial College London
London, UK

Cameron C. Weber
School of Science
Auckland University of Technology
Auckland, New Zealand

Section I

Ionic Liquids in Organic Catalysis

1 Task-Specific Ionic Liquids

Ahmed Ali Hullio

CONTENTS

1.1 INTRODUCTION

Ionic liquids have turned out to be curious class of compound having unique combination of properties, such as being liquids at room temperature, undetectable vapor pressure, higher thermal stability, nonflammability, and interesting solubility trends [1]. Further, all of physical features like polarity, solubility, miscibility, viscosity, boiling, and melting points can be tuned or adjusted as per requirement of applications through simple variation in its structure. Thus, we can prepare a liquid with properties appropriate for any chemical application, therefore, ionic liquids are called designer solvents. In addition, in many cases, ionic liquids are recyclable, thus avoiding the generation of chemical waste and thus are termed as "green solvents". Recycling is especially beneficial both economically and environmentally in the case of ionic liquid supported catalytic system [2].

All these amazing properties and features associated with ionic liquids attracted the attention of chemists across the globe. Various chemists employed them for different versatile chemical applications including as green reaction media achieving different chemical reactions, organic synthesis, biphasic separation, metal catalysis, and so on.

Ionic liquids, being an unusual liquid with an unprecedented bunch of properties became focus of attention for exploration of its application. Apart from its common use as reaction media, ionic liquids have profoundly affected the biphasic processes by making it possible to recycle the homogenous metal catalysts. The brilliant performance of transition metal catalysis tempted the chemist to apply the same methodology in organo-catalysis. However, the dismal solubility of organo-catalysts in ionic liquids turns out to be the major impediment in that procedure [3], since many useful organic reactions are promoted by organo-catalysts. Some of the organic chemists thought that the performance of the organo-catalyzed reaction can be further enhanced by making the organo-catalyst the part of ionic liquid. By doing so, many of the drawbacks of organo-catalyzed reaction can be overcome developing their ionic liquid versions, for instance, clean synthesis, reducing reaction time, improving product yield, and recycling of catalyst system [4].

Task specific ionic liquids are special kind of imidazole-based ionic liquids, which are synthesized by attaching the specific organo-functional group to the structure of imidazolium cation ring [5]. Such special class of ionic liquids containing structurally and functionally complex side chains attached with their cationic part is functionalized ionic liquids or task-specific ionic liquids (TSILs). This ionic liquid technology has been successfully applied in several classical organic processes and other procedures where TSILs act as both reagent and medium coupled [6].

The incorporation of functional groups into the ionic liquid represents one of the simplest and most effective strategies for the modification of the liquid's properties, and a large number of salts with functional groups attached to the cation have now been reported. These include, for example, amines [7], amides [8], nitriles [9], ethers and alcohols [10], acids [11], urea and thiourea [12], and fluorous chains [13]. In addition, functional anions have also received some attention; those giving rise to low melting salts include amino acids [14], alkene substituted anions [15], triazole anions [16], selenium-based anions [17], functionalized borate anions [18], carboranes [19], and transition metal-carbonyl anions. This line of thought gave birth to a revolutionary idea of a new class of reagents designed as TSIL, an IL-type part is attached with an extra functional group designed for the specific property. Many organo-catalysts have been converted into their ionic liquids versions, leading to excellent combinations of advantages.

Inherited drawbacks associated with the classical methods have always kept chemists on track of hot pursuit of new and better methodologies. Task-specific ionic liquid is the hunt and is focus of intense research and has become very large field because different ionic liquids can be designed to perform various intended tasks [20]. The recent examples include homo and heterogeneous transition metal catalyzed reactions, asymmetric synthesis [21], and solvent extractions. In this part, we present a brief discussion of various task-specific ionic liquids conceived, designed, and applied for organo-catalyzed reactions. There is large number of task-specific ionic liquids that have been designed, synthesized, and employed to generate various kinds of improvements and many review articles have been published.

The positive outcome of this methodology has triggered the ever increasing growth in the field of task-specific ionic liquids. It is interesting to note that we can synthesize the specific ionic liquids for any particular task, for example, chemical transformations. This fact justifies their name as TSILs or designer solvents that contain specific functionalities and are capable of carrying out specific tasks [22].

Brief account of designing, synthesis, and application of task-specific ionic liquids, especially for key organic transformation are presented here.

1.1.1 BRÖNSTED ACIDIC IONIC LIQUIDS CATALYZED REACTION

Keeping in view the importance of strong Brönsted acids in chemical catalysis, various methodologies have been developed from time to time. Every technique is found to be associated with certain drawbacks. In continuation of trend of developing better catalytic systems, Cole et al. reported the first series of ionic liquids that are designed to be strong Brönsted acids [23]. In each of the new Brönsted acid task-specific ionic liquids, an alkane sulfonic acid group is covalently tethered to the ionic liquid cation, Figure 1.1.

These inexpensive Brönsted acidic TSILs consist of alkane sulfonic acid group attached to suitable cation obtained from triphenyl phosphine, trialkylamine, 1-methyl imidazole, and pyridine. These are SO_3H-functional halogen-free acidic ionic liquids which exhibit catalytic activity for acid-catalyzed reactions and have proved their potential for some well-known acid-catalyzed organic reactions, such as Fischer esterification, ether formation by alcohol dehydrodimerization, pinacol-benzopinacole rearrangement, Mannich reaction, and synthesis of chalcones via the Claisen–Schmidt condensation. Cole et al. tested the potential of first imidazolium **1a-** and triphenyl phosphonium **1b**-based Brönsted task-specific ionic liquids for Fischer esterification, alcohol dehydrodimerization, and pinacol/benzopinacole rearrangement and got excellent results, Figure 1.2 [23].

Sahoo et al. have investigated the potential of other SO_3H functional Brönsted acidic ionic liquids **2a** and **2b** for Mannich reaction, Figure 1.3 [24]. The ionic liquids bearing triphenyl phosphonium sultone with *p*-toluene sulfonate anion **2a** or methyl imidazolium sultone with *p*-toluene sulfonate anion **2b** were used as Brönsted acid task-specific ionic liquid.

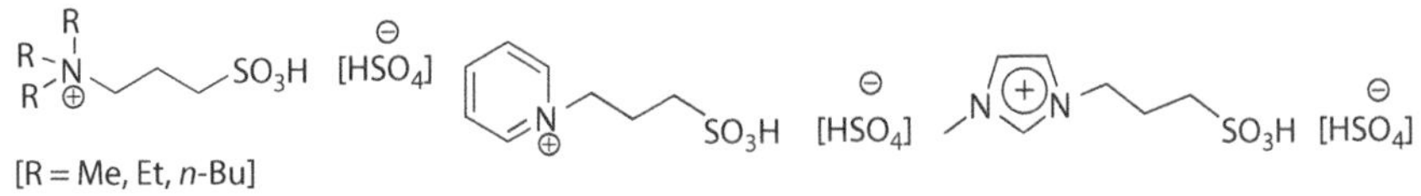

FIGURE 1.1 Structures of Brönsted acidic task specific ionic liquids.

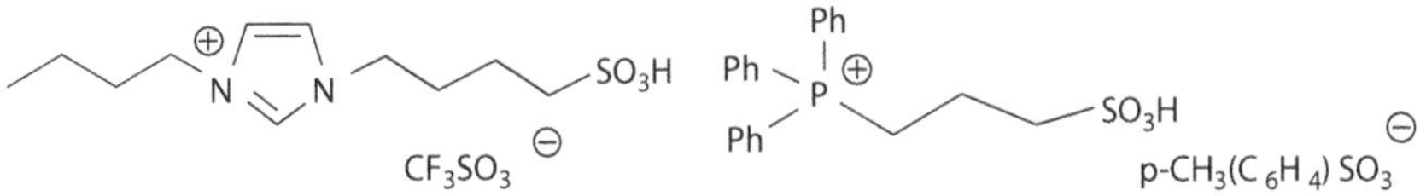

FIGURE 1.2 Brönsted acidic task-specific ionic liquids.

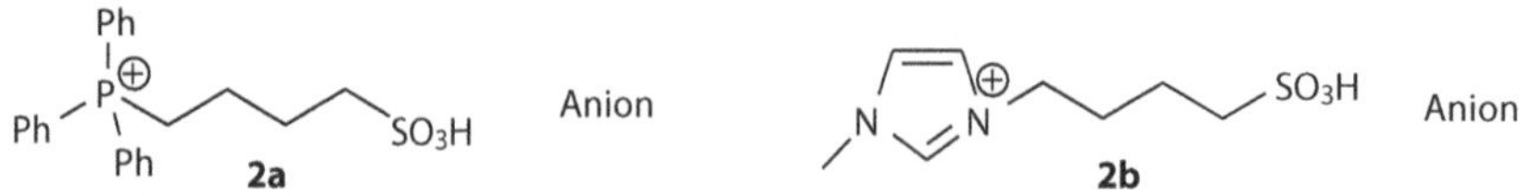

2a

Triphenyl phosphine sultone

Anion = PTSA, TFA

2b

N-methyl imidazolium sultone

FIGURE 1.3 Brönsted acidic task-specific ionic liquids.

Using catalytic amount of these ionic liquids for Mannich reaction of different aldehydes, ketones, and amines afforded corresponding β-amino carbonyl compounds. The reactions proceeded very fast with high yield of the desired Mannich base using catalytic amount of ionic liquid, Scheme 1.1.

Dong et al. synthesized a series of acyclic SO_3H-functionalized halogen-free acids and tested the efficacy of Brönsted-acidic TSILs on the performance of Claisen–Schmidt condensation, Figure 1.4 [25]. The best results were achieved with **3a**.

Few other derivatives of benzaldehydes and acetophenones were also subjected to the Claisen–Schmidt condensation to form corresponding α,β-unsaturated carbonyls under the optimized reaction conditions in presence of [TMPSA] [HSO$_4$], Scheme 1.2.

Wang et al. reported some pyridinium-based Brönsted-acidic task-specific ionic liquids containing sulphonic groups attached to pyridinium cation through alkyl chain of three carbons, Figure 1.5 [26]. Several versions of this ionic liquid were prepared using different anions prepared by deprotonation of some inorganic and

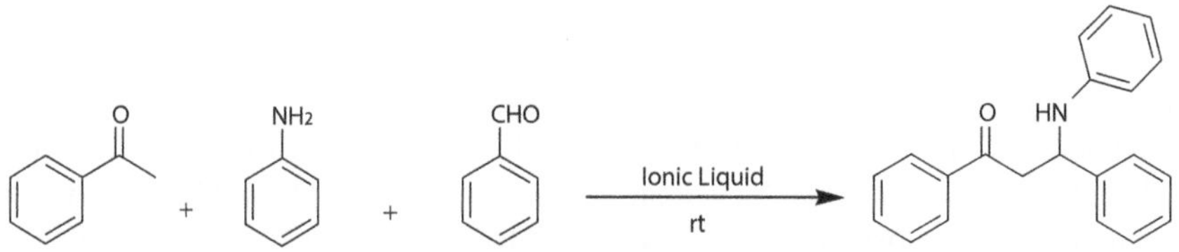

SCHEME 1.1 Mannich reaction in Brönsted acid task-specific ionic liquid.

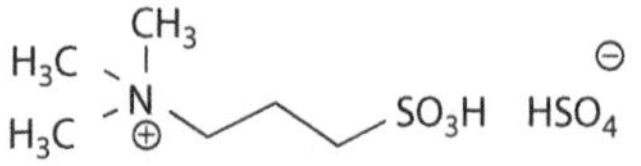

3a   [TMPSA][HSO$_4$]

FIGURE 1.4 Brönsted acidic task-specific ionic liquid.

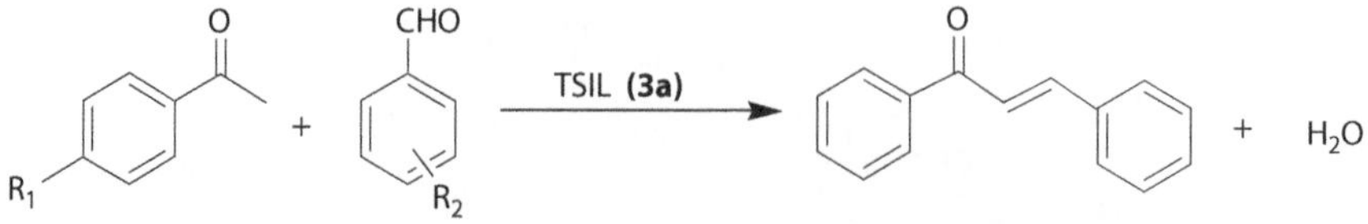

SCHEME 1.2 Claisen–Schmidt condensation catalyzed by the TSIL **3a**.

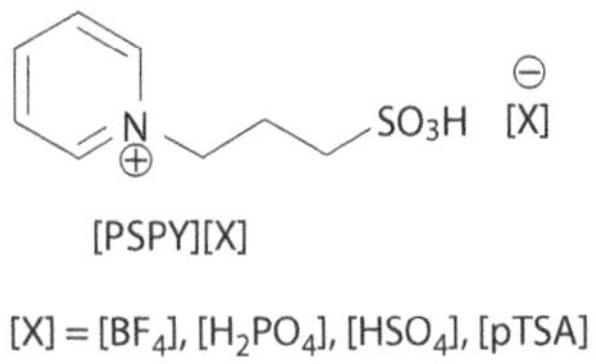

[PSPY][X]

[X] = [BF$_4$], [H$_2$PO$_4$], [HSO$_4$], [pTSA]

FIGURE 1.5 *N*-Propane sulfone pyridinium (PSPy).

organic acids. These acidic task-specific ionic liquids demonstrated their acidic activity in achieving the esterification of benzoic acid with different alcohols like methanol, ethanol, and butanol.

The acidic with bisulphate anion [PSPy][HSO$_4$] exhibited maximum acidic activity in the esterifications. Furthermore, the high degree of partial immiscibility of TSILs with the produced esters facilitates the esterification reaction equilibrium, shifting it to the product side. The different esters products were separated simply by decantation, thus allowing the recycling of pyridinium-based Brönsted acidic task-specific ionic liquid.

Jiang et al. have shown applications of 1,1,3,3-tetramethylguanidinium (TMG)-based Brönsted acidic task-specific ionic liquid to catalyze the Henry reactions of variety of nitroalkanes and carbonyl compounds to form corresponding 2-nitroalcohols [27]. Different kinds of aliphatic and aromatic carbonyl compounds and cyclic ketones proceeded to form corresponding products with good yields at room temperatures, moreover, the catalyst could be recyclable repeatedly. Two different guanidine-based ionic liquids were prepared by neutralizing tetramethylguanidine with trifluoroacetic acid (TFA) and lactic acid to give 1,1,3,3-TMG trifluoroacetate [TMG][F3Ac] and TMG lactate [TMG][Lac], respectively. Both of these ionic species are liquids under normal conditions, Scheme 1.3.

The successful results were obtained with [TMG][Lac], it catalyzed Henry reactions of aromatic and aliphatic aldehydes, giving higher yields than ([TMG][F3Ac]), Scheme 1.4.

SCHEME 1.3 Preparation of guanidinium anion salt [TMG][Lac].

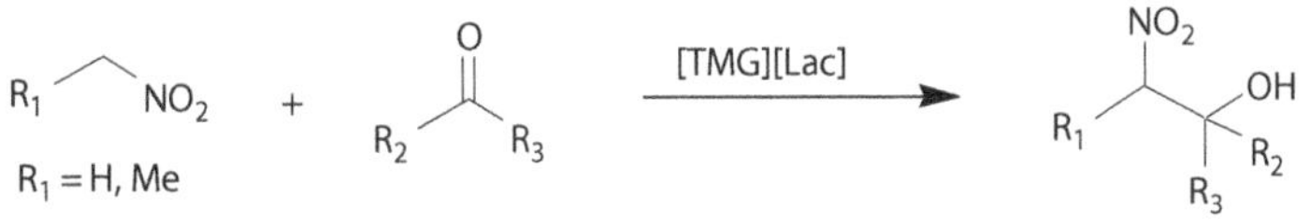

SCHEME 1.4 [TMG][Lac] ionic liquid catalyzed Henry reactions.

FIGURE 1.6   Task-specific ionic liquids possessing two Brönsted acid sites.

Liu et al. developed a series of air and moisture stable nonchloroaluminate Brönsted acidic ionic liquids [28]. These Brönsted acidic task-specific ionic liquids had two Brönsted acid sites. One Brönsted acidic site –COOH was attached through cationic ring one and two carbons units and the other existed as anion HSO_4^-, or $H_2PO_4^-$, Figure 1.6. These new task-specific ionic liquids have been used for the esterification of isopropanol by chloroacetic acid under mild conditions and without any additional organic solvent with good yields. There was no need of tedious work-up because the ester products were simply decanted as they formed an immiscible layer. The recycling of ionic liquid was possible after its drying. These Brönsted acidic ionic liquids can find their role as catalysts or as a media-cum-catalyst in large number of acid promoted chemical changes, such as rearrangements, dehydrations, polymerizations, etherifications, Friedel-Crafts reactions, and many more.

The esterification of isopropanol and chloroacetic acid was tested for using all different task-specific ionic liquids as the solvent-catalyst in a batch type process, Scheme 1.5. The yields of isopropyl chloroacetate esters were considerably higher (94.5% and 87.8%) in **6a** and **6b** ionic liquid versions.

1.1.2   DISULFONYL CHLORIDE-BASED TASK-SPECIFIC IONIC LIQUIDS FOR ACETALIZATION OF CARBONYL COMPOUNDS

Li et al. reported imidazole-based mono and disulfonyl chloride task-specific ionic liquid **7a** and **7b** for the acetalization of aldehydes and ketones with good catalytic performance under mild reaction conditions, Figure 1.7 [29].

Acetalization of butyl aldehyde with isoamyl alcohol was tested in **7a** and **7b** gave excellent results, Scheme 1.6.

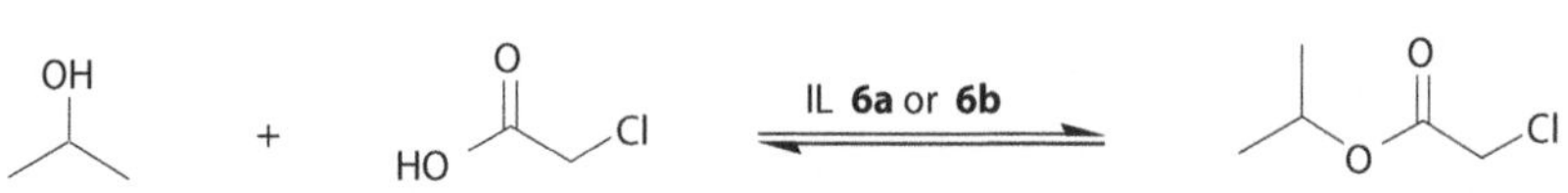

SCHEME 1.5   Esterification of isopropanol by chloroacetic acid.

FIGURE 1.7   Imidazole-based mono and disulfonyl chloride task-specific ionic liquid.

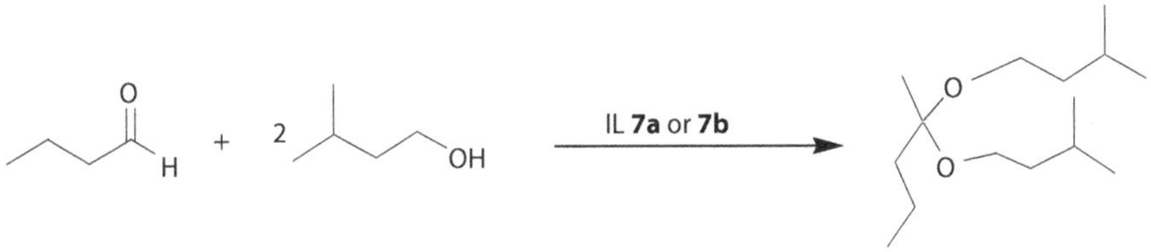

SCHEME 1.6 Acetalization of butyl aldehyde with isoamyl alcohol.

The catalytic activity was found to be excellent with **7a**, by achieving 94.0% conversion and 98.9% selectivity. However, **7b** demonstrated lower activity (81.8%), which may be attributed to the steric hindrance of this ionic liquid.

1.1.3 SULFONYL CHLORIDE-BASED TASK-SPECIFIC IONIC LIQUID FOR BECKMANN REARRANGEMENT

The Beckmann rearrangement is the acidic reagent promoted rearrangement of a different kind of ketoximes to corresponding amides. It is used in industry preparation of ε-caprolactam from cyclohexanone oxime. Sun et al. have reported a novel task-specific ionic liquid consisting of sulfonyl chloride specifically designed to promote Beckmann rearrangement, Figure 1.8 [30].

The sulfonyl chloride-based task-specific ionic liquid **8a** demonstrated best potential for smooth preparation of ε-caprolactam from cyclohexanone oxime in high yield at ca. 80°C. Other varieties of ketoximes exhibited similar results, Scheme 1.7.

1.1.4 HIGH TEMPERATURE TASK-SPECIFIC IONIC LIQUIDS

Armstrong et al. have reported synthesis and application of several thermally stable ionic liquids with thermal stability range lying from 330°C to 400°C [31]. These ionic liquids were based on two five-membered rings connected with each other through alkyl chain of nine carbons, and each ring possessed positive charge and

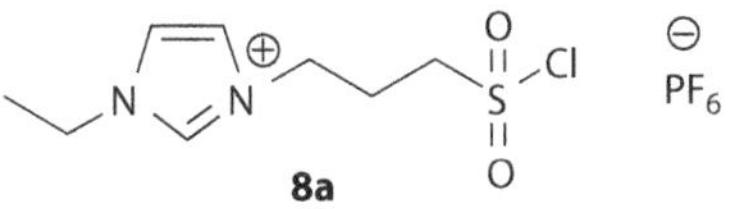

FIGURE 1.8 Imidazole-based sulfonyl chloride ionic liquid.

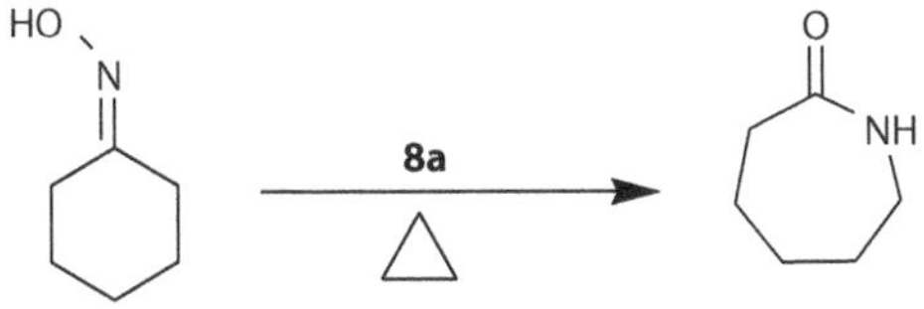

SCHEME 1.7 Task-specific ionic liquid-promoted Beckmann reaction.

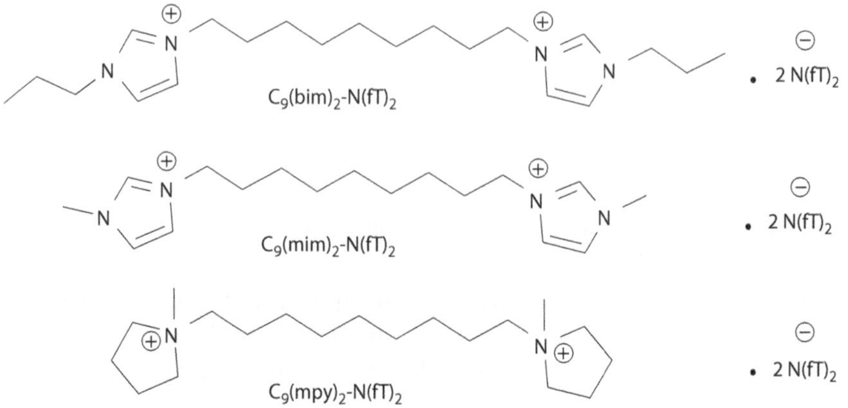

FIGURE 1.9 Structure of the high temperature dicationic ionic liquid.

were generally named as geminal dicationic ionic liquids, Figure 1.9. These thermally stable ionic liquids were designed as a solvent for organic transformations occurring at higher temperatures, for example, the isomerization reaction, the Claisen rearrangement, and the thermally induced Diels-Alder reaction.

The yield of isomerization of carvone to carvacrol at 300°C was higher in $C_9(mpy)_2$-NTf$_2$. The Claisen rearrangement of allyl phenyl ether at 250°C produced better yields of products in the $C_9(mim)_2$-NTf$_2$ and $C_9(bim)_2$-NTf$_2$ ionic liquids. The heat-promoted Diels-Alder reaction was checked in the $C_9(mim)_2$-NTf$_2$ (51%) and $C_9(mpy)_2$-NTf$_2$ (47%) ionic liquids where anthracene on reaction with diethyl fumarate produced Diels-Alder adducts in suitable yields at 220°C in 10 min.

1.1.5 QUINUCLIDINE-BASED TASK-SPECIFIC IONIC LIQUID FOR MORITA-BAYLIS-HILLMAN REACTIONS

The Morita-Baylis-Hillman reaction involves the tertiary amine-mediated reaction between activated alkenes and aldehydes to form corresponding allylic alcohols. Mi et al. have designed and synthesized quinuclidine-based task-specific ionic liquid to catalyze Baylis-Hillman reactions, Figure 1.10 [32]. This catalysis was performed under homogeneous reaction and product formed was separated by simple extraction. The efficiency of ionic liquid-supported quinuclidine as a Baylis-Hillman catalyst

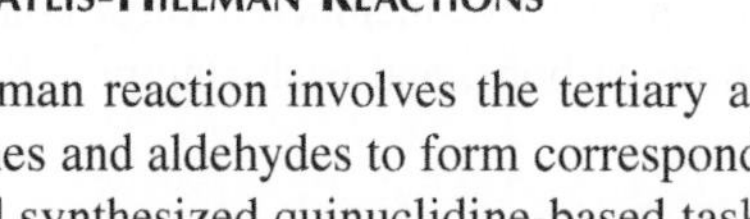

10a

FIGURE 1.10 Quinuclidine-based TSIL.

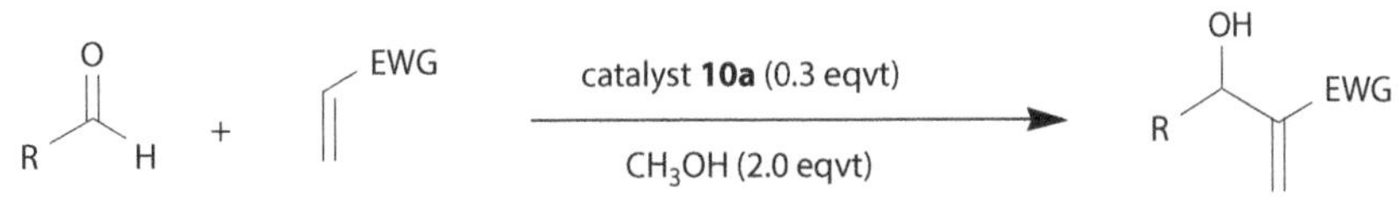

SCHEME 1.8   Ionic liquid-supported quinuclidine catalyzed Baylis-Hillman reaction.

found equally good as its nonimmobilized counterpart. The different allylic alcohol products were formed in moderate to excellent yields. The quinuclidine-based task-specific ionic liquid was easily recovered and recycled at least six times without much loss of catalytic efficiency.

Under the optimum conditions, methyl acrylate and acrylonitrile underwent qui-nuclidine-based TSIL-mediated reaction with different types of aldehydes to form corresponding allylic alcohols at 25°C. Various derivatives of aliphatic and aromatic aldehydes gave efficient response to 0.3 equivalent of Baylis-Hillman reagent and reacted with acrylates, producing the relevant Baylis-Hillman adducts in reasonable to best amounts, for example, 60%–96%, Scheme 1.8.

1,3-Dilakylimdazolium cation has acidic hydrogen at C-2 which undergoes depro-tonation under strongly basic conditions, as in the case of Baylis-Hillman reaction which forms carbene which undergoes addition reaction with aldehydes. Jeong et al. solved this problem by reporting novel 1,3-dialkyl-1,2,3-triazolium ionic liquids which are chemically inert under basic conditions and more suitable media for the reactions involving bases like Baylis-Hillman reaction than the common 1,3-dialkyl-imidazolium ionic liquids, Figure 1.11 [33].

The stereotype of novel 1,2,3-triazolium ionic liquid is shown in Figure 1.11, in which the problematic acidic C-2 proton of 1,3-dialkylimidazolium cation has been replaced by the nitrogen of the 1,3-dialkyl-1,2,3-triazolium cation to obtain the stability under basic conditions. The reported novel ionic liquids could be used as an efficient reagent for the Baylis-Hillman reaction.

The reaction rates of Baylis-Hillman reaction in these novel ionic liquids were compared using different aldehydes (1 mmol), acrylates (2 mmol), and 1,4-diazabicyclo[2.2.2]octane (DABCO) (2 mmol) in the presence of 1,2,3-triazolium ionic liquids (0.1 mL). The results clearly demonstrated that the Baylis-Hillman reaction can be greatly accelerated in [bmTr][NTf$_2$], [bmTr][PF$_6$] and [dbTr][NTf$_2$], Scheme 1.9.

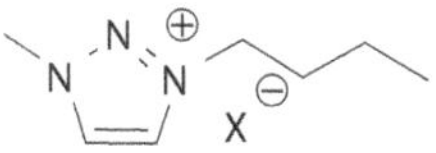

11a   [bmTr][X]

X = I, NTf$_2$, OTf, PF$_6$, BF$_4$

FIGURE 1.11   Novel 1,2,3-triazolium ionic liquid.

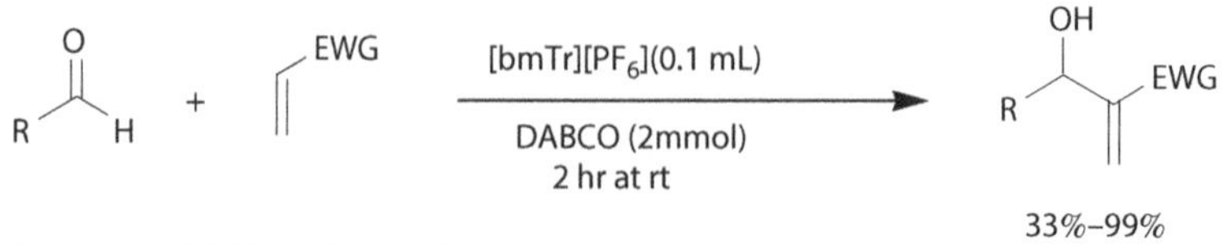

R = alkyl, aryl, **EWG** = CO$_2$CH$_3$, CN

X = I, NTf$_2$, OTf, PF$_6$, BF$_4$

SCHEME 1.9   Baylis-Hillman reaction in 1,2,3-triazolium-based task-specific ionic liquid.

1.1.6 Nicotine-Based Task-Specific Ionic Liquids

The different kinds of liquid compounds when used as solvent act both as reaction media as well as reagent. For instance, pyridine and hexamethylphosphoramide are used as nucleophilic solvents in synthetic schemes as catalytic solvents for some specific reactions like acylation reactions. However, immense toxicity of such solvents has often been a barrier in their widespread use, therefore, search of their nontoxic substitutes has been a hot pursuit. Pyridine is the key compound which is used both as solvent and as a coreagent, but its toxic effects on human health are well documented. Handy et al. developed the ionic liquid substitute of pyridine and investigated its use as a better nucleophilic solvent [34]. They converted nicotine into room temperature ionic liquid and demonstrated its use as a better green alternate to all pyridine dependent reaction, Figure 1.12.

Nicotine-based ionic liquids (NBILs) were simply prepared by *N*-alkylation nitrogen atom in nicotine providing two possible versions **12a** and **12b** of nicotine-based ionic liquid. Both of these were used in standard acylation reactions. In contrast to nicotine-based ionic liquid **12b**, the **12a** lacks nucleophilic pyridine, thus was not expected to catalyze acetylations, Scheme 1.10. However, it did effectively mediate the acylation of 2-phenylethanol, as well as a more hindered secondary alcohol, 1-phenylethanol with acetic anhydride, and no reaction was observed with a tertiary alcohol at room temperature.

12a   **12b**

FIGURE 1.12   Nicotine-based nucleophilic ionic liquids.

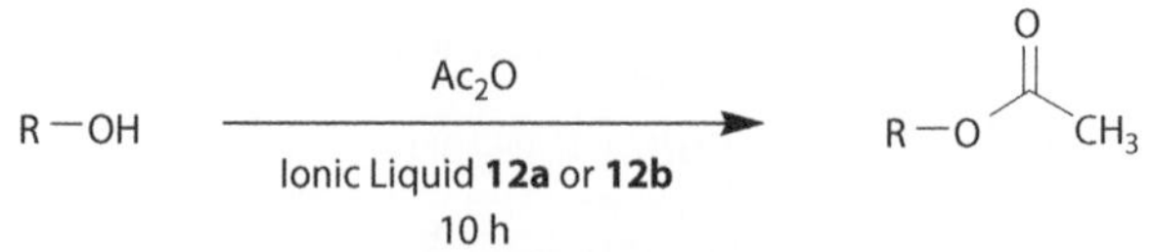

SCHEME 1.10   Ionic liquid catalyzed acetylation of alcohol.

The NBIL **12b** displayed similar reactivity to pyridine itself. Both of the solvents, **12a** and **12b** proved competent in acylating various structural variants of primary alcohols at 25°C. However, in the case of secondary alcohols, the response of both of the solvents was dismally poor. Neither of the solvents completely acylated the secondary alcohol even within 20 hours. The acylation of relatively hindered alcohols like 1-phenylethanol and 2-phenyl-2-propanol was achieved by just heating the NBIL, **12b**, reaction mixture to higher temperatures as 70°C providing significant yields of corresponding esters. The NBIL **12b** was compatible with some other acylating agents like acid chlorides which successfully acylated both secondary and tertiary alcohols to relevant esters in best yields. Catalysts **12a** and **12b** were recycled after removal of products.

Handy et al. also reported synthesis and applications of fructose-based ionic liquid, Figure 1.13 [35]. Although, fructose cannot be converted into cations due to lack of quaternizable elements, it can be converted into hydroxymethylene imidazole **13a** ionic liquid by adopting Darby-Trotter methodology. By this method, ionic liquid **13b** is obtainable after two successive alkylations followed by appropriate anionic exchange.

This protic ionic liquid proved its worth for Pd-catalyzed Heck reaction and for homogenous phase synthesis. The Pd-catalyzed Heck reaction using ionic liquid **13b** gave yields more than 95%, without any side reaction, Scheme 1.11.

As a second application, ionic liquid **13b** was used as a support for homogeneous supported-phase synthesis, Scheme 1.12.

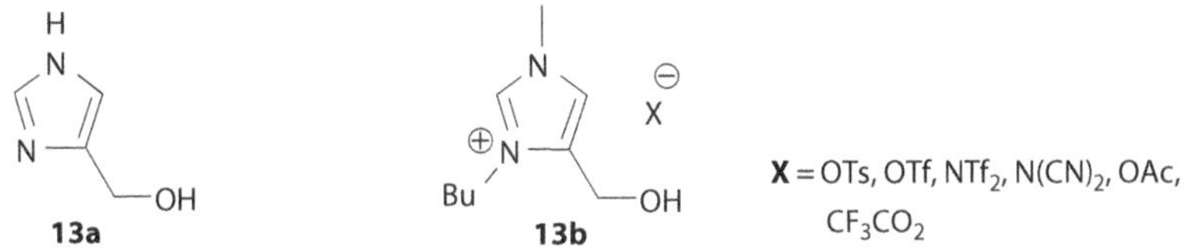

FIGURE 1.13   Fructose-derived ionic liquid synthesis.

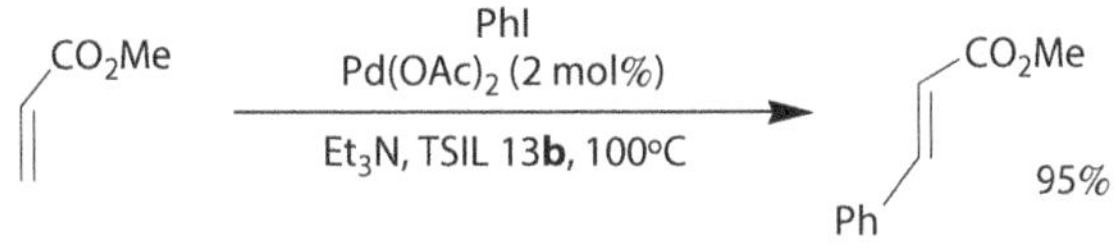

SCHEME 1.11   Heck reactions in fructose-derived RTIL.

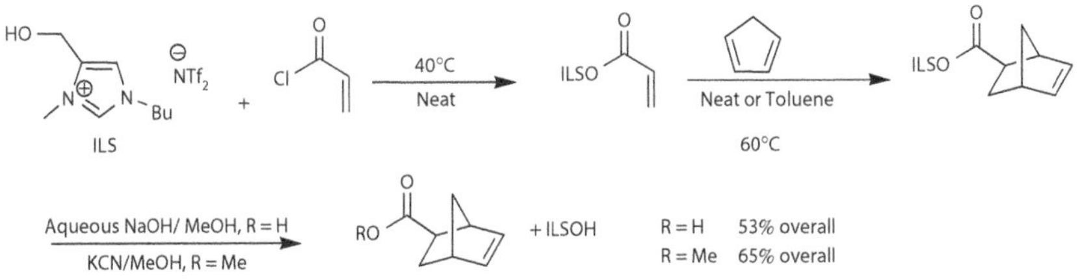

SCHEME 1.12   Homogeneous-supported synthesis using fructose-derived IL.

1.1.7 PYRROLIDINES-BASED CHIRAL TASK-SPECIFIC IONIC LIQUIDS

Chiral molecules are often used as chiral organo-catalyst to achieve asymmetric versions of different reactions. Pyrrolidine is one of such powerful chiral organo-catalysts, it is cyclic five-membered secondary amine which is now regarded as one of the "privileged" backbones for asymmetric catalysis [36]. Its high potential for achieving enantioselective synthesis has inspired many chemists to explore its further potential. Some of the chemists have developed the pyrrolidine-based chiral task-specific ionic liquids. The pyrrolidine is anchored on imidazolium cation *via* various alkyl chains to use it under ionic liquid conditions. Some of the important contributions from different chemists are as follows.

Luo et al. achieved successful direct asymmetric aldol reaction through pyrrolidine-based chiral TSILs **14a** and **14b**. The reaction occurred via an enamine intermediate, Figure 1.14, and the procedure maintained the biphasic properties of ionic liquid, thereby ensuring good recyclability and reusability [37].

Chiral TSILs **14a** and **14b** catalyzed the model reaction of *p*-nitrobenzaldehyde and acetone. The reaction provided best yields in presence of water and acetic acid, providing the required direct aldol products along with some dehydration by-products. Other aldehydes and ketones also provided good results, Scheme 1.13.

Similarly, other carbonyls were treated with different aldehydes. The two diastereomeric products of aldol products were obtained in reasonable yield, Scheme 1.14.

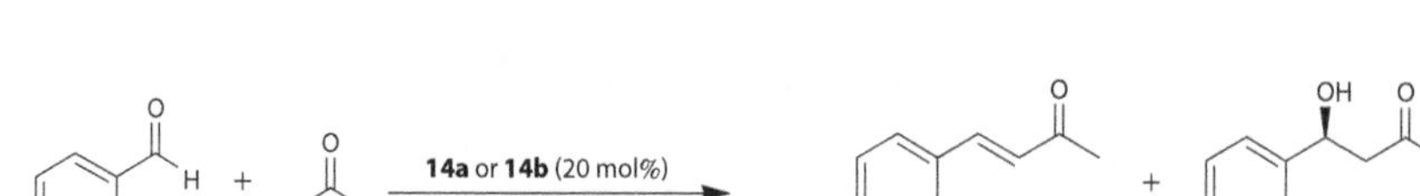

FIGURE 1.14 Best chiral task-specific ionic liquids.

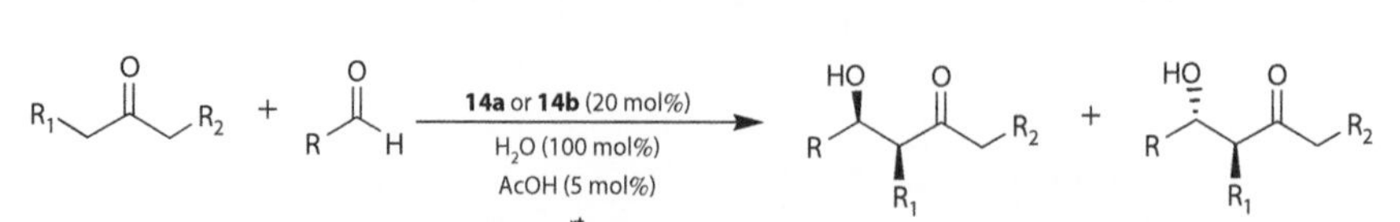

SCHEME 1.13 Chiral TSIL **14a** and **14b** catalyzed direct aldol reaction.

SCHEME 1.14 Pyrrolidine-based TSIL promoted the direct aldol reaction.

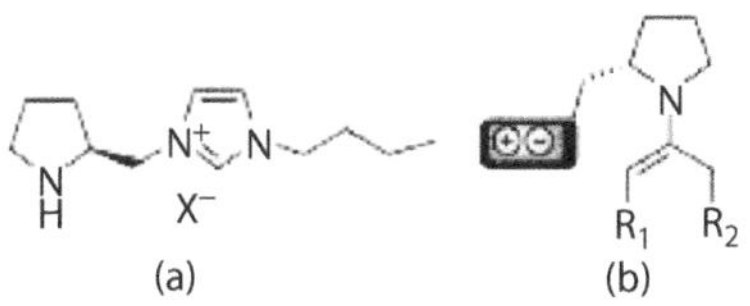

FIGURE 1.15   (a) Pyrrolidine-type functional ionic liquids and (b) the enamine intermediate.

These experimental observations can be rationalized by assuming that reaction occurs through *syn*-enamine intermediate, and the ionic liquid part provides shielding effect for the aldehyde acceptors leading to observed enantioselectivity, Figure 1.15.

Luo et al. also investigated the potential of same pyrrolidine-type chiral TSILs **14a** and **14b** for the Michael addition of cyclohexanone to *trans*-β-Nitrostyrene [38]. The proposed chiral TSILs provided substantial yields and best diastereoselectivities (*syn/anti* = 99:1) and enantioselectivities (98% Enantiomeric excess [*ee*]).

The asymmetric Michael addition of cyclohexanone to nitrostyrene furnished 75% yield after 60 hours (Diastereomeric ratio (***dr***) = 95:5 and 75% *ee* for the syn diastereomer). The best performances were quantitative yields with high diastereoselectivity (*syn/anti* = 99:1) and enantioselectivity (98% *ee*), Scheme 1.15.

Both electron-rich and electron-deficient nitrostyrenes were excellent Michael acceptors for cyclohexanone and cyclopentanone, showing reasonable diastereoselectivities and enantioselectivities. In the presence of **14a**, acetone formed Michael adducts with moderate yields and enantioselectivities. **14a**-mediated Michael reaction of acetone with cyclic nitroolefins provided products with good yields and enantioselectivities (*syn*: 76% *ee*, *anti*: 80% *ee*).

The chiral TSILs **14a** and **14b** also catalyzed the Michael addition of aldehydes. Under the optimized conditions, the addition of isobutyraldehyde to *trans*-β-Nitrostyrene provided the corresponding Michael adduct in best yields and 89% *ee*. The best diastereo- and enantioselectivities achieved through chiral pyrrolidine-based ionic liquid catalysis arose from an acyclic synclinal transition state. Where the ionic liquid part serves to shield the Si face of the enamine double bond in the ketone donor and directs the reaction to proceed via a Re-Re approach Figure 1.16. The highly polar imidazolium ring also stabilizes the transition state.

Ni et al. have the designed and synthesized new type of pyrrolidine-based chiral ionic liquid by attaching the pyrrolidine moiety on imidazolium cation, Figure 1.17 [39]. This chiral task-specific ionic liquid serve as chiral organo-catalyst, and chirality is controlled by pyrrolidine which contains N–H bond to control the stereoselectivities by hydrogen bonding. This chiral ionic liquid was found to catalyze the

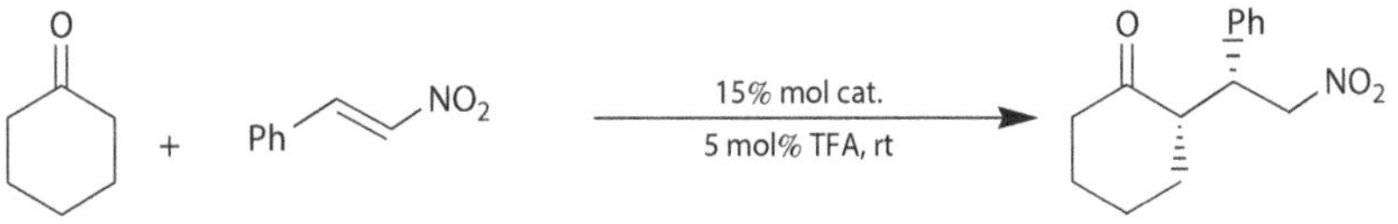

SCHEME 1.15   The Michael reaction of cyclohexanone with nitroalkenes.

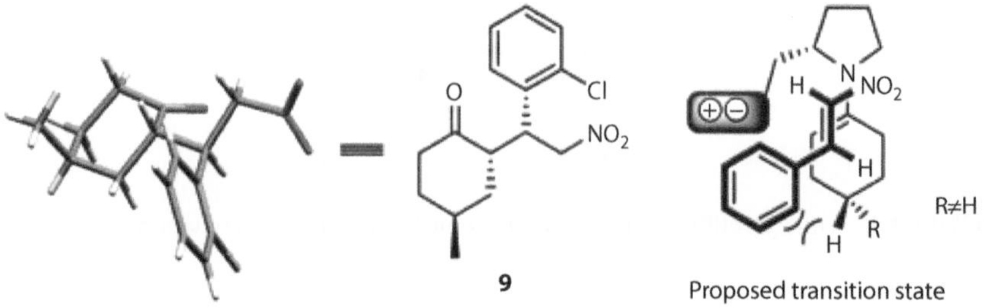

9

Proposed transition state

R≠H

FIGURE 1.16 Crystal structure of **9** and the proposed transition state.

17a

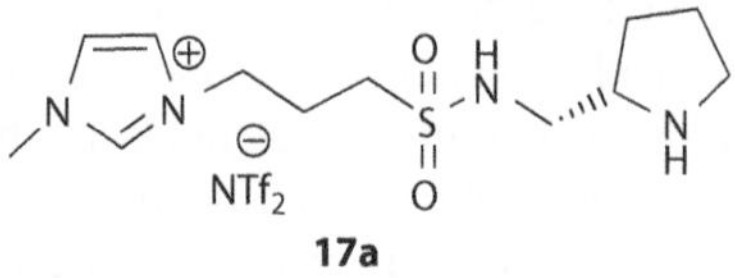

FIGURE 1.17 Pyrrolidine-based chiral ionic liquid.

Michael addition reaction of aldehydes and nitrostyrenes to give moderate yields, good enantioselectivities, and high diastereoselectivities.

The chiral TSIL catalyst **17a** catalyzed a series of asymmetric Michael addition reaction of different aldehydes with nitroolefins. The Michael reaction between isobutyraldehyde and nitrostyrene was achieved at room temperature using chiral TSILcatalyst **17a** as a catalyst in MeOHand *i*-PrOH. The reaction proceeded smoothly to give the desired Michael adduct in good yields (62%–80%) and enantioselectivities (66%–67% *ee*), Scheme 1.16.

Then various aldehydes underwent successful Michael reactions with different aryl-substituted nitrostyrenes in the presence of 20 mol% of **17a** in Et₂O at 4°C, giving the corresponding Michael adducts in moderate yields (29%–64%), with good enantioselectivities (64%–82% *ee*), and high diastereoselectivities (*syn/anti* ratio up to 97:3).

The N–H bond in pyrrolidine affects the course of reaction by its hydrogen bonding to nitrostyrene in such a way that C–C bond formation occurs preferably by enamine addition to relatively less hindered *Si* face of the nitrostyrene, Scheme 1.17. In addition, pyrrolidine-sulfonamide-based chiral TSIL catalyst **17a**, being a bifunctional catalyst, is expected to stabilize the transition state and make the selectivity possible.

Pyrrolidine-type chiral imidazolium TSILs, catalyzes the asymmetric synthesis with high diastereo- and enantioselectivities. However, it suffers from formation of side products resulting from the deprotonation of acidic C2 hydrogen of the

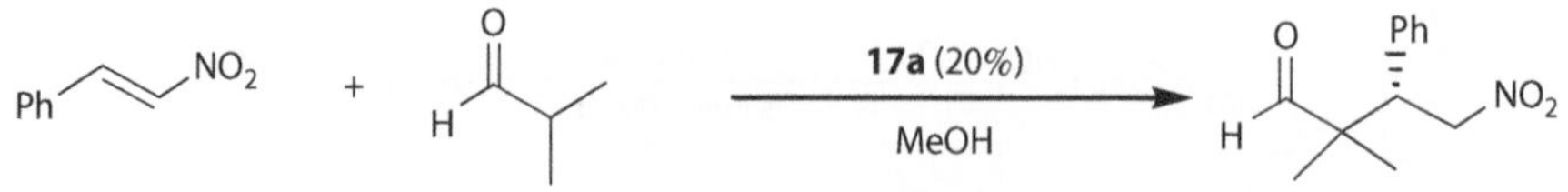

SCHEME 1.16 Asymmetric Michael addition.

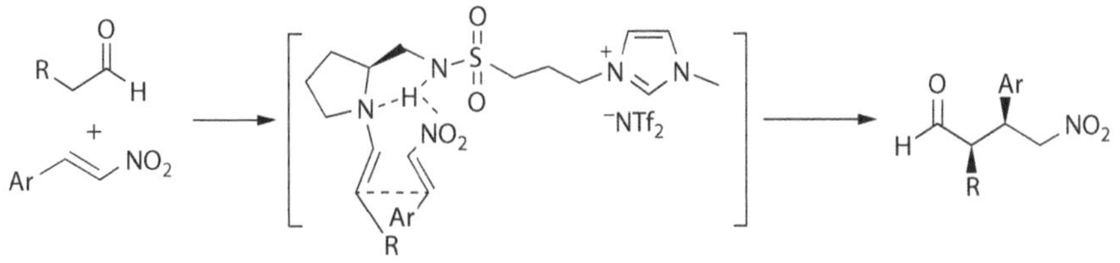
O
R
H
+
Ar
NO2
N
N--H
O
S
O
NO2
N
N
⁻NTf2
Ar
R
O
Ar
H
NO2
R

SCHEME 1.17 Mechanism for the Michael addition using catalyst **17a**.

imidazolium cation ring. Pyridinium-ionic liquids can serve as a suitable substitute to prevent such drawbacks associated with use of imidazolium-based ionic liquids. Ni et al. reported the new group of chiral ionic liquid containing pyrrolidine moiety attached to pyridinium ring, Figure 1.18 [40]. These chiral ionic liquids served as an effective green organo-catalyst to achieve better enantioselective Michael addition of various cyclic ketones to nitroalkenes.

All the chiral TSILs synthesized were soluble in polar solvent, such as MeOH, CH_3CN, dimethylformamide (DMF), and dimethyl sulfoxide, but were immiscible in Et_2O, EtOAc, and hexane. Chiral TSILs **18a–c** were soluble in H_2O, while the anion NTf_2, **18d**, was immiscible in H_2O. The different solubility allows them to be easily extracted for reuse.

All chiral pyridinium TSILs **18a–d** catalyzed the asymmetric Michael reaction of cyclohexanone with nitrostyrene. The nature of anion was found to have dramatic effects on the catalytic, as well as enantioselective activities of chiral ionic liquids. Ionic liquid **18a** with chloride anion demonstrated the best diastereo- and enantioselectivity with reasonable yield at 25°C, Scheme 1.18.

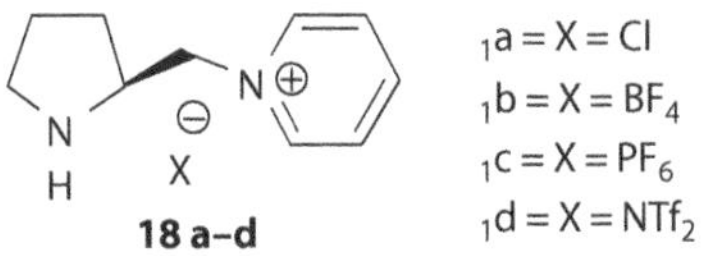
N
H
N⊕
⊖
X
18 a–d
₁a = X = Cl
₁b = X = BF4
₁c = X = PF6
₁d = X = NTf2

FIGURE 1.18 Pyrrolidine-based chiral pyridinium TSILs.

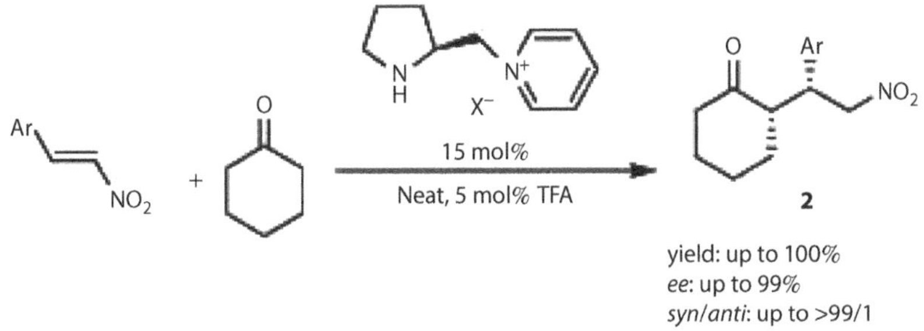
Ar
NO2
+
O
N
H
N+
X⁻
15 mol%
Neat, 5 mol% TFA
O
Ar
NO2
2
yield: up to 100%
ee: up to 99%
syn/anti: up to >99/1

SCHEME 1.18 Asymmetric Michael reaction of cyclohexanone with nitrostyrene.

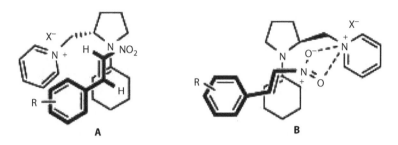

FIGURE 1.19 Transition state for Michael additions.

The origin of observed enantio- and diastereoselecivity shown by pyridinium-based chiral ionic liquids lies in an acyclic synclinal transition state **A**, where pyridinium ring shields the *si*-face of enamine double bond. The contribution observed enantioselectivity may arise from electrostatic attraction between pyridinium cation and nitro group of the substrate (transition state **B** in Figure 1.19).

Zhuo et al. have already used same ionic liquid-supported proline to catalyze direct asymmetric aldol reaction, Figure 1.20 [41]. This time they used it for Knoevenagel condensation reactions. The Knoevenagel condensation reaction of aldehydes with malononitrile was found to be catalyzed by ionic liquid-supported proline efficiently [42]. The method represented a better alternative to the classical synthesis strategies and exhibited the advantage of performing in ionic liquids.

The [Promim][CF₃CO₂] (**20a**) was used to catalyze the Knoevenagel condensation of malononitrile with varying aldehydes and carbonyl derivatives, Scheme 1.19. The reactions were performed with 30 mol% of the catalyst for 24 hours at 80°C, resulting in expected Knoevenagel condensation products in good yield.

Dong et al. used pyrrolidine amide-based task-specific ionic liquid to formulate and arrange standard homogeneous conditions for Claisen–Schmidt condensation reaction, Figure 1.21 [43].

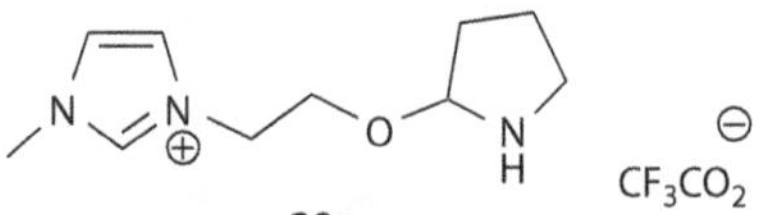

20a

FIGURE 1.20 Ionic liquid-supported proline.

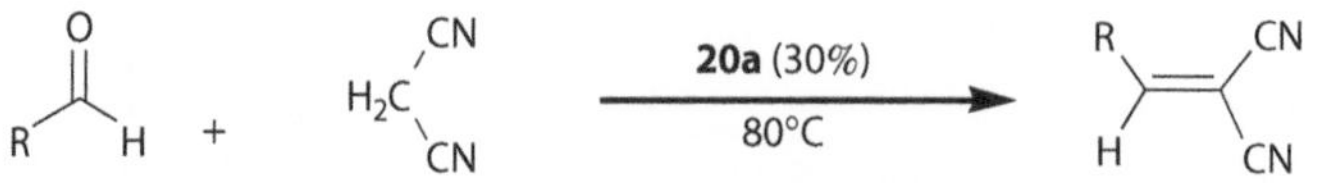

SCHEME 1.19 Knoevenagel condensation of malononitrile with varying aldehydes.

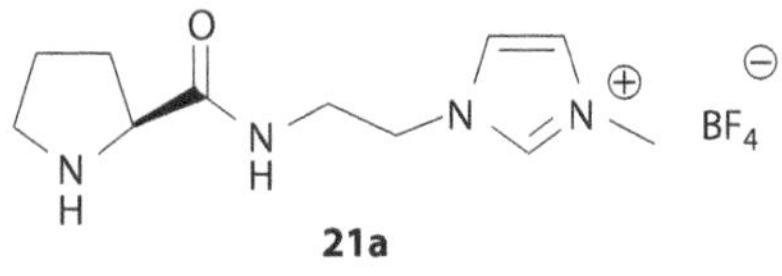

FIGURE 1.21 Pyrrolidine amide-based task-specific ionic liquid.

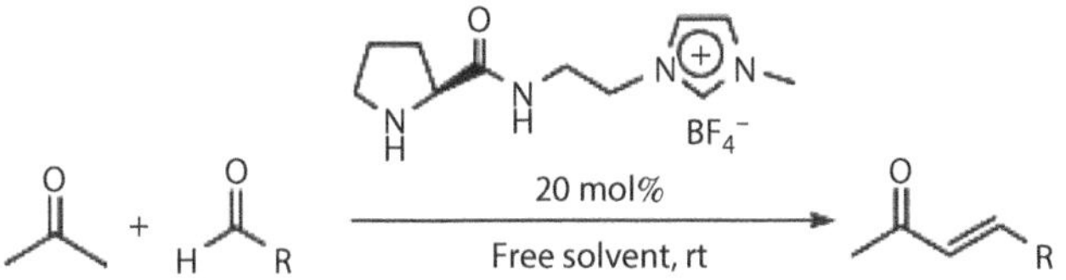

SCHEME 1.20 Task-specific ionic liquid catalyzed Claisen–Schmidt condensation reaction.

The Claisen–Schmidt reaction of acyclic and cyclic ketones with different aromatic aldehydes gave (E)-α,β-unsaturated ketones under solvent-free conditions at 25°C and led to best yields (Scheme 1.20).

21a-Catalyzed Claisen–Schmidt reaction between various ketones and aldehydes exhibited desirable results under softer reaction conditions. Initially, **21a** was applied to the aldol condensation reaction of 4-nitro benzaldehyde with acetone with 30 mol% of **21a** in dimethyl sulfoxide at 25°C. Instead of β-hydroxy ketone, it gave dehydration product, α,β-unsaturated ketone with (E)-configuration in excellent yield, Scheme 1.21. The results proved the catalytic capacity of **21a** for convenient synthesis of the α,β-unsaturated ketone.

1.1.8 SULFOXIDE-BASED TASK-SPECIFIC IONIC LIQUID FOR SWERN OXIDATION

Swern oxidation is most powerful method of converting almost every type of alcohol into corresponding carbonyl compounds quantitatively under milder conditions. However, Swern oxidation involves use of sulphur compounds which are difficult to handle due to their toxicity and pungent smell. Chan et al. developed a sulfide-based task ionic liquid clean and odorless method for Swern oxidation, Figure 1.22 [44].

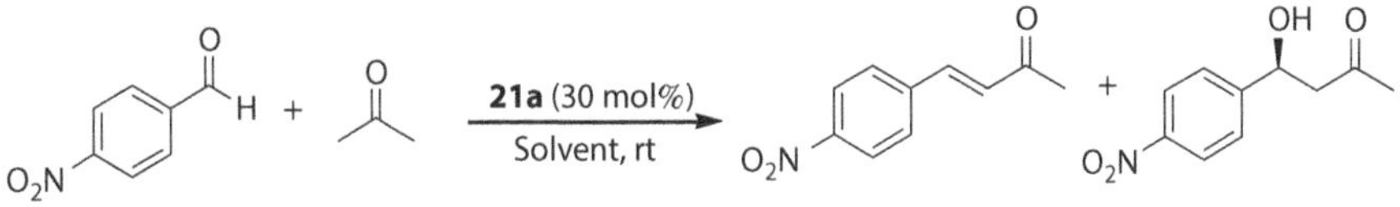

SCHEME 1.21 Aldol condensation reaction of 4-nitrobenzaldehyde with acetone in the presence of **21a**.

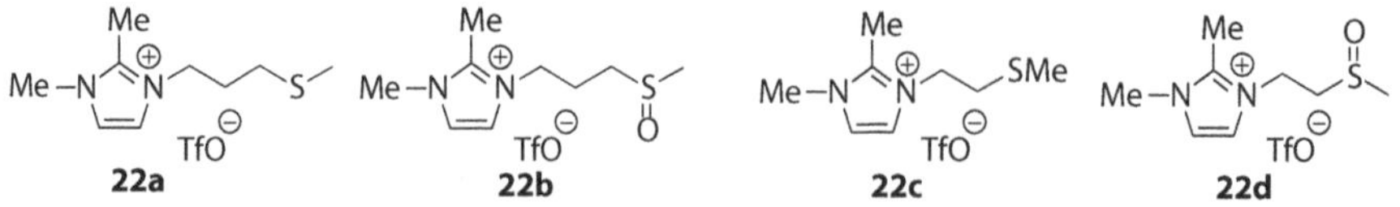

FIGURE 1.22   Ionic liquid-grafted sulfides and sulfoxides.

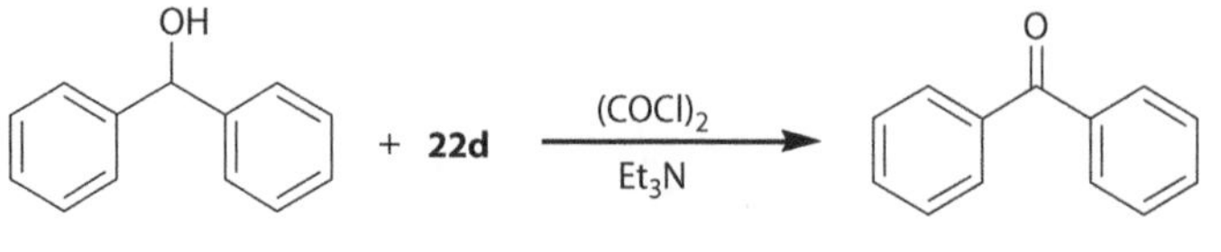

SCHEME 1.22   Oxidation of different types of alcohols under new protocol.

After optimization of condition, several kinds of primary and secondary alcohols were subjected to sulfoxide-based ionic liquid-catalyzed Swern oxidation. The two versions of sulfoxide-based ionic liquids, for example, **22d** or **22b** exhibited better activity. In addition to this, oxalyl chloride and triethylamine were used as activator and base, respectively, in a binary solvent system $CH_3-CN: CH_2Cl_2$ at lower temperature, Scheme 1.22. All kinds 1° and 2° alcohols furnished relevant carbonyl compounds in best yields.

All ionic liquid-based reagents due to their appropriate miscibility behavior are recoverable from reaction media and are recycled numbers of times. After extraction of products with immiscible ether, the sulfide-based ionic liquids **22c** or **22a** were recovered by initially treating the mixture with aq: K_2CO_3, then extracting in a binary solvent system $CH_3-CN: CH_2Cl_2$. The recovered sulfide **22c or 22a** were again oxidized with HIO_4 and reused for the Swern oxidation, Scheme 1.23.

1.1.9   NHPI COMPLEX-BASED TASK-SPECIFIC IONIC LIQUID

N-hydroxy phthalimide (NHPI) is well-known organo-catalyst for catalytic power to oxidize different compounds to higher oxidation states under soft conditions. It can

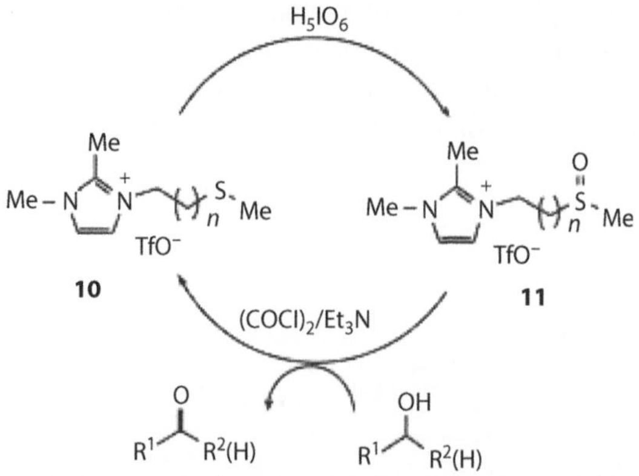

SCHEME 1.23   Recycling profile of sulfoxide-based ionic liquid.

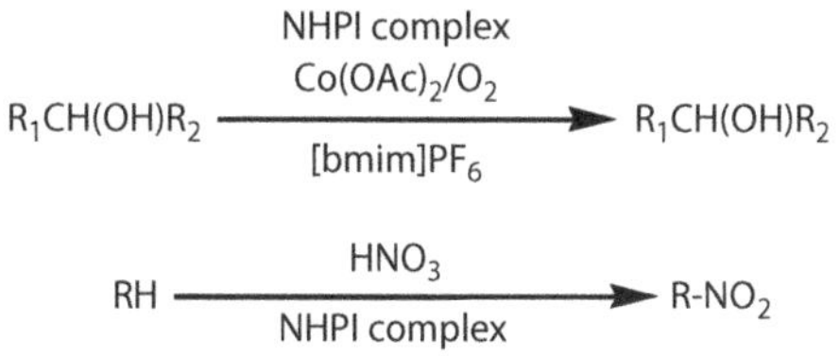

SCHEME 1.24 NHPI-promoted oxidation of alcohols to carbonyls and nitration of alkanes.

oxidize suitable alcohols to relevant ketones and carboxylic acids and can nitrate the C–H bond, Scheme 1.24. It can also oxidize carbohydrates and other carbinols in form of stable phthalimide N-oxy [45].

NHPI as a catalyst has manifested its better oxidizing strength when used in ionic liquids as reaction media. The 1-phenyl ethanol was oxidized completely to acetophenone by NHPI-Co(OAc)$_2$-O$_2$ system using ionic liquids like [BMIM][BF$_4$] [BMIM] [PF$_6$] and [BMIM][CF$_3$SO$_3$], Scheme 1.25. However, the recycling efficiency was quite dismal and yields first cycle 93% were reduced to 80% in the second and to 26% in third one.

These results show that NHPI is separated smoothly from ionic liquid with organic solvents, and that it is impossible to construct the reusable reaction system because of the disappearance of NHPI as a catalyst. Kitazume et al. have reported the catalytic potential of NHPI complexes-based ionic liquid, Figure 1.23, as a better alternative to ordinary NHPI organo-catalyst. Like most of task-specific ionic liquids, it provided the convenience in procedure and easy recycling of catalytic system for the oxidations [46].

The oxidation reaction of 1-phenyl ethanol to acetophenone shows that the system using by the IL-NHPI (10 mol%)-Co(OAc)$_2$-O$_2$ in ionic liquid ([BMIM] [PF$_6$]) is reusable. In the same system, various types of carbinols are transformed into the corresponding aldehydes and/or ketones in good yield and secondary hydroxy group is selectively oxidized in this system. Products of oxidation in

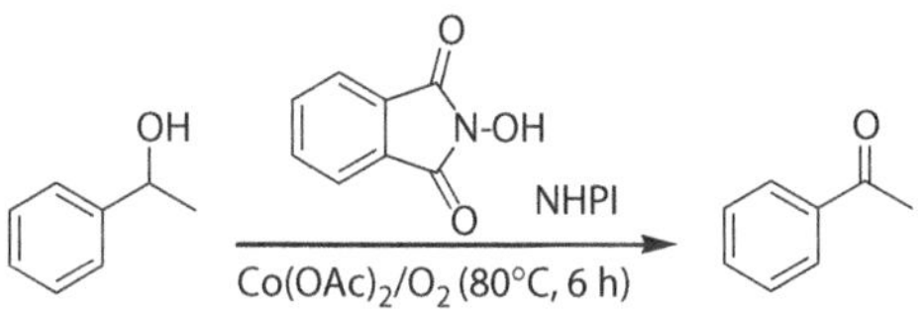

SCHEME 1.25 Oxidation of 1-phenyl ethanol in the NHPI-Co(OAc)$_2$-O$_2$ system.

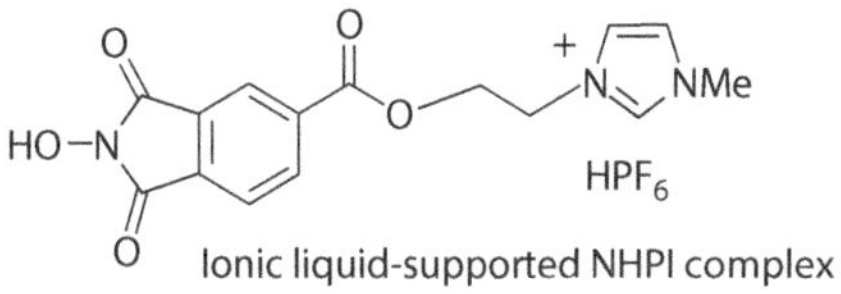

Ionic liquid-supported NHPI complex

FIGURE 1.23 NHPI complex-based ionic liquid.

IL-supported NHPI complex-ionic liquid [BMIM][PF$_6$] were extracted with supercritical carbon dioxide.

Direct nitration of carbon-hydride bond was also examined in this novel catalytic system. Nitration of alkanes requires as high as 250°C to 400°C temperature and use of toxic chemicals like nitric acid or any other source of NO$_2$ gas. Thus, search for greener conditions is necessary for this useful reaction. The nitration of alkanes with HNO3 as nitrating agent was investigated in 25 mol% NHPI complex-based ionic liquid as a catalyst in [BMIM][PF$_6$] as a reaction media. The nitrations occurred smoothly at relatively lower 80°C giving best conversion yields.

1.1.10 TEMPO-BASED TASK-SPECIFIC IONIC LIQUIDS FOR OXIDATION OF ALCOHOL

The TEMPO abbreviation stands for 2,2,6,6-tetramethylpiperidine-1-oxyl, which is oxidizing organo-catalyst known due to its ability to selectively oxidize the alkyl hydroxyl group to carbonyl group. It offers many advantages like metal-free oxidation under nontoxic condition and, more importantly, it exhibits reversible redox behavior which makes it recyclable. According to Anelli et al. protocol, for these oxidations, 1 mole equivalent of alcohol is treated with 1 mol% of TEMPO and equimolar ratio of any suitable terminal oxidant, such as meta-chloroperbenzoic acid, NaClO$_3$, bleach, and N-chlorosuccinimide [47]. However, this method suffers from many procedural inconveniences, such as cumbersome work-up and tedious products separation from tedious isolation of products from TEMPO.

Qian et al. have reported the green ionic liquid version of TEMPO as TEMPO-based task-specific ionic liquid, Figure 1.24, and it was investigated for its competence to ordinary TEMPO for chemoselective oxidation of 1° and 2° carbinols to aldehydes and ketones in aqueous-[BMIM]PF$_6$ biphasic conditions, respectively [48].

The oxidation of variety of alcohols was achieved according to the Aneli et al. protocol using TEMPO-based ionic liquid radical as catalyst. The resulting aldehydes and ketones products were obtained in better yields, Scheme 1.26 [47].

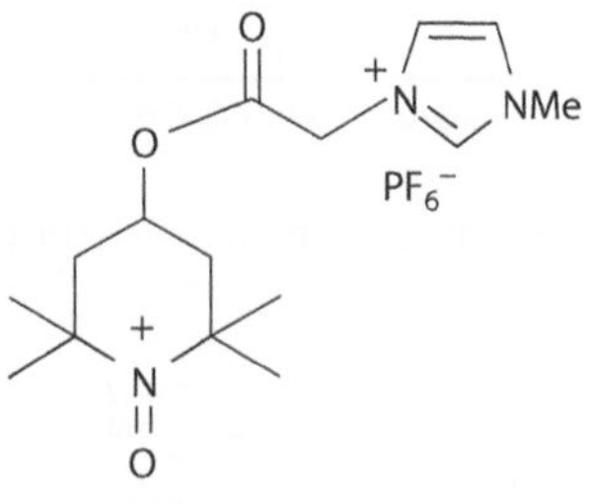

25a

FIGURE 1.24 TEMPO-derived task-specific ionic liquid.

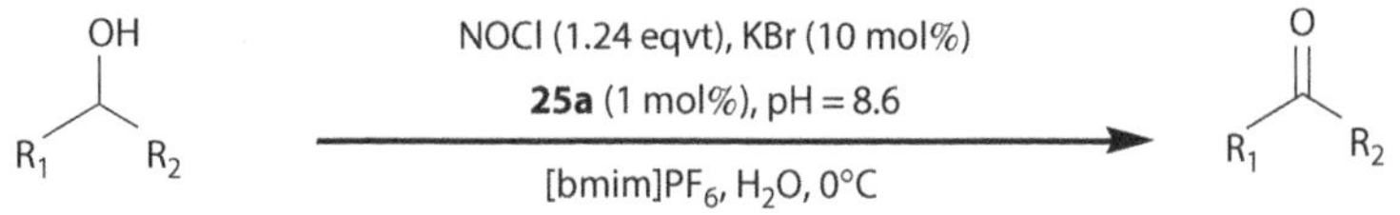

SCHEME 1.26 Oxidation of alcohols by novel IL-supported TEMPO.

1.1.11 HYPERVALENT IODINE III REAGENT-BASED TASK-SPECIFIC IONIC LIQUID

The use of hypervalent iodine organo-reagents for highly selective oxidation of alcohols to carbonyl compounds is not uncommon. The pentavalent iodine reagents like Dess-Martin periodane and o-iodoxybenzoic acid are often used as oxidizing agents to achieve efficient and selective oxidation of alcohols to carbonyl compounds in dimethyl sulfoxide, CH_2Cl_2, and acetone. However, the use of these iodine (V) oxidants is not safer because they can explode. Many reaction protocols have been reported. Qian et al. incorporated the hypervalent iodine reagent to ionic liquid, Figure 1.25, and have reported some additional advantages over the routine procedure [49].

The oxidation of alcohols was conducted in the ionic liquid [EMIN][BF₄] using [diBMIM][BF₄] as an oxidant in the presence of a low concentration of bromide ions under mild conditions (30°C, a 1:1.4 ratio of substrate **2**:oxidant **1**). A variety of primary and secondary alcohols were oxidized to carbonyl compounds in moderate to excellent yields at room temperature Under these conditions, primary alcohols were oxidized in less than 4 hours to the corresponding aldehydes in 57%–95% yields without any noticeable over oxidation to the carboxylic acids. Secondary alcohols were oxidized to the corresponding ketones over longer reaction times, Scheme 1.27.

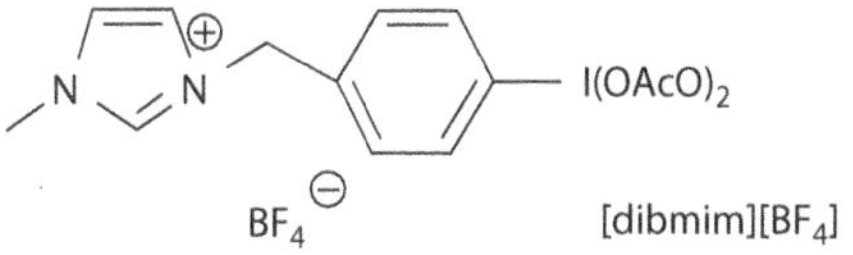

FIGURE 1.25 IL-supported hypervalent iodine reagent.

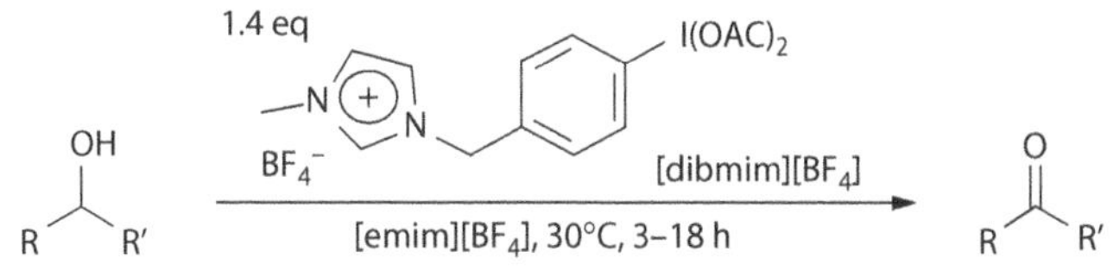

SCHEME 1.27 Selective oxidation of alcohols by [diBMIM][BF₄] in [EMIM][BF₄].

1.1.12 Task-Specific Ionic Liquid for Stille Cross-Coupling Reaction

Stille cross-coupling reaction deals with palladium-mediated reaction between any alkyl stannanes with aromatic halides, which is mostly employed to achieve alkylation of aromatic compounds or biarylation. Reaction suffers from some procedural inconveniences and dismal reactivity of organo-stannanes. All these shortcomings of this useful reaction can be surmounted by conducting this reactions under ionic liquids conditions. Vitz et al. developed a new ionic liquid-supported tin reagent, Figure 1.26, that was synthesized as an ionic liquid substitute of organic compounds for easy and efficient ligand transfer [50]. This method was found to have many edges over the classical method, for example, lowered reaction temperature and recycling ionic liquid version of the Stille reagent.

With these tin reagents supported on ionic liquids, the catalysts $Pd_2dba_3 \cdot CHCl_3$ and $Pd(dba)_2$ provided good yields of the desired biaryl products from aryl iodides, and other cross-coupling reactions were successful even at low reaction temperatures without addition of copper salts or ligands. In all cases of Stille reactions of aryl iodides, very high levels of conversion could be achieved with 2 equivalent of **27a** ($n = 2$, iodide) within six hours in the presence of $Pd_2dba_3 \cdot CHCl_3$, at the temperature of 35°C to afford a product in 94% yield, Scheme 1.28.

In the case of substituted aryl iodides, alkyl- and methoxyiodobenzenes took relatively longer time duration as compared to simple iodobenzenes. Similarly 3-iodo pyridine also afforded smooth Stille coupling.

The catalyst efficiency on recycling of organotin-palladium reagent was investigated, Scheme 1.29. On completion of reaction, all organic compounds were extracted by using nonpolar solvents like hexatane, which was not miscible with the tin containing ionic liquid, and the halogenotin-supported ionic liquid **10** was restored by addition of phenyl lithium. In this way, the ionic liquid-supported tin reagent was recycled five times with no significant loss of catalytic potential.

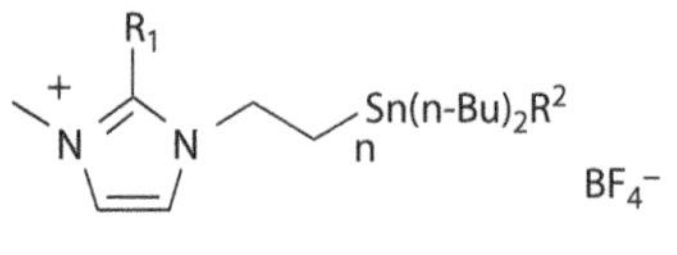

FIGURE 1.26 Synthesis of tin-supported task ionic liquid R_1 = H (**27a** $n = 2$) (**27b** $n = 5$).

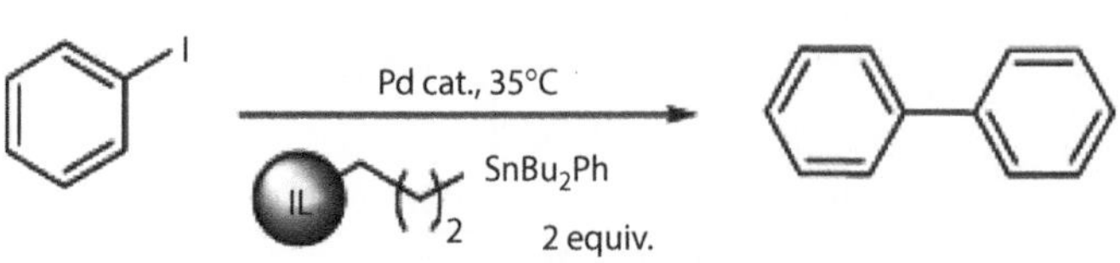

SCHEME 1.28 Stille coupling reaction with tin containing ionic liquid.

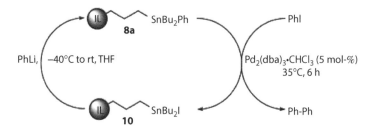

SCHEME 1.29   Recycling of the tin compounds.

1.1.13   DMF-Like Multifunctional Task-Specific Ionic Liquid

DMF is commonly used as polar aprotic solvents in chemical labs. However, it has been used as coreagent with many organic, as well as inorganic reagents. DMF is also reacted with phosphoryl chloride or another relevant compound to prepare Vilsmeier reagent, a wonderful reactive reagent used to achieve key organic transformations. Since task-specific ionic liquids are reported with efficient performance, as compared to ordinary counterparts, therefore, it was conceived to develop its ionic liquid version to make its use convenient, efficient, and recyclable. Hullio et al. introduced new term by reporting world's first multifunctional task ionic liquid [51]. An ionic liquid with DMF functionality was anchored on imidazolium cation with two carbon spacings, Figure 1.27.

The DMF-like task-specific ionic liquids were declared to be multifunctional because it can be used to achieve variety of diverse chemical transformations. Whereas all reported task-specific ionic liquids were essentially designed for "specific" reaction only. The DMF-like Multipurpose Task Specific Ionic Liquid (MTSIL) was applied to various DMF-dependent reactions like nucleophilic substitutions, sodium periodate-DMF catalyzed direct oxidation of alkyl halides into corresponding carbonyl compounds, Scheme 1.30.

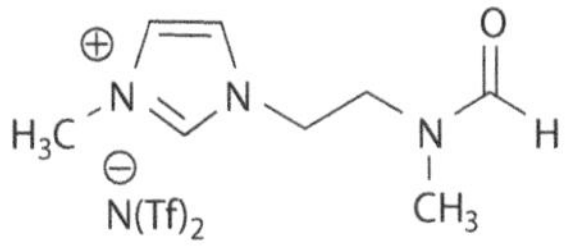

FIGURE 1.27   DMF-like multifunctional task-specific ionic liquid.

R = R, H

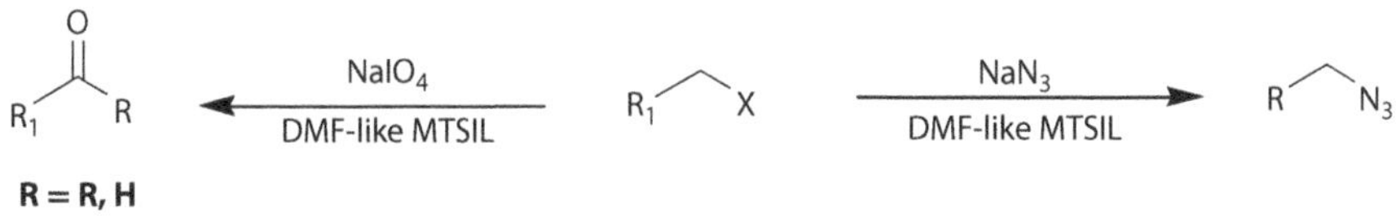

SCHEME 1.30   Oxidation and azidation of alkyl halides in multipurpose DMF-like TSIL.

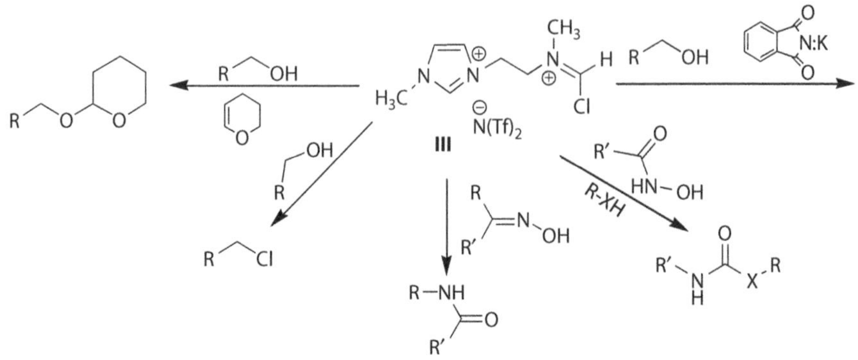

SCHEME 1.31 TCT-DMF-like ionic liquid catalyzed reactions.

Same group is further investigating the multipurpose nature of the DMF-like task-specific ionic liquid by developing the ionic liquid version of Vilsmeier reagent from TCT (Trichlorotriazine)-DMF-like task-specific ionic liquid complex. All Vilsmeier reagent-mediated reactions will be tested to prove the multipurpose nature of proposed DMF-like task-specific ionic liquid, Scheme 1.31. In addition to this, the new ionic liquid system is likely to be efficient, versatile accompanied by easy procedure, and easy recovery of product and recycling of the ionic system.

REFERENCES

1. (a) Welton, T. Room-temperature ionic liquids: Solvents for synthesis and catalysis. *Chem. Rev.*, **1999**, *99*, 2071–2083. (b) Wasserscheid, P. and Keim, W. Ionic liquids-new "solutions" for transition metal catalysis. *Angew. Chem., Int. Ed.* **2000**, *39*, 3772–3789. (c) Dupont, J., de Souza, R. F., and Suarez, P. A. Z. Ionic liquid (molten salt) phase organometallic catalysis. *Chem. Rev.*, **2002**, *102*, 3667–3692.
2. Freemantle, M. Green solutions for ionic liquids. *C&E News*, August 24, **1998**, 32–37.
3. For recent examples, see: (a) Kotrusz, P., Kmentova, I., Gotov, B., Toma, S., and Solcaniova, E. Proline-catalysed asymmetric aldol reaction in the room temperature ionic liquid [bmim][PF$_6$]. *Chem. Commun.*, **2002**, *21*, 2510–2511. (b) Loh, T. P., Feng, L. C., Yang, H. Y., and Yang, J. Y. L-Proline in an ionic liquid as an efficient and reusable catalyst for direct asymmetric aldol reactions. *Tetrahedron Lett.*, **2002**, *43*, 8741–8743. (c) Chowdari, N. S., Ramachary, D. B., and Barbas III, C. F. Organocatalysis in ionic liquids: Highly efficient L-proline-catalyzed direct asymmetric Mannish reactions involving ketones and aldehyde nucleophiles. *Synlett.*, **2003**, *35*, 1906–1909.
4. For selected studies with aiming at the recovery and reuse of organocatalysts, see: (a) Chowdari, N. S., Ramachary, D. B., and Barbas III, C. F. Organocatalysis in ionic liquids: Highly efficient L-proline-catalyzed direct asymmetric Mannich reactions involving ketone and aldehyde nucleophiles. *Synlett.*, **2003**, *35*, 1906–1909. (b) Bensa, D., Constantieux, T., and Rodriguez, J. P. BEMP: A new efficient and commercially available user-friendly and recyclable heterogeneous organocatalyst for the Michael addition of 1,3-dicarbonyl compounds. *Synth.*, **2004**, *6*, 923–927.
5. (a) Sawant, A. D., Raut, D. G., Darvatkar, N. B., and Salunkhe, M. M. Recent developments of task-specific ionic liquids in organic synthesis. *Green Chem. Lett. Rev.* **2011**, *4*, 41–54. (b) Sadaf Khan, S., Hanelt, S., and Liebscher, J. Versatile synthesis of 1,2,3-triazolium-based ionic liquids. *ARKIVOC*, **2009**, *XII*, 193–208.

6. Lee, S. Functionalized imidazolium salts for task-specific ionic liquids and their applications. *Chem. Commun.*, **2006**, *37*, 1049–1063.

7. Davis, J. H. Task-specific ionic liquids. *Chem. Lett.*, **2004**, *33*, 1072–1077.

8. Zhao, D., Fei, Z., Scopelliti R., and Dyson, P. Synthesis and characterization of ionic liquids incorporating the nitrile functionality. *J. Inorg. Chem.*, **2004**, *43*, 2197.

9. Branco, L. C., Rosa, J. N., Moura Ramos, J. J., and Alfons, C. A. M. Preparation and characterization of new room temperature ionic liquids. *Chem. Eur. J.*, **2002**, *8*, 3671.

10. Holbrey, J. D., Reichert, W. M., Tkatchenko, I., Bouajila, E., Walter, O., and Tommasi, I. 1,3-Dimethylimidazolium-2-carboxylate: The unexpected synthesis of an ionic liquid precursor and carbene-CO_2 adduct. *Chem. Commun.*, **2003**, *347*, 28–29.

11. Visser, A. E., Swatloski, R. P., Reichert, W. M., Mayton, R., Sheff, S., Wierzbicki, A., Davis, J. H., and Rogers, R. D. Task-specific ionic liquids for the extraction of metal ions from aqueous solutions. *Chem. Commun.*, **2001**, *1*, 135–136.

12. Merrigan, T. L., Bates, E. D., Dorman, S. C., and Davis, J. H. New fluorous ionic liquids function as surfactants in conventional room temperature ionic liquids. *Chem. Commun.*, **2000**, *20*, 2051–2052.

13. Bicak, N. A new ionic liquid: 2-hydroxy ethylammonium formate. *J. Mol. Liquids*, **2005**, *116*, 15–18.

14. Yoshizawa, M., Ogihara, W., and Ohno, H. Novel polymer: Electrolytes prepared by copolymerization of ionic liquid. monomers. *Polym. Adv. Technol.*, **2002**, *13*, 589–594.

15. (a) Ogihara, W., Yoshizawa, M., and Ohno, H. Novel ionic liquids composed of only azole ions. *Chem. Lett.*, **2004**, *33*, 1022. (b) Xue, H., Gao, Y., Twamley, B., and Shreeve, J. M. New energetic salts based on nitrogen-containing heterocycles. *Chem. Mater.*, **2005**, *17*, 191–198. (c) Katritzky, A. R., Singh, S., Kirichenko, K., Holbrey, J. D., and Smiglak, M. 1-Butyl-3-methylimidazolium 3,5-dinitro-1,2,4-triazolate: A novel ionic liquid containing a rigid, planar energetic anion. *Chem. Comm.*, **2005**, *36*, 868–870.

16. Kim, H. S., Kim, Y. J., Lee, H., Park, K. Y., Lee, C., and Chin, C. S. Ionic liquids containing anionic selenium species: Applications for the oxidative carbonylation of aniline. *Angew. Chem., Int. Ed.*, **2002**, *41*, 4300–4303.

17. (a) Zhao, D., Fei, Z., Ohlin, C. A., Laurenczy, G., and Dyson, P. J. Dual-functionalised ionic liquids: Synthesis and characterisation of imidazolium salts with a nitrile-functionalized anion. *Chem. Commun.*, **2004**, *21*, 2500–2501.

18. Larsen, A. S., Holbrey, J. D., Tham, F. S., and Reed, C. A. Designing ionic liquids: Imidazolium melts with inert carborane anions. *J. Am. Chem. Soc.*, **2000**, *122*, 7264–7272.

19. Bates, E. D., Mayton, R. D., Ntai, I., and Davis, J. H. CO_2 capture by a task-specific ionic liquid. *J. Am. Chem. Soc.*, **2002**, *124*, 926–927.

20. Gordon, C. M. New developments in catalysis using ionic liquids. *Appl. Catal. A: Gen.*, **2001**, *222*, 101–117.

21. Sheldon, R. A., Lau, R. M., Sorgedrager, M. J., van Rantwijk, F., and Seddon, K. R. Biocatalysis in ionic liquids. *Green Chem.*, **2002**, *4*, 147–151.

22. Pastre, J. C., Génisson, Y., Saffon, N., Dandurand, J., and Carlos R. D. Synthesis of novel room temperature chiral ionic liquids: Application as reaction media for the heck arylation of aza-endocyclic acrylates. *J. Braz. Chem. Soc.*, [online]. **2010**, *21*, 821–836.

23. Cole, A. C., Jensen, J. L., Ntai, I., Tran, K. L. T., Weaver, K. J., Forbes, D. C., and Davis, J. H. Novel Brönsted acidic ionic liquids and their use as dual solvent-catalysts. *J. Am. Chem. Soc.*, **2002**, *124*, 5962–5963.

24. Sahoo, S., Joseph T., and Halligudi, S. B. Mannich reaction in Brönsted acidic ionic liquid: A facile synthesis of D, α-amino carbonyl compounds. *J. Mol. Cat. A: Chemical*, **2006**, *244*, 179–182.

25. Dong, F., Jian, C., Zhenghao, F., Kai, G., and Zuliang, L. Synthesis of chalcones via Claisen-Schmidt condensation reaction catalyzed by acyclic acidic ionic liquids. *Catal. Comm.*, **2008**, *9*, 1924–1927.

26. Wang, T., Xing, H., Zhou, Z., and Dai Y. Novel Brönsted-acidic ionic liquids for esterifications. *Ind. Eng. Chem. Res.*, **2005**, *44*, 4147–4150.

27. Jiang, T., Gaoa, H., Han, B., Zhaoa, G., Changa, Y., Wua, W., Gaoa, L., and Yanga, G. Ionic liquid catalyzed Henry reactions. *Tetrahedron Lett.*, **2004**, *45*, 2699–2701.

28. Liu, D., Guib, J., Zhu, X., Song, L., and Sun, Z. Synthesis and characterization of task specific ionic liquids. *Synth. Commun.*, **2007**, *37*, 759–765.

29. Li, D., Shi, F., Peng, J., Guo, S., and Deng, Y. Application of functional ionic liquids possessing two adjacent acid sites for acetalization of aldehydes. *J. Org. Chem.*, **2004**, *69*, 3582–3585.

30. Sun, Z., Gui, J., Deng, Y., and Hu, Z. A novel task-specific ionic liquid for Beckmann rearrangement: A simple and effective way for product separation. *Tetrahedron Lett.*, **2004**, *45*, 2681–2683.

31. Han, X., Armstrong, D. W. Using geminal dicationic ionic liquids as solvents for high temperature organic reactions. *Org. Lett.*, **2005**, *7*, 4205–4208.

32. Mi, X., Luo, S., and Cheng, J. P. Ionic liquid-immobilized quinuclidine-catalyzed Morita-Baylis-Hillman reactions. *J. Org. Chem.*, **2005**, *70*, 2338–2341.

33. Jeong, Y. and Ryu, J. S. Synthesis of 1,3-dialkyl-1,2,3-triazolium ionic liquids and their applications to the Baylis-Hillman reaction. *J. Org. Chem.*, **2010**, *75*, 4183–4191.

34. Handy, S. T. Greener solvents: Room temperature ionic liquids from bio-renewable sources. *Chemistry—A Eur. J.*, **2003**, *9*, 2938–2944.

35. Handy, S. T., Okello, M., Dickenson, G., and Egrie, C. Ionic liquids from bio-renewable resources: Nicotine and fructose. In *Thirteenth International Symposium on Molten Salts* (Eds. P. Trulove, T. S. H. DeLong), Electrochemical Society, Pennington, NJ, **2002**.

36. (a) Trotter, J. and Darby, W. 4(5)-Hydroxymethylimidazole hydrochloride. *Org. Synth. Coll. Vol. III*, Wiley, New York, **1973**, 460–461. (b) Dalko, P. I. and Moisan, L. Chiral amine catalyzed direct aldol reactions. *Angew. Chem., Int. Ed.*, **2004**, *43*, 5138–5176.

37. Luo, S., Mi, X., Zhang, L., Liu, S., Xu, H., and Cheng, J.-P. Functionalized ionic liquids catalyzed direct aldol reactions. *Tetraheron*, **2007**, *63*, 1923–1930.

38. Luo, S., Mi, X., Zhang, L., Liu, S., Xu, H., and Cheng, J. P. Functionalized chiral ionic liquids as highly efficient asymmetric organocatalysts for Michael addition to nitroolefins. *Angew. Chem., Int. Ed.*, **2006**, *45*, 3093–3097.

39. Zhang, Q., Ni, B., and Headley, A. D. Asymmetric Michael addition reactions of aldehyde with nitrostyrenes catalyzed by functional chiral ionic liquids. *Tetrahedron*, **2008**, *64*, 5091–5097.

40. Ni, B., Zhang, Q., and Headley, A. D. Pyrrolidine-based chiral pyridinium ionic liquids as recyclable and highly efficient organocatalysts for the asymmetric Michael addition reactions. *Tetrahedron Lett.*, **2008**, *49*, 1249–1252.

41. Chen, Z., Li, Y., Xie, H., Hu, C. G., and Dong, X. A novel and efficient ionic liquid-supported proline catalyzed Knoevenagel condensation. *Russian J. Org. Chem.*, **2008**, *44*, 1807–1810.

42. Zhuo, C., Xian, D., Jian-wei, W., and Hui, X. An efficient and recyclable ionic liquid-supported proline catalyzed Knoevenagel condensation. *ISRN Org Chem.*, **2011**, Article ID 676789, 5.

43. Yang, S. D., Wu, L. Y., Yan, Z. Y., Pan, Z. L., and Liang, Y. M. A novel ionic liquid supported organocatalyst of pyrrolidine amide: Synthesis and catalyzed Claisen-Schmidt reaction. *J. Mol. Catal. A: Chemical*, **2007**, *268*, 107–111.

44. He, X. and Chan, T. H. New non-volatile and odorless organosulfur compounds anchored on ionic liquids. Recyclable reagents for Swern oxidation. *Tetrahedron*, **2006**, *62*, 3389–3394.

45. (a) Bragd, P. L., Bekkum, H. V., and Besemer, A. C. TEMPO-mediated oxidation of polysaccharides: Survey of methods and applications. *Topics Catal.*, **2004**, *27*, 49–66. (b) Ishii, Y. Innovation of hydrocarbon oxidation with molecular oxygen and related reactions. *J. Synth. Org. Chem., Jpn.*, **2003**, *61*, 1056–1063. (c) Ishii, Y., Matsunaka K., and Sakaguchi, S. The first catalytic sulfoxidation of saturated hydrocarbons with SO_2/ O_2 by a vanadium species. *J. Am. Chem. Soc.*, **2000**, *122*, 7390–7391. (d) Isozaki, S., Nishiwaki, Y., Sakaguchi, S., and Ishii, Y. Nitration of alkanes with nitric acid catalyzed by *N*-hydroxyphthalimide. *Chem. Commun.*, **2001**, *15*, 1352–1353.

46. Koguchi, S. and Kitazume T. Synthetic utilities of ionic liquid supported NHPI complex. *Tetrahedro Lett.*, **2006**, *47*, 2797–2801.

47. (a) Anelli, P. L., Biffi, C., Montanari, F., Quici, S. Fast and selective oxidation of primary alcohols to aldehydes or to carboxylic acids and of secondary alcohols to ketones mediated by oxoammonium salts under two-phase conditions. *J. Org. Chem.*, **1987**, 52, 2559–2562. (b) Anelli, P. L., Biffi, C., Montanari, F., Quici, S. Oxidation of diols with alkali hypochlorites catalyzed by oxammonium salts under two-phase conditions. *J. Org. Chem.*, **1989**, 54, 2970–2972.

48. Qian, W., Jin, E., Bao, W., and Zhang, Y. Clean and selective oxidation of alcohols catalyzed by ion-supported TEMPO in water. *Tetrahedron*, **2006**, *62*, 556–562.

49. Qian, W., Jin, E., Bao, W., and Zhang, Y. Clean and highly selective oxidation of alcohols in an ionic liquid by using an ion-supported hypervalent iodine (III) reagent. *Angew. Chem., Int. Ed.*, **2005**, *44*, 952–955.

50. Pham, P. D., Vitz, J., Chamignon, C., Martel, A., and Legoupy, S. Stille cross-coupling reactions with tin reagents supported on ionic liquids. *Eur. J. Org. Chem.*, **2009**, *19*, 3249–3257.

51. Hullio, A. A., Mastoi, G. M., and Khan, K. M. First multipurpose task specific ionic liquids: Designing and synthesis of novel dimethyl formamide-like ionic liquid and its application as a green solvent alternative to dimethyl formamide dependent reactions. *Asian J. Chem.*, **2011**, *23*, 5411–5418.

2 Ionic Liquid-Supported Organocatalysts for Asymmetric Organic Synthesis

Allan D. Headley

CONTENTS

2.1 INTRODUCTION AND BACKGROUND

The use of specific catalysts for the synthesis of desired asymmetric compounds has generated tremendous interest from an academic, as well as industrial perspective, specifically in the synthesis of pharmaceutical compounds. There are several factors that must be carefully considered in order to determine appropriate solvents, reaction conditions, and especially the selection of suitable catalysts for asymmetric reactions. The factors that must be considered for the selection of a suitable catalyst include: catalyst stability; recyclability of the catalyst and, hence, environmental concerns; cost involved in catalyst synthesis; and, equally important, the efficiency of the catalyst to transform reactants to the desired asymmetric products. Catalysts that are widely used for organic transformations, especially in asymmetric synthesis, can be divided into three general categories: (a) biocatalysts, (b) metal ion catalysts, and (c) organocatalysts. Biocatalysts are typically very efficient, but their isolation and, hence, purification after the completion of a reaction are often difficult and very laborious. In addition, the substrate scope that these catalysts can efficiently catalyze is typically narrow. Metal ion catalysts have proven to be very useful for the synthesis of various asymmetric compounds, but they are often expensive, moisture- and air-sensitive, and they typically require very harsh reaction conditions for optimum efficiency. In addition, the metals used are typically toxic and are of environmental concern. On the other hand, there are many advantages for the use of organocatalysts. Organocatalysts typically have low molecular weights, their synthesis is typically low-cost and straightforward; most are easily made from readily available starting compounds, such as naturally occurring amino acids; they are typically stable in most solvents and air; and they do not involve the use of toxic metals. The field of organocatalysis can be divided loosely into the following categories: *N*-heterocyclic carbene,[1] ammonium enolate,[2] enamine,[3] iminium,[4] hydrogen bonding,[5] and strong Brønsted acid.[6] A major challenge faced in the use of organocatalysts, however, is that relatively high catalyst loading is required; typically, 10 mol%–30 mol% is required in order to complete transformations in reasonable timescales. In addition, there have been challenges associated with catalyst recovery and recyclability. Based on the development of more efficient organocatalysts over the years, the efficiencies and recovery of most organocatalysts have been greatly improved and most are applicable to a wide variety of fundamental organic transformations.[7]

Most organocatalyzed processes are typically carried out in organic solvents, and, as a result, there are challenges associated with the separation of the catalyst from the reaction mixture. There are obvious benefits for catalyst recovery, including catalyst recycling and environmental benefits. In order to address these challenges, the design and synthesis of efficient, easily recoverable, and reusable catalysts are active areas of research. Over the years, various approaches have been utilized in an effort to improve the efficiency of catalyst recovery. For example, aqueous[8] and fluorous biphase catalysis[9,10] have been utilized and reactions have been carried out in super-critical media.[11]

Another technique that was discovered to address the challenges associated with the use of conventional organocatalysts is that of attaching catalysts to supports in order to make them more efficient and at the same time recyclable. As a result, a new generation of supported catalysts, which were attached to various supports typically via a covalent bond, was developed.[12] A major advantage in the use of catalysts that are supported is that they are easily recovered for reuse. An added advantage of supported catalysts is that high catalyst loading is typically not required, compared to unsupported catalysts.[13] As a result of the potential advantages in the use of supported catalysts, much effort has focused on the design and development of new types of homogeneous-supported organocatalysts that are easily recyclable; require low catalyst loading; and are effective for a wide range of asymmetric reactions. Immobilized supports typically include silica and polymeric materials, but recently there has been growing interest in the use of ionic liquids as support.

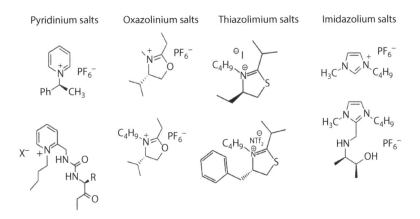

FIGURE 2.1   Examples of common types of ionic liquids.

2.2   IONIC LIQUIDS

In 1914, Paul Walden reported the first synthesis of the room temperature ionic liquid, ethylammonium nitrate.[14] Ever since that report, the field of ionic liquids has grown exponentially and, through this growth, there have been extensive environmental and technological benefits.[15] Growing interest in ionic liquids started with the discovery of a new class of ionic liquids based on alkyl-substituted imidazolium and pyridinium cations with halide or tetrahalogenoaluminate anions. Ionic liquids are generally defined as organic salts with melting points below some arbitrary temperature, such as 100°C. Some ionic liquids can maintain their liquid state at temperatures as high as 200°C.[16,17] Ionic liquids are non-flammable, lack measurable vapor pressure, and have high thermal and chemical stabilities. A number of different types of ionic liquids have been synthesized and examples are shown in Figure 2.1.

There are numerous applications of ionic liquids: they have been used as electrolytes for batteries,[18,19] in the production of aluminum,[20] and in the nuclear industry.[21] In addition, since they are polar compounds, they are perfectly suited to serve as solvents for polar compounds[22] and as solvents for a wide range of organic reactions.[23] They have been used in other areas of chemistry as well, including catalysis, synthesis, and gas absorption.[24] A major advantage in the use of ionic liquids is that their solubility in various reaction media can be controlled and easily fine-tuned by modifying the structures of their cations and anions. Similarly, variations of the structures of their cations or anions can result in modified properties. Owing to the property diversity of ionic liquids, they are ideal to serve as reusable homogenous supports for catalysts.[25]

2.3   MAGNETIC IONIC LIQUIDS

A type of support that has been used in catalysis is magnetic nanoparticles (MNPs). A major advantage in the use of MNPs is that they can be easily removed from the reaction mixture due to their magnetic properties.[26] Ionic liquids that have magnetic

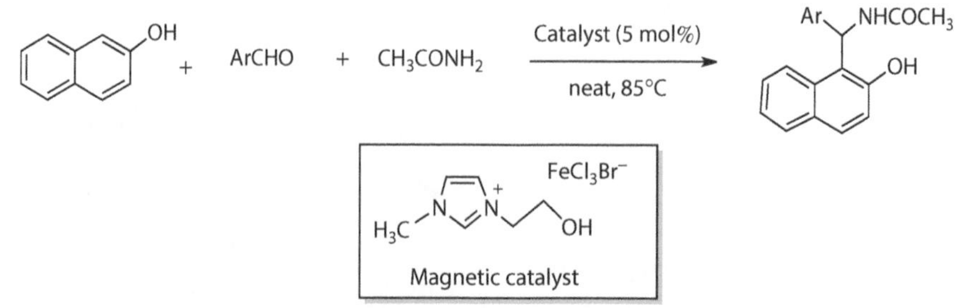

FIGURE 2.2 An example of ionic liquid with magnetic properties.

properties form a unique category of compounds that have gained widespread application.[27] Recently, the ionic liquid shown in Figure 2.2 was used for the synthesis of 1-amidoalkyl-2-napthols. For these reactions, short reaction times and high yields (>80%) were obtained.[28] A very attractive feature of this ionic liquid is that it could be separated from the reaction medium by means of its magnetic property.

Other ionic liquids with magnetic properties have been synthesized and evaluated, for example, the Warner group synthesized magnetic ionic liquids from amino acids.[29] For these magnetic ionic liquids, the Warner group investigated their chiral discrimination properties, but they were not used as catalysts for enantioselective reactions. Owing to their magnetic property, this category of ionic liquids should serve as ideal supports.

2.4   IONIC LIQUIDS AS SOLVENTS FOR ORGANIC REACTIONS

Ionic liquids are ideal as reusable reaction media for asymmetric organic synthesis since they provide a polar medium in which reactions can take place.[30] Compared to conventional solvents, the ionic liquid [BMIM][BF$_4$] was found to be an effective reaction medium for the asymmetric aminohalogenation of functionalized alkenes (Figure 2.3).[31] A similar observation was made for the metal-catalyzed regio- and stereoselective aminohalogenation of cinnamic esters.[32]

More recently, Ishikawa et al. examined the stereoselectivity of the Diels-Alder reaction in imidazolium ionic liquids, which contain various delocalized anions (Figure 2.4).[33] It was shown that the acidity of the hydrogens on the imidazolium

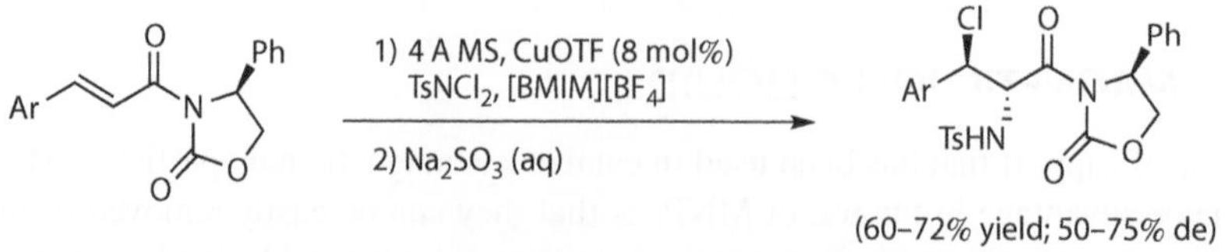

(60–72% yield; 50–75% de)

FIGURE 2.3 Asymmetric aminohalogenation of functionalized alkenes in an ionic liquid.

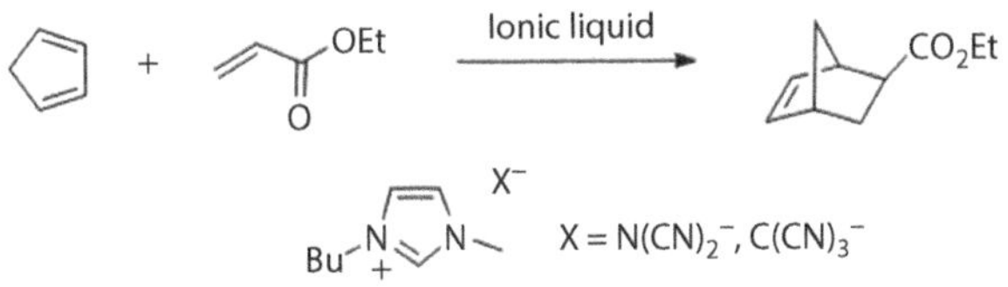

FIGURE 2.4   Diels-Alder reaction carried out in imidazolium ionic liquid.

ring interacted with the anions of the ionic liquids, which in turn played a role in the solvation and stability of the solutes for this reaction. High endo/exo and percent conversion were obtained in these imidazolium media.

In another example, the synthesis of cellulose-2,3-bis(3,5-dimethylphenlycarbamate), which are important stationary phases that are used for the separation of enantiomers, was carried out in the ionic liquid, 1-allyl-3-methylimidazolium chloride.[34]

In an effort to produce a more environmentally friendly category of ionic liquids, the incorporation of various types of anions has been envisioned. Recently, the artificial sweetener, saccharin, was used as the anion for a new type of ionic liquid (Figure 2.5). Saccharin was chosen due to the low toxicity and the resulting ionic liquid was a viscous liquid, which could be used as a solvent medium for reactions. For the reaction shown in Figure 2.5, the ionic liquid plays an essential role in substrate activation by directly facilitating the "solvent-free" Michael addition of the thiol to ferrocenyl enone.[35]

Davis et al. were one of the first researchers to report that ionic liquid derivatives could not only be used as solvents for various reactions, but they can be used to catalyze reactions.[36] Figure 2.6 shows one of the earlier ionic liquids used to catalyze the esterification reaction of ethanol and acetic acid.[37]

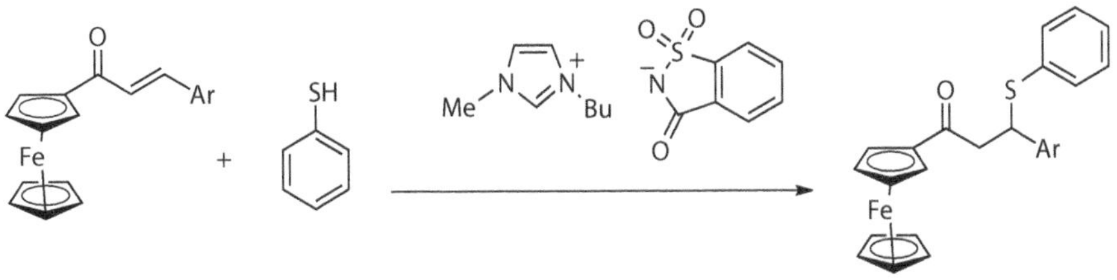

FIGURE 2.5   Saccharin ionic liquid for the Michael addition of thiol to ferrocenyl enone.

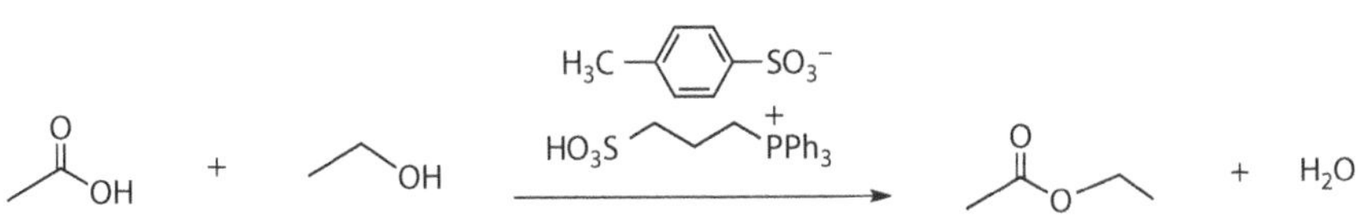

FIGURE 2.6   Ionic liquid used for the esterification reaction of ethanol and acetic acid.

2.5   CHIRAL IONIC LIQUIDS

Ionic liquids have not only been used as solvents for different reactions, but as co-solvents and co-catalysts for various organic transformations.[38] Ionic liquids that contain chiral groups are another category that has been synthesized. One such category, which contains two stereogenic centers, was synthesized by the Headley group and is shown in Figure 2.7.[39] The synthesis of these chiral ionic liquids is easy and straightforward.

Another class of chiral ionic liquid designed and synthesized contained a fused-ring structure.[40] Even though these ionic liquids were not tested as chiral solvents for asymmetric reactions, the chiral environment that they present should serve to positively affect the outcomes of such reactions. Other examples of chiral ionic liquids are shown in Figure 2.8, these were synthesized from natural amino acids as the starting material. Chiral ILs **1** were shown to be stable compounds, with low melting points;[41] chiral IL **2** was used to catalyze the asymmetric Michael addition of aldehydes to nitrostyrenes; and for these reactions, enantioselectivity (*ee*) up to 99% and *dr* (*syn/anti*) up to 97:3 were obtained.[42] Chiral IL **3** was synthesized from both

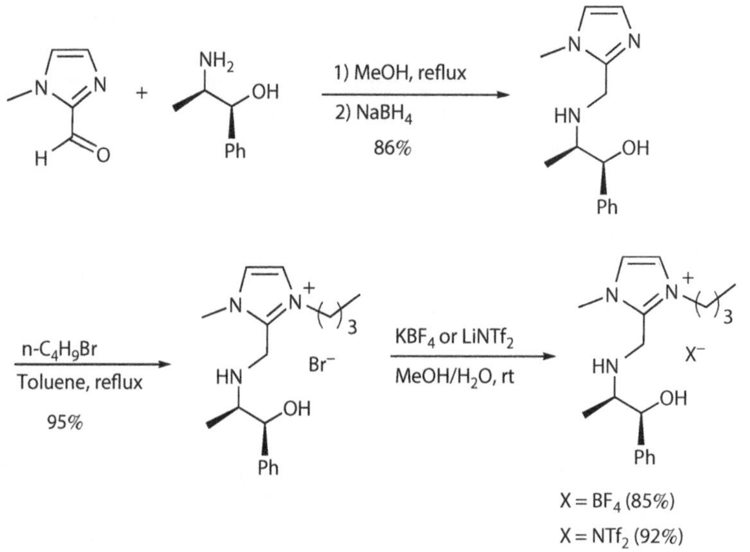

FIGURE 2.7   Synthesis of ionic liquid with two sterogenic centers.

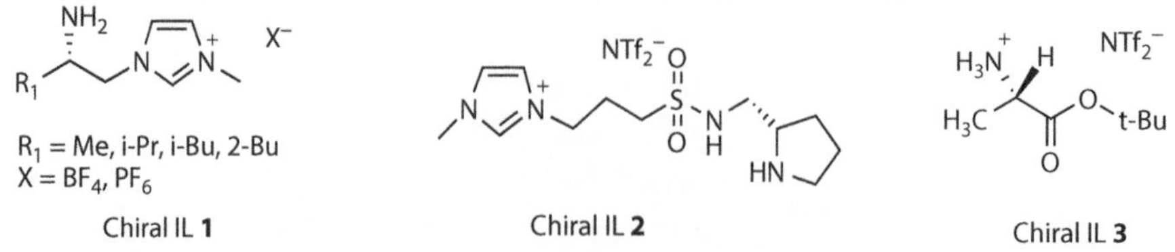

FIGURE 2.8   Chiral ionic liquids synthesized from amino acids.

L and D alanine and is a stable compound and a liquid at room temperature.[43] Even though chiral IL **3** was not tested as solvent or catalyst for enantioselective reactions, it demonstrated enantiomeric recognition abilities.

2.6 SILICA AND POLYMER-SUPPORTED IONIC LIQUIDS

Anchoring catalysts to supports has many advantages, including the reduction of the amount of catalyst used, and equally important, the ease of catalyst recovery and recyclability. Merrifield first proposed the use of insoluble polymer resin as a support.[44] In an effort to improve the efficiency and recovery of these ionic liquids, especially when used as catalysts, researchers have anchored ionic liquids to solid supports, such as silica gel, as shown in Figure 2.9.[45] Catalyst **4** was used to catalyze a hydroformylation reaction;[46] for the reactions studied, it was shown that there was an improvement in the efficiency of the catalyst, compared to stand alone catalysts, and also that the catalyst could be easily recycled.

Another type of covalently anchored ionic liquid catalyst is shown in Figure 2.10, in which the ionic liquid is linked onto a highly cross-linked polystyrene resin. These were used for the synthesis of cyclic carbonates via cycloaddition reactions of CO_2 with epoxides. High yields, up to 99%, and excellent selectivity, up to 99%, were achieved using these supported catalysts.[47] In addition, it was shown that catalyst **5** could be reused for as many as six times without loss of catalytic activity.

In 2012, a series of polyvinylidene chloride-immobilized ionic liquids were synthesized and were derived from 4-dimethylaminopyridine. These catalysts were used in conjunction with L-proline and provided an efficient method for the asymmetric catalysis of the asymmetric aldol reactions between cyclohexanone and aromatic aldehydes in the presence of water. High yields of up to 99%, as well as diastereoselectivities of up to 6:94, and excellent enantioselectivities of up to 98% were obtained.[48] In addition, these catalytic systems could be recycled and reused up to seven times (Figure 2.11).

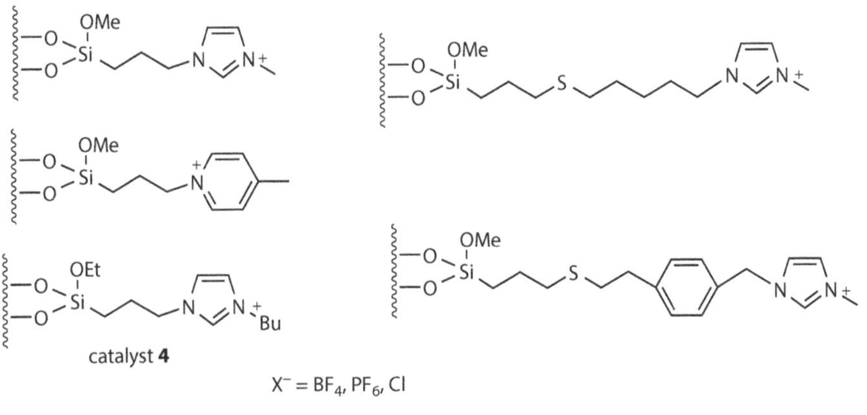

catalyst **4**

$X^- = BF_4, PF_6, Cl$

FIGURE 2.9 Various silica-gel-immobilized ionic liquid catalysts.

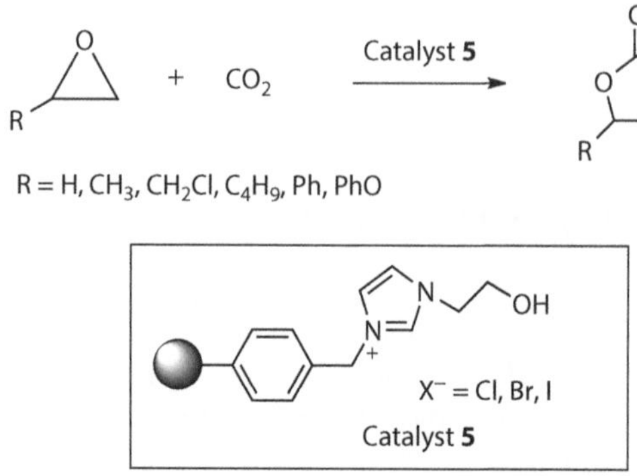

$R = H, CH_3, CH_2Cl, C_4H_9, Ph, PhO$

$X^- = Cl, Br, I$

Catalyst **5**

FIGURE 2.10   Covalently anchored onto a highly cross-linked polystyrene resin.

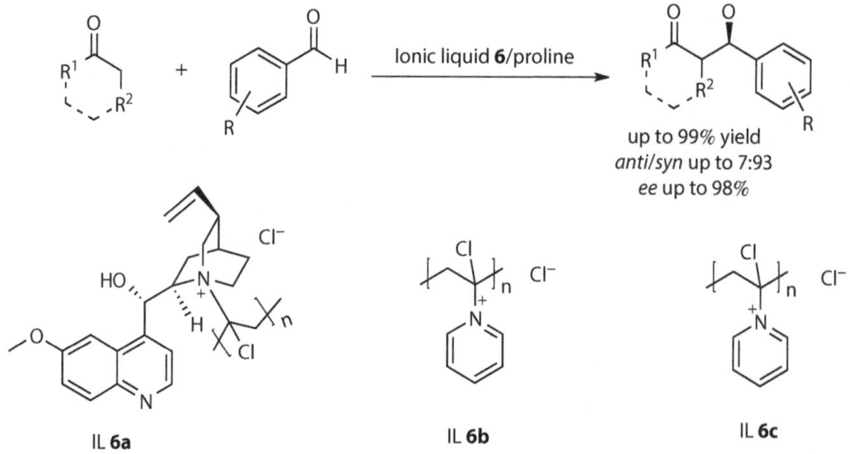

Ionic liquid **6**/proline

up to 99% yield
anti/syn up to 7:93
ee up to 98%

IL **6a**   IL **6b**   IL **6c**

FIGURE 2.11   Asymmetric aldol reactions catalyzed by polyvinylidene chloride-immobilized ionic liquids/proline.

2.7   ORGANOCATALYSTS SUPPORTED ON IONIC LIQUIDS

In the previous sections, it was shown that ionic liquids have been used as solvents and as catalysts for various reactions by simply adding them to the reaction media. An alternate approach is the use of ionic liquids, as we have seen in Section 2.6, as supports for catalysts: ionic liquid-supported (ILS) organocatalysts.[49] The use of ionic liquids to generate homogeneous-immobilized organocatalysts for organic synthesis is one of the preeminent achievements in the field in recent history.[50,51] This design involves covalently anchoring the catalyst to ionic liquids, which is illustrated in Figure 2.12.

Compared to other solid-support and fluorous catalysts, ionic liquid-supported organocatalysts have several advantages: (1) they are not only effective catalysts

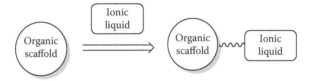

FIGURE 2.12   Basic design of ionic liquid-supported (ILS) organocatalysts.

for asymmetric homogeneous reactions, but can be easily recycled based on their solubility in different reaction media; (2) their high polarity enhances reaction rates synergistically[52]; (3) they offer better loading capacity than other solid supports; and (4) they are economical, compared to most other supported organocatalysts that are presently being used. A major advantage in the use of ionic liquid-supported catalysts is that when they are used to induce the outcomes of organic reactions, they can be easily separated from the organic phase. After the reaction is complete, a less polar solvent, such as diethyl ether, is typically added and the ionic liquid-immobilized catalyst is easily isolated due to solubility differences. As a result, these organocatalysts can be regenerated for additional use.

2.8   CHIRAL ORGANOCATALYSTS SUPPORTED BY IONIC LIQUIDS

Over the years, there have been different types of ionic liquids used as supports for organocatalysts. Also, there have been numerous different types of chiral scaffolds used, but amines, and specifically, proline derivatives appear to be the category that is most widely used. Ever since it was demonstrated that proline could be used to catalyze the asymmetric aldol reaction, it has been one of the most widely used scaffolds for organocatalysts.[53]

2.8.1   PYRROLIDINE ORGANOCATALYSTS

The use of secondary amines for the catalysis of carbonyl type reactions via enamine and iminium ion intermediates has emerged as a powerful synthetic method for transforming carbonyl containing compounds into asymmetric compounds. Among the various classes of aminocatalysts used, chiral cyclic secondary amines and, in particular, (S)-proline[54] and its derivatives,[55] as well as MacMillan's imidazolidinones have proven to be the most effective.[56] These type organocatalysts have been shown to be very effective for asymmetric transformations in organic synthesis to produce high yields, high enantioselectivities, and diastereoseletivities.[57] (S) proline has been used to catalyze various reactions, including aldol reaction[58,59] and the Mannich reaction.[60] Since proline was shown to be an effective asymmetric catalyst, various derivatives have been developed and some examples are shown in Figure 2.13.[61]

As proline gained widespread popularity for the catalysis of various reactions, more robust variations of proline derivatives have emerged over the years, as shown in Figure 2.14. These have been used to catalyze the enantioselective aldol and Michael reactions.[62]

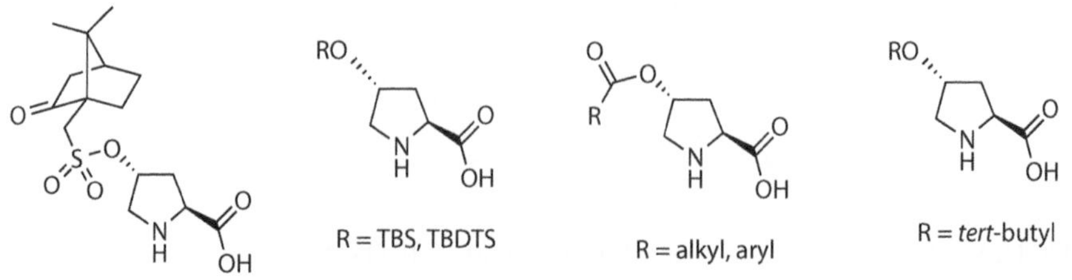

FIGURE 2.13   Examples of proline organocatalyst derivatives.

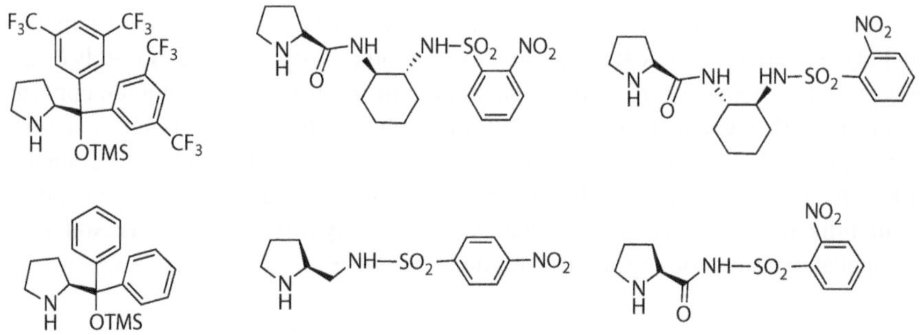

FIGURE 2.14   Examples of organocatalysts derived from pyrrolidine.

2.8.2   PYRROLIDINE IMIDAZOLIUM-IMMOBILIZED ORGANOCATALYSTS

As the use of different types pyrrolidine organocatalysts began to gain widespread use, they were also being anchored to ionic liquids and successfully used to catalyze various essential carbon-carbon forming reactions. The catalysis of the Michael reaction using catalyst 7 is shown in Figure 2.15.[63]

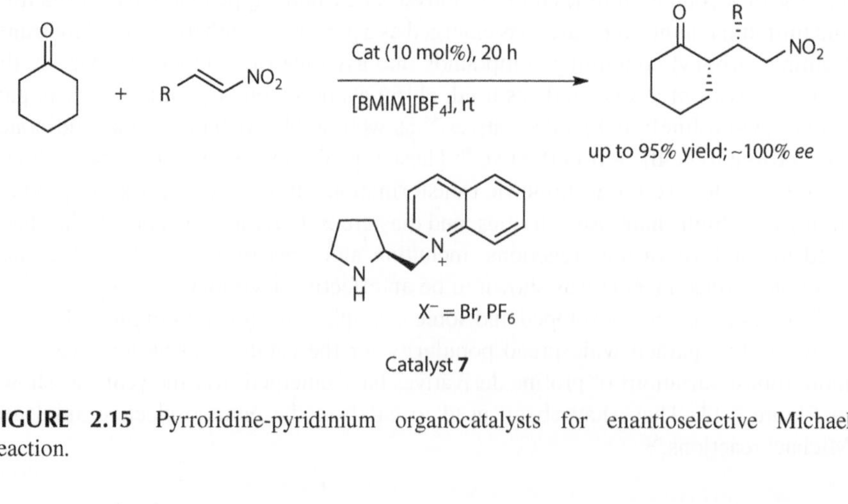

Catalyst **7**

FIGURE 2.15   Pyrrolidine-pyridinium organocatalysts for enantioselective Michael reaction.

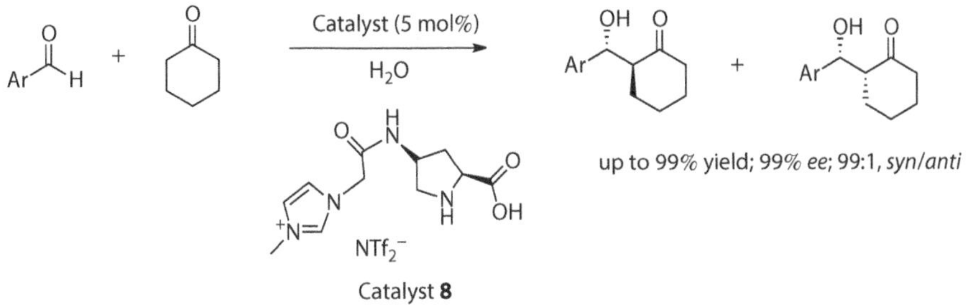

FIGURE 2.16 Ionic liquid-tagged proline catalysts used for the asymmetric aldol reaction.

Recently, a new type of ionic liquid-tagged proline catalyst (catalyst **8**, Figure 2.16) was developed and proven to be very efficient at catalyzing the asymmetric aldol reaction.[64] The synthesis of this catalyst is straightforward in which protected hydroxyl proline is used as a starting material. Of interest, this reaction could be carried out in water, and it was also effective in different ionic liquids as solvents, including [Bmim][NTf$_2$].

In 2007, the surfactant-type IL-immobilized organocatalyzed asymmetric Michael reaction of nitrostyrene and cyclohexanone in water was reported, in which catalysts **9–12** were used.[65] These catalysts are easily synthesized by exchange of the anion with a surfactant sulfonate anion by anion metathesis or neutralization. Catalyst **9** (X = DS) was found to be the optimal catalyst, affording the desired product in 93% yield, 97:3 *syn/anti*, and 97% *ee* (Figure 2.17).

In 2008, Zlotin et al.[66] designed and synthesized a new type of IL-immobilized proline catalyst, catalyst **12**, which contains a long-chain hydrocarbon group (Figure 2.18). Owing to the amphiphilic nature of this category of catalysts, excellent diastereoselectivities (*anti/syn* ratio up to 97:3) and enantioselectivities (up to 99% *ee*) of the aldol reaction between cyclic ketones and aromatic aldehydes in water were achieved. In addition, replacement of the long-chain hydrocarbon group with

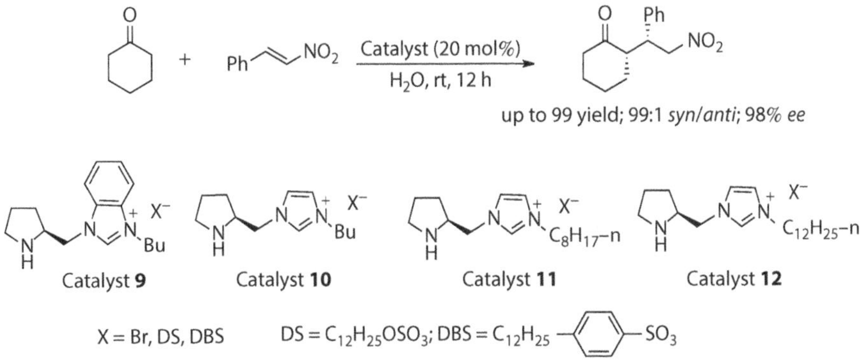

FIGURE 2.17 Surfactant-type IL-supported organocatalysts for the asymmetric Michael reaction.

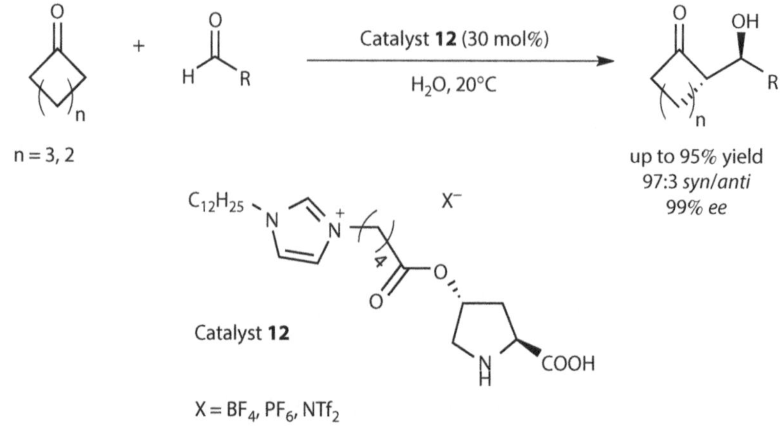

Catalyst 12

X = BF$_4$, PF$_6$, NTf$_2$

FIGURE 2.18   ILS pyrrolidinium organocatalysts used for the asymmetric aldol reaction.

a shorter chain resulted in the catalyst being inactive in an aqueous environment, highlighting the importance of the long-chain hydrocarbon group in the catalyst for catalytic activity.

In 2008, Lombardo et al.[67] utilized the ionic tag concept and attached the imidazolium ionic liquid moiety into *trans*-4-hydroxy-L-proline to prepare the IL-immobilized organocatalyst **13** (Figure 2.19), which was applied to the enantioselective aldol reactions; high enantioselectivity; and diastereoselectivity were achieved under aqueous biphasic conditions.

Lombardo et al.[68] prepared another type of novel water-soluble organocatalysts **14a** (Figure 2.20) and used it as a catalyst for the aldol reaction of cyclic ketones and aromatic aldehydes. It was found that the hydrophilic portions of the catalysts made them water soluble, and the highly hydrophobic anion allowed the catalyst to be transported into the organic phase of the biphasic ketone-water system where the aldol reaction takes place. As a result, high enantioselectivities and *anti*-diastereoselectivities (up to ≤99%) were obtained.

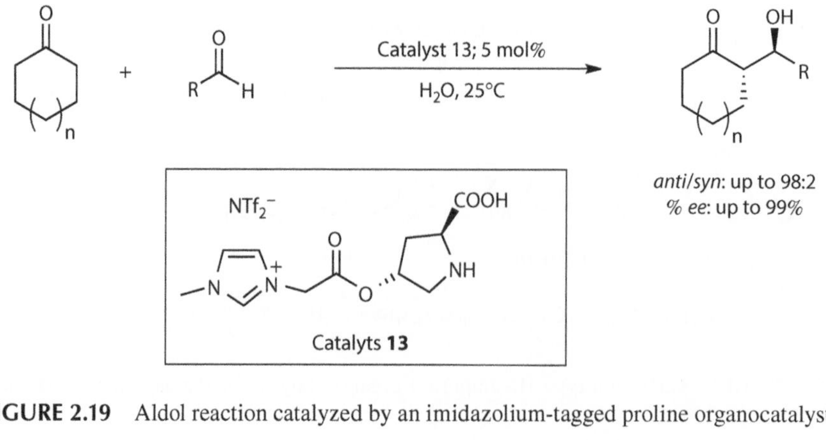

Catalyts 13

FIGURE 2.19   Aldol reaction catalyzed by an imidazolium-tagged proline organocatalyst.

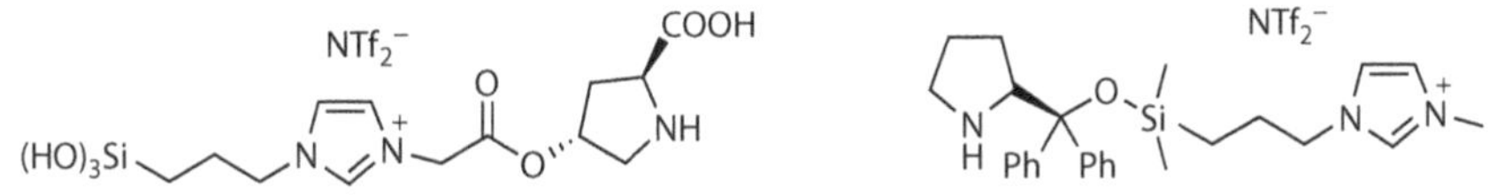

Catalyst **14a** Catalyst **14b**

FIGURE 2.20 Different kinds of ionic liquid-immobilized organocatalysts.

Later, that same group reported an ion-tagged diphenylprolinol silyl ether **14b** (Figure 2.20), which was used to catalyze the Michael addition of aliphatic aldehydes to nitroalkenes.[69] The authors observed that catalyst loading could be decreased to 0.25 mol%–5 mol% in the presence of only a slight excess of aldehyde (1.2–2 equivalent) to achieve a high 99.5% *ee*. Both CH_2Cl_2 and water could be used as the solvent for the reaction, but CH_2Cl_2 was a much better solvent compared to water for the reactions of propanal with aromatic nitroalkenes. Water was a better reaction medium, however, when less reactive aliphatic nitroalkenes or longer chain aldehydes were used. This phenomenon can be ascribed to poor solubility in water when the high molecular weight aldehydes are used. When sterically hindered nitroalkenes or aldehydes were used, an acid as additive was necessary to ensure high conversions.

Other types of ILS organocatalysts developed by the Headley group are shown in Figure 2.21.[70] It was shown that **15c** had superior catalytic activity, compared to catalysts **15a** and **15b**. The justification given is that a much tighter transition state could be maintained by that catalyst, compared to the others.

2.8.3 Pyrrolidinium Ammonium-Immobilized Organocatalysts

A novel idea of using biorenewable non-toxic raw materials through a simple and green route to synthesize IL [Choline][Pro] **16** (Figure 2.22) was developed by Hu and co-workers.[71] They found that this catalyst could be used to catalyze the direct

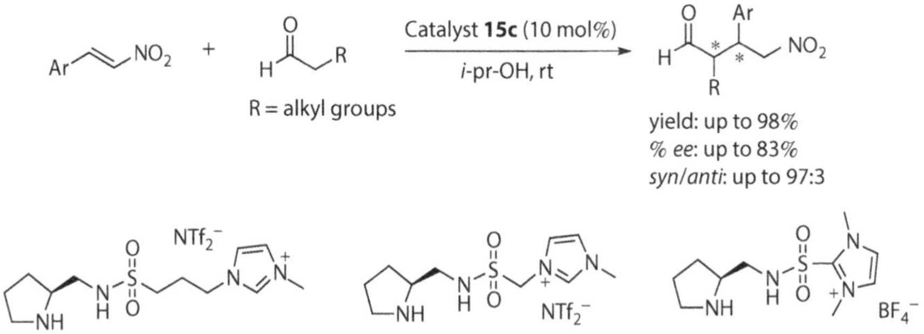

Catalyst **15a** Catalyst **15b** Catalyst **15c**

FIGURE 2.21 ILS organocatalysts used for the Michael asymmetric reactions.

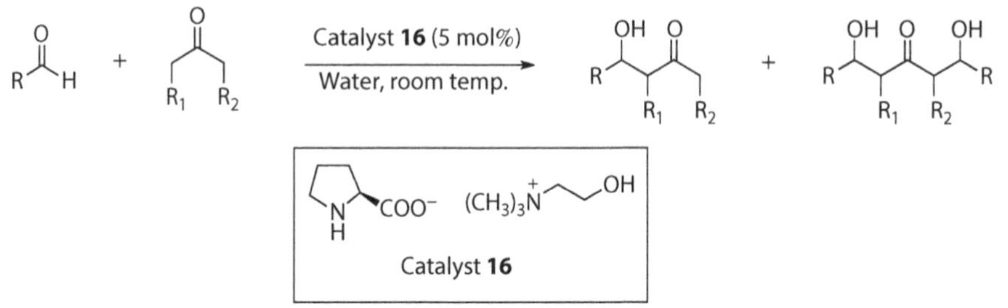

FIGURE 2.22 Direct aldol reaction of aromatic aldehydes and ketones.

aldol reaction between ketones and aromatic aldehydes efficiently in water with good yields. In addition, the authors observed that the reaction mixtures separated into an aqueous phase and an organic phase after the reaction was completed, and the catalyst is soluble in the aqueous phase. As a result, the catalyst could be recycled up to three times without obvious decrease in yields.

The chiral quaternary ammonium ionic liquid (catalyst **17**, Figure 2.23) was shown to be effective at catalyzing the asymmetric Michael addition reaction of ketones and aldehydes to nitroolefins. Excellent yields, enantioselectivities, and diastereoselectivities were obtained with this catalyst. In addition, the catalyst could be reused up to five times without a significant loss in catalytic activity.[72]

The pyrrolidine-based organocatalyst **18** was another quaternary ammonium type ionic liquid-immobilized catalyst developed by the Headley group[73] (Figure 2.24). They discovered that catalyst **18** in the presence of a newly developed ionic liquid-supported benzoic acid as co-catalyst was shown to be highly effective in the asymmetric Michael reaction in water.[74] For the Michael reactions studied, excellent diastereo- and enantioselectivities were obtained using only 5 mol% of catalyst loading. Another advantage of the system was that after extraction of the product, the catalyst remained in the aqueous layer and could be reused in subsequent reactions for ten times without significant loss of enantioselectivity.

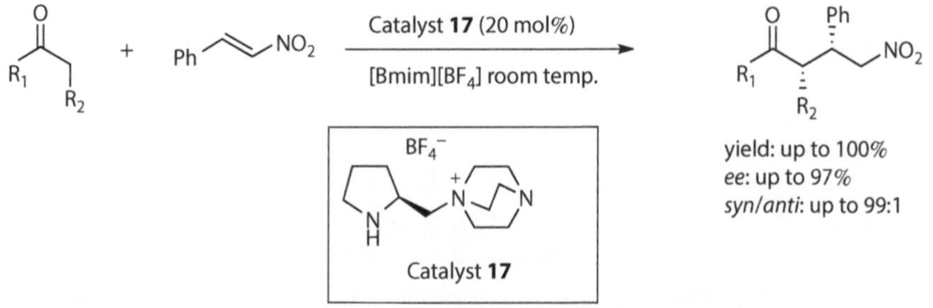

FIGURE 2.23 Chiral quaternary ammonium ionic liquid-immobilized organocatalyst.

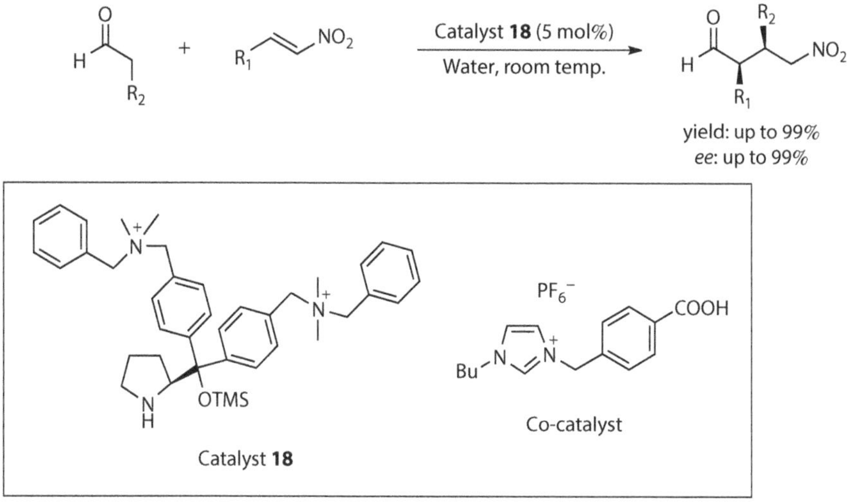

FIGURE 2.24 Organocatalytic asymmetric Michael reaction using aldehydes and nitroalkenes.

2.8.4 PYRIDINIUM ION-IMMOBILIZED ORGANOCATALYSTS

A novel category of ionic liquids that contains both Lewis and Brønsted acid functionalities was synthesized and used to catalyze various Michael addition reactions.[75] It was shown that catalyst **19** was stable to air and water and could be used in aqueous solutions and gave excellent yields for the reactions studied (Figure 2.25).

The Headley group synthesized a novel type of pyrrolidine-based chiral pyridinium ionic liquid-supported organocatalyst.[76] Catalyst **20** was shown to be recyclable and highly efficient organocatalysts for the asymmetric Michael addition reactions. For the Michael reaction shown in Figure 2.26, catalyst **20** gave excellent stereochemical outcome (Figure 2.27).

Organocatalysts **21**, **22**, and **23** were investigated as recyclable oganocatalysts for the aldol reaction.[77] It was found that organocatalysts **23** could effectively catalyze the aldol reaction in water with high diastereoselectivity (up to 98:2) and

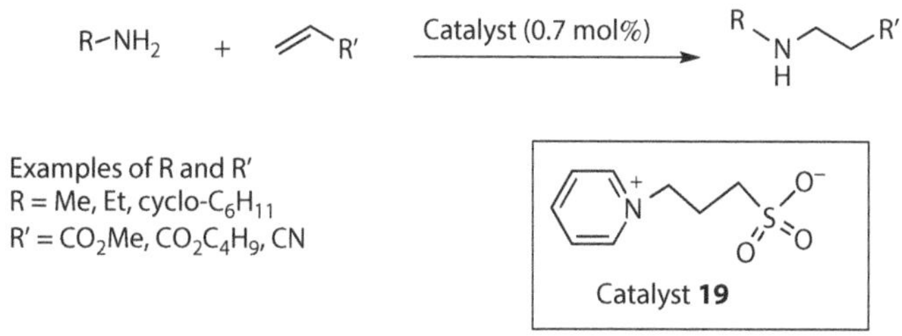

Examples of R and R'
R = Me, Et, cyclo-C_6H_{11}
R' = CO_2Me, $CO_2C_4H_9$, CN

Catalyst **19**

FIGURE 2.25 Michael conjugate addition using catalyst **24**.

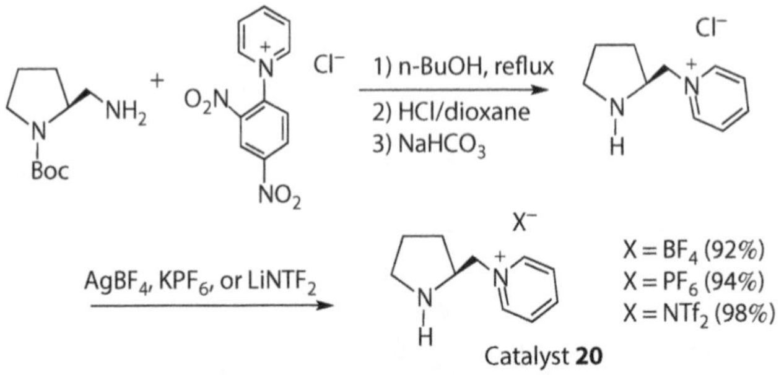

FIGURE 2.26 Synthesis of pyridinium ILS organocatalysts.

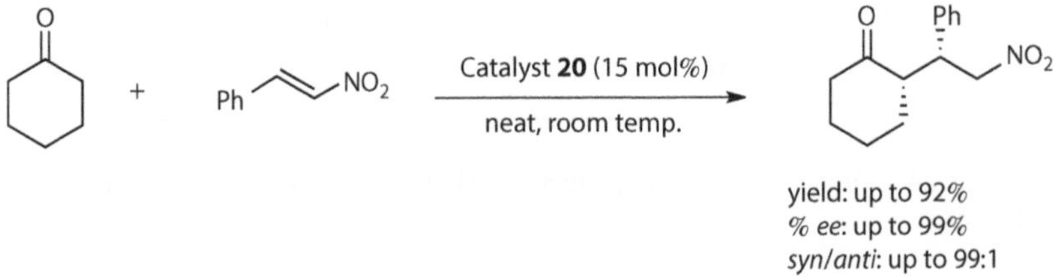

FIGURE 2.27 Pyridinium ILS organocatalysts for the Michael reaction.

enantioselectivity (up to 99%). In addition, it was discovered that organocatalysts **21** could be reused up to eight times without significant loss on catalytic activity.

2.8.5 OTHER HETEROCYCLIC IONIC LIQUID-IMMOBILIZED ORGANOCATALYSTS

Two novel thiazolium-based organocatalysts were prepared and used for the etherification reaction.[78,79] Figure 2.28 shows the etherification reaction involving catalyst **22**, it was demonstrated that the methyl group had a negative effect on the etherification

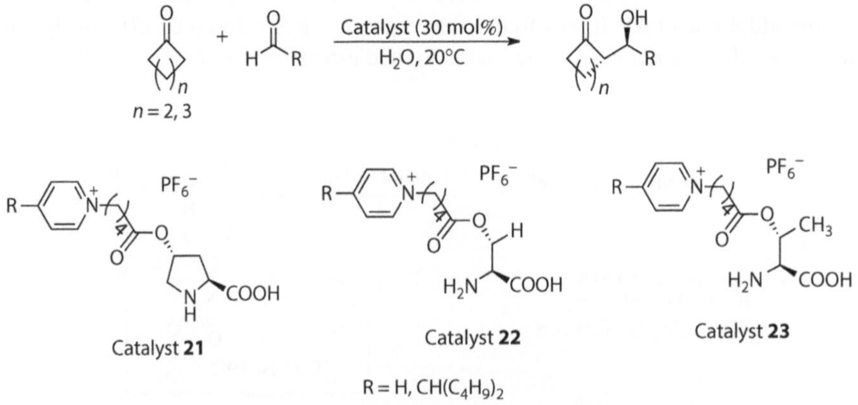

FIGURE 2.28 Pyridinium ILS organocatalysts used for the aldol reaction.

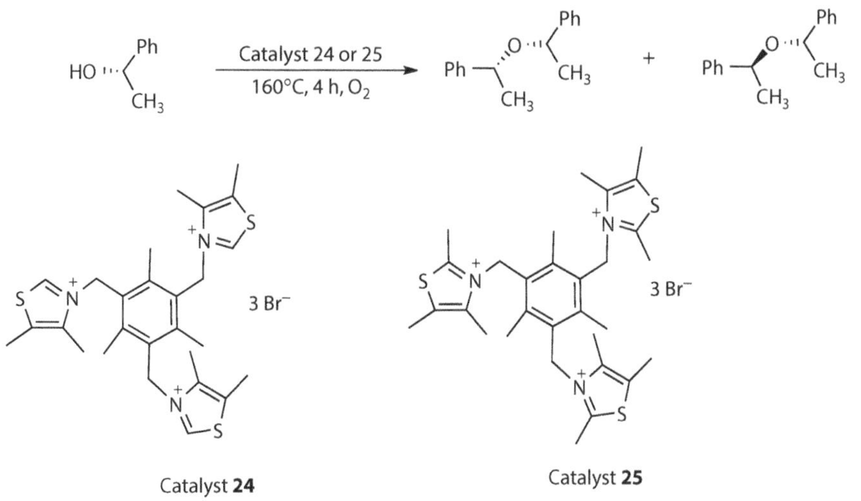

FIGURE 2.29   Etherification reaction of (*S*)-1-phenylethanol using thiazolium organocatalysts.

reaction. The percent conversion using catalyst **24** by far exceeded that using catalyst **25**, which has a methyl group in the two positions. For catalyst **24**, high conversion and selectivities up to 92% and 82%, respectively, were obtained (Figure 2.29).[80]

2.9   RECYCLABILITY OF IONIC LIQUID-SUPPORTED ORGANOCATALYSTS

A major challenge faced in the use of catalysts, especially organocatalysts, is catalyst recovery and reuse. As pointed out earlier, this aspect has been addressed by covalently bonding the organocatalysts to various supports, such as silica and resins. Organocatalysts that are bonded to ionic liquid supports through covalent bonds offer unique advantages owing to the exceptional properties of the ionic liquid support. An ideal situation exists when catalysts are used to catalyze reactions, especially asymmetric reactions, and then recycled for additional use; Figure 2.30 illustrates the strategy for catalyst recovery and reuse.

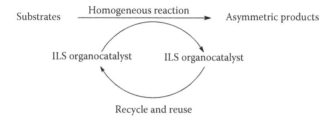

FIGURE 2.30   Strategy for recovery and reuse of ILS organocatalysts.

The reality is that after the reaction is complete, the catalyst, along with products and unwanted side products and unreacted starting compounds, also exist in the reaction mixture. Thus, the major challenge becomes the effective separation of the catalyst from the reaction mixture, so that it can be reused for additional reactions.

The stability of the catalyst plays a major role in it being able to be effective after being recovered. ILS organocatalysts have proven to be extremely stable under various reaction conditions and temperature ranges. An important aspect of consideration for catalyst recovery is catalyst leaching. Owing to the organic nature of these catalysts, in addition to their ionic properties, catalyst leaching[81] has been a major challenge for the recyclability of these type organocatalysts. Leaching typically occurs when the catalyst crosses boundaries. For ILS organocatalysts, leaching is possible typically during catalyst separation, in which the catalyst remains in the reaction mixture phase and not extracted effectively into the aqueous phase. Even with this challenge, there are many reports of successful recycling of ILS organocatalysts using this method.

The process that is typically used for ILS organocatalysts isolation is illustrated in Figure 2.31. The difference is solubility of ILS ionic liquids, compared to that of the organic products in solvents of different polarity, is utilized for effective separation. The typical method used to evaluate the effectiveness of catalyst recovery and reuse is to determine percent yield and stereochemical outcome after each cycle. Michael reaction of isovaleraldehyde with *trans*-β-Nitrostyrene, chiral IL **2** (Figure 2.8) could be effectively used after five cycles; the yield went from 97% to 86%, the % *ee* from 69 to 68, and *syn/anti* went from 97/3 to 96/4. The surfactant-type IL, catalyst **9** (Figure 2.17), when used for the asymmetric Michael reaction of nitrostyrene and cyclohexanone in water, catalyst **9** could be recycled at least five times in water

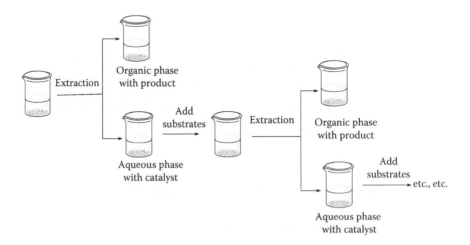

FIGURE 2.31 Illustration of catalyst recovery and reuse process.

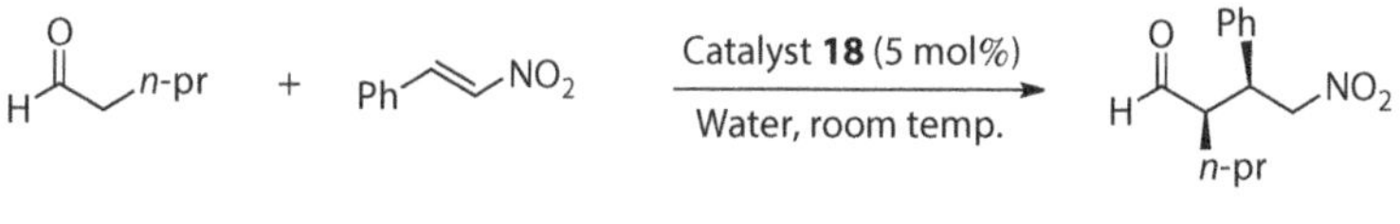

FIGURE 2.32 Catalyst **18**, along with ILS benzoic acid as co-catalyst, for recyclability studies.

without significant loss of activities and selectivities. In 2009, Zlotin et al.[82] reported that proline, serine, and threonine-derived chiral-supported ionic liquid organocatalysts could be effectively recycled for the aldol reactions in which various ketone donors with aromatic aldehydes are used.

For catalyst **18** (Figure 2.24), it was demonstrated that the catalyst could be used up to ten times without significant loss in enantioselectivity or diastereoselectivity for the Michael reaction shown in Figure 2.32. It was reported that the *syn:anti* went from 97:3 to 86:14 after the ten cycles, and for the same number of cycles, the enantioselectivity went from >99% to 98%. Another ionic liquid-supported organocatalyst was developed by anchoring it to MNP coated with silica, which was used for the aldol reaction (Figure 2.33).

A major advantage of this organocatalyst is that it could be recovered from the reaction mixture by simple magnet extraction and could be reused effectively up to five times for the aldol reaction. It was also shown that there was no substantial leaching of the catalyst on extraction and that it was a stable and robust catalyst.[83] In addition, when it was used as a catalyst for the reaction between cyclohexanone and 2-nitrobenzaldehyde, high yields (up to 92%), diastereoselectivity (*dr*; 88/12) and enantioselectivity (85%) were obtained using 10 mol% of a catalyst. Another method that was recently used for the extraction of organocatalysts is to take advantage of the magnetic properties of MNP. A hydrophobic magnetic room temperature ionic liquid, trihexyltetradecylphosphonium tetrachloroferrate (III) ([$3C_6PC_{14}$][$FeCl_4$]), was synthesized from trihexyltetradecylphosphonium chloride and $FeCl_3 \cdot 6H_2O$. This magnetic room temperature ionic liquid has a novel magnetic ionic liquid design and was investigated as a possible separation agent. The separation strategy is based on the strong paramagnetism of [$3C_6PC_{14}$][$FeCl_4$], which responds to an external neodymium magnet.[84]

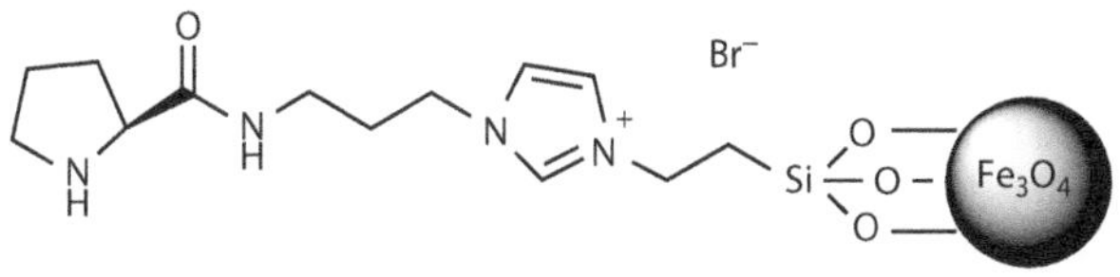

FIGURE 2.33 Ionic liquid organocatalysts supported by MNP coated with silica.

2.10 CONCLUSIONS

Catalysts have played a major role in chemistry over the years, but only recently has there been an increased interest in the use of supported organocatalysts, especially ionic liquid-supported organocatalysts. Their use has become essential in organic synthesis, especially in the asymmetric synthetic; they are very effective for a wide range of asymmetric reactions, including many cornerstone carbon-carbon forming reactions in organic chemistry. A major advantage in the use of these supported organocatalysts is that they are typically readily recyclable, which assists in meeting the environmental challenges faced by industry and research. The main advantage of these types of catalysts is that they are environmentally benign and eco-friendly, and they can be easily recovered and reused many times without loss of activity and selectivity. Owing to the ionic nature of these compounds, they are very effective in aqueous media as a result, can be separated easily from organic solvents, as well as aqueous phases by simply adjusting the polarity of the media. In addition, the amount of ILS catalyst used is typically less than other types of catalysts. There is a bright future for further design of more efficient and practical ionic liquid-immobilized organocatalysts that are functional under mild catalytic conditions and for a wider range of reactions.

REFERENCES

1. (a) Biju, A. T., Kuhl, N., and Glorius, F. Extending NHC-catalysis: Coupling aldehydes with unconventional reaction partners, *Acc. Chem. Res.* 2011, 44, 1182–1195. (b) Patil, N. Merging metal and N-heterocyclic carbene catalysis: On the way to discovering enantioselective organic transformations, *Angew. Chem. Int. Ed.* 2011, 50, 1759–1761.
2. Gaunt, M. J. and Johansson, C. C. C. Recent developments in the use of catalytic asymmetric ammonium enolates in chemical synthesis, *Chem. Rev.* 2007, 107, 5596–5605.
3. Mukherjee, S., Yang, J. W., Hoffmann, S., and List, B. Asymmetric enamine catalysis, *Chem. Rev.* 2007, 107, 5471–5569.
4. (a) Erkkila, A., Majander, I., and Pihko, P. M. Iminium catalysis, *Chem. Rev.* 2007, 107, 5416–5470. (b) Lelais, G. and MacMillan, D. W. C. Modern strategies in organic catalysis: The advent and development of iminium activation, *Aldrichimica Acta* 2006, 39, 79–87.
5. (a) Doyle, A. G. and Jacobsen, E. N. Small-molecule H-bond donors in asymmetric catalysis, *Chem. Rev.* 2007, 107, 5713–5743. (b) Taylor, M. S. and Jacobsen, E. N. Asymmetric catalysis by chiral hydrogen-bond donors, *Angew. Chem. Int. Ed.* 2006, 45, 1520–1543.
6. (a) Terada, M. Chiral phosphoric acids as versatile catalysts for enantioselective carbon-carbon bond forming reactions, *Synthesis* 2010, 1929–1982. (b) Akiyama, T. Stronger Brønsted acids, *Chem. Rev.* 2007, 107, 5744–5758.
7. List, B (Guest Editor), *Chem. Rev.* 2007, 107(12), 5413 Special issue on Organocatalysis: 5883, Special issue on Organocatalysis.
8. Arhancet, J. P., Davis, M. E., Merola, J. S., and Hanson, B. E. Hydroformylation by supported aqueous-phase catalysis: A new class of heterogeneous catalysts, *Nature* 1989, 339, 454–455.
9. Corlils, B. and Kuntz, E. G. *Aqueous-Phase Organometallic Catalysis*, Cornils, B. and Herrmann, W. A., (Eds.), Wiley-VCH, Weinheim, Germany, 1988.

10. Horvath, I. T. and Rabai, J. Facile catalyst separation without water: Fluorous biphase hydroformylation of olefins, *Science* 1994, 266, 72–75.

11. Jessop, P. G., Ikariva, T., and Movori, R. Homogeneous catalysis in supercritical fluids, *Chem. Rev.* 1999, 99, 475–494.

12. Hartley, F. R. *Supported Metal Complexes. A New Generation of Catalysis* Kluwer: Dordrecht, the Netherlands, 1985.

13. For some selected examples, see: (a) Enders, D. and Seki, A. Proline-catalyzed enantioselective Michael additions of ketones to nitrostyrene, *Synlett* 2002, 26–28. (b) Mase, N., Thayumanavan, R., Tanaka, F., and Barbas III, C. F. Direct asymmetric organocatalytic Michael reactions of α,α-disubstituted aldehydes with β-Nitrostyrenes for the synthesis of quaternary carbon-containing products, *Org. Lett.* 2004, 6, 2527–2530. (c) Wang, W., Wang, J., and Li, H. Direct, highly enantioselective pyrrolidine sulfonamide catalyzed Michael addition reactions of aldehydes to nitrostyrenes, *Angew. Chem. Int. Ed.* 2005, 44, 1369.

14. Walden, P. Molecular weights and electrical conductivity of several fused salts. *Bull. Acad. Imper. Sci.* (St. Petersburg.) 1914, 1800, 405–422.

15. (a) Welton, T. Room-temperature ionic liquids. solvents for synthesis and catalysis, *Chem. Rev.* 1999, 99, 2071–2084. (b) Sheldon, R. Catalytic reactions in ionic liquids, *Chem. Commun.* 2001, 23, 2399–2407. (c) Dupont, J., de Souza, R. F., and Suarez, P. A. Z. Ionic liquid (molten salt) phase organometallic catalysis, *Chem. Rev.* 2002, 102, 3667–3692. (d) Weingaertner, H. Understanding ionic liquids at the molecular level: Facts, problems, and controversies, *Angew. Chem. Int. Ed.* 2008, 47, 654–670. (e) Lee, J. W., Shin, J. Y., Chun, Y. S., Jang, H. B., Song, C. E., and Lee, S.-G. Toward understanding the origin of positive effects of ionic liquids on catalysis: Formation of more reactive catalysts and stabilization of reactive intermediates and transition states in ionic liquids, *Acc. Chem. Res.* 2010, 43, 985–994. (f) Dupont, J. From molten salts to ionic liquids: A "nano" journey, *Acc. Chem. Res.* 2011, 44, 1223–1231.

16. Diaw, M., Chagnes, A., Carre, B., Willmann, P., and Lemordant, D. Mixed ionic liquid as electrolyte for lithium batteries, *J. Power Sour.* 2005, 146, 682–684.

17. Chiappe, C. and Pieraccini, D. Ionic liquids: Solvent properties and organic reactivity, *J. Phys. Org. Chem.* 2005, 18, 275–297.

18. Chum, H. L., Koch, V. R., Miller, L. L., and Osteryoung, R. A. An electrochemical scrutiny of organometallic iron complexes and hexamethylbenzene in a room temperature molten salt, *J. Am. Chem. Soc.* 1975, 97, 3264–3265.

19. Wilkes, J. S., Levisky, J. A., Wilson, R. A., and Hussey, C. L. Dialkylimidazolium chloroaluminate melts: A new class of room-temperature ionic liquids for electrochemistry, Spectroscopy, and synthesis, *Inorg. Chem.* 1982, 21, 1263–1264.

20. Zhang, M., Kamavarum, V., and Reddy, R. G. New electrolytes for aluminum production: Ionic liquids, *JOM* 2003, 55(11) 54–57.

21. Ha, S. H., Mencchavez, R. N., and Koo, Y.-M. Reprocessing of spent nuclear waste using ionic liquids, *Korean J. of Chem. Eng.* 2010, 27(5) 1360–1365.

22. (a) Hunt, P. A., Gould, I. R., and Kirchner, B. The structure of imidazolium-based ionic liquids: Insights from ion-pair interactions, *Aust. J. Chem.* 2007, 60, 9–14. (b) Wilkes, J. S. Properties of ionic liquid solvents for catalysis, *J. Mol. Catal. A* 2004, 214, 11–17. (c) Anderson, J. L., Ding, J., Welton, T., and Armstrong, D. W. Characterizing ionic liquids on the basis of multiple solvation interactions, *J. Am. Chem. Soc.* 2002, 124, 14247–14254. (d) Huddleston, J. G., Visser, A. E., Reichert, W. M., Willauer, H. D., Broker, G. A., and Rogers, R. D. Characterization and comparison of hydrophilic and hydrophobic room temperature ionic liquids incorporating the imidazolium cation, *Green Chem.* 2001, 3, 156–164.

23. (a) Song, C. E. Enantioselective chemo- and bio-catalysis in ionic liquids, *Chem. Commun.* 2004, 1033–1043. (b) Lee, S. Functionalized imidazolium salts for task-specific ionic liquids and their applications, *Chem. Commun.* 2006, 1049–1063. (c) Pârvulescu, V. I. and Hardacre, C. Catalysis in ionic liquids, *Chem. Rev.* 2007, 107, 2615–2665. (d) Martins, M. A. P., Frizzo, C. P., Moreira, D. N., Zanatta, N., and Bonacorso, H. G. Ionic liquids in heterocyclic synthesis, *Chem. Rev.* 2008, 108, 2015–2050. (e) Hallett, J. P. and Welton, T. Room-temperature ionic liquids: Solvents for synthesis and catalysis, *Chem. Rev.* 2011, 111, 3508–3576.

24. Priv, R. G. Task-specific ionic liquids, *Angewandte Chemie Intl. Ed.*, 2010, 49(16), 2834–2839.

25. (a) Ding, J. and Armstrong, D. W. Chiral ionic liquids: Synthesis and applications, *Chirality* 2005, 17, 281–292. (b) Headley, A. D. and Ni, B. Chiral imidazolium ionic liquids: Their synthesis and influence on the outcome of organic reactions, *Aldrichimica Acta* 2007, 40, 107–117. (c) Wasserscheid, P. and Welton, T. *Ionic Liquids in Synthesis* 2nd ed., Wiley-VCH, Weinheim, Germany, 2008. (d) Toma, S., Meciarová, M., and Sebesta, R. Are ionic liquid suitable media for organocatalytic reactions, *Eur. J. Org. Chem.* 2009, 321–327.

26. Mrówcaynski, R., Nan, A., and Leibscher, J. Magnetic nanoparticle-supported organocatalysis: An efficient way of recycling and reuse, *RSC Adv.* 2014, 4, 5927–5952.

27. (a) Hayashi, S. and Hamaguchi, H. Discovery of a magnetic ionic liquid [bmim]FeCl$_4$, *Chem. Lett.* 2004, 33, 1590–1591. (b) Santos, E., Albo, J., and Irabien, A. Magnetic ionic liquids: Synthesis, properties and applications, *RSC Adv.* 2014, 4, 40008–40018. (c) Wang, J., Yao, H., Nie, Y., Bai, L., Zhang, X., and Li, J. Application of iron-containing magnetic ionic liquids in extraction process of coal direct liquefaction residues, *Ind. Eng. Chem. Res.* 2012, 51 (9), 3776–3782.

28. Sayyahi, S., Azin, A., and Saghanezhad, S. J. Synthesis and characterization of a novel paramagnetic functionalized ionic liquid as a highly efficient catalyst in a one-pot synthesis of 1-amidoalkyl-2-napthols, *J. Mol. Liq.* 2014, 198, 30–36.

29. Li, M., De Rooy, S. L., Bwambok, D. K., El-Zahab, B., DiTusa, J. F., and Warner, I. M. Magnetic chiral ionic liquids derived from amino acids, *Chem. Commun.* 2009, 45, 6922–6924.

30. (a) Welton, T. Room-temperature ionic liquids. solvents for synthesis and catalysis, *Chem. Rev.* 1999, 99, 2071–2084. (b) Shelton, R. Catalytic reactions in ionic liquids, *Chem. Commun.* 2001, 23, 2399–2407.

31. Xu, X., Kotti, S. R. S. S., Liu, J., Cannon, J. F., Headley, A. D., and Li, G. Ionic liquid media resulted in the first asymmetric aminohalogenation reaction of alkenes, *Org. Lett.* 2004, 6, 26, 4881–4884.

32. Kotti, S. R. S. S., Xu, X., Wang, Y., Headley, A. D., and Li, G. Ionic liquid media resulted in more efficient regio- and stereoselective aminohalogenation of cinnamic esters, *Tetrahedron Lett.* 2004, 45, 7209–7212.

33. Nobuoka, K., Kitaoka, S., Yanagisako, A., Maki, Y., Harran, T., and Ishikawa, Y. Stereoselectivity of the Diels-Alder reaction in ionic liquids with cyano moieties: Effect of the charge delocalization of anions on the relation of solvent–solvent and solute–solvent interactions, *RSC Adv.* 2013, 3, 19632–19638.

34. Liu, R., Zhang, Y., Bai, L., Huang, M., Chen, J., and Zhang, Y. Synthesis of cellulose-2,3-bis(3,5-dimethylphenylcarbamate) in an ionic liquid and its chiral separation efficiency as stationary phase, *Int. J. Mol. Sci.* 2014, 15(4), 6161–6168.

35. Kumar, A., Srivastava, S., Gupta, G., Kumar, P., and Sakar, J. Functional ionic liquid [bmim][Sac] mediated synthesis of ferrocenyl thiopropanones via the "dual activation of the substrate by the ionic liquid", *RSC Adv.* 2013, 3, 3548–3552.

36. Davis, J. H., and Forrester, K. J. Thiazolium-ion based organic ionic liquids (oil). novel OILs which promote the benzoin condensation, *Tetrahedron Lett.* 1999, 40, 1621–1622.

37. Davis, J. H. Task-specific ionic liquids, *Chem. Lett.* 2004, 33, 1072–1077.
38. Dominquez de Maria, P. (Ed.), *Ionic Liquids in Biotransformations and Organocatalysis, Solvents and Beyond*, John Wiley & Sons, 2012, Hoboken, NJ.
39. Garre, S., Parker, E., Ni, B., and Headley, A. D. Design and synthesis of bistereogenic chiral ionic liquids and their use as solvents for asymmetric Baylis-Hilman reactions, *Org. Biomol. Chem.* 2008, 6, 3041–3043.
40. Ni, B., Garre, S., and Headley, A. D. Design and synthesis of fused-ring chiral ionic liquids from amino acid derivatives, *Tetrahedron Lett.* 2007, 48, 1999–2002.
41. Luo, S.-P., Xu, D.-Q., Yue, H.-D., Wang, L.-P., Yang, W.-L., and Xu, Z.-Y. Synthesis and properties of novel chiral-amine-functionalized ionic liquids, *Tetrahedron: Asym.* 2006, 17, 2028–2033.
42. Ni, B., Zhang, Q., and Headley, A. D. Functionalized chiral ionic liquid as recyclable organocatalyst for asymmetric Michael addition to nitrostyrenes, *Green Chem.* 2007, 9, 737–739.
43. Bwambok, D. K., Marwani, H. M., Fernand, V. E., Fakayode, S. O., Lowry, M., Negulescu, I., Strongin, R. M., and Warner, I. M. Synthesis and characterization of novel chiral ionic liquids and investigation of their enantiomeric recognition properties, *Chirality* 2008, 20(2), 151–158.
44. Merrifield, R. B. Solid phase peptide synthesis. 1. The synthesis of a tetrapeptide. *J. Am. Chem. Soc.* 1963, 85, 2149–2154.
45. (a) Gruttaduria, M., Riela, S., Aparile, C., Meo, P. L., D'Anna, F. D., and Noto, R. Supported ionic liquids. New recyclable materials for the L-proline-catalyzed aldol reaction, *Ad. Syn. Catal.* 2006, 348, 82–92. (b) Aprile, C., Giacalone, F., Gruttaduria, M., Maculescu, A. M., Noto, R., Revell, J. D., and Wennemers, H. New ionic liquid-modified silica gels as recyclable materials for L-Proline or H-Pro-Pro-Asp-NH2-catalyzed reaction, *Green Chem.* 2007, 9, 1328–1334.
46. Mehnert, C., Cook, R. A., Dispenziere, C., and Afework, M. Supported ionic liquid catalysis- A new concept for homogeneous hydroformylation catalysis, *J. Am. Chem. Soc.* 2002, 124, 12932–12933.
47. Sun, J., Chen, W., Fan, W., Wang, Y., Meng, Z., and Zhang, S. Reusable and efficient polymer-supported task-specific ionic liquid catalyst for cycloaddition of expide with CO_2, *Catal. Today* 2009, 148, 361–367.
48. Zhang, X., Zhao, W., Qu, C., Yang, L., and Cui, Y. Efficient asymmetric aldol reaction catalyzed by polyvinylidene chloride-supported ionic liquid/l-proline catalyst system, *Tetrahedron: Asym.* 2012, 23, 468–473.
49. Mehnert, C. P. Supported ionic liquid catalysis, *Chem. Eur. J.* 2005, 11, 50–56.
50. Miao, W. and Chan, T. H. Ionic-liquid-supported synthesis: A novel liquid-phase strategy for organic synthesis. *Acc. Chem. Res.* 2006, 39, 897–908.
51. Handy, S. T. and Okello, M. Homogeneous supported synthesis using ionic liquid supports: Tunable separation properties. *J. Org. Chem.* 2005, 70, 2874–2877.
52. Handy, S. T. Room temperature ionic liquids: Different classes and physical properties, *Curr. Org. Chem.* 2005, 9, 959–988.
53. Lattanzi, A. Alpha, alpha-diarylprolinols: Bifunctional organocatalysts for asymmetric synthesis, *Chem. Commun.* 2009, 12, 1452–1463.
54. List, B., Proline-catalyzed asymmetric reactions, *Tetrahedron* 2002, 58, 5573–5590.
55. Jensen, K. L., Dickmeiss, G., Jiang, H., Albrecht, L., and Jorgensen, K. A. The diarylprolinol silyl ether system: A general organocatalyst, *Acc. Chem. Res.* 2012, 45, 248–264.
56. (a) Jen, W. S., Wiener, J. J. M., and MacMillan, D. W. C. New strategies for organic catalysis: The first enantioselective organocatalytic 1,3-dipolar cycloaddition, *J. Am. Chem. Soc.* 2000, 121, 9874–9875. (b) Paras, N. A. and MacMillan, D. W. C. New strategies in organic catalysis: The first enantioselective organocatalytic Friedel-Crafts alkylation, *J. Am. Chem. Soc.* 2001, 123, 4370–4371.

57. For some selected reviews, see: (a) Mukherjee, S., Yang, J. W., Hoffmann, S., and List, B. Asymmetric enamine catalysis, *Chem. Rev.* 2007, 107, 5471–5569. (b) List, B. (Ed.). *Asymmetric Organocatalysis*, Topics in Current Chemistry, 2010, Vol. 291, Springer, Heidelberg, Germany.

58. List, B., Pojarliev, P., and Castello, C. Proline-catalyzed asymmetric aldol reactions between ketones and α-unsubstituted aldehydes, *Org. Lett.* 2001, 3(4), 573–575.

59. List, B., Lerner, R. A., and Barbas, C. F. III. Proline-catalyzed direct asymmetric aldol reactions, *J. Am. Chem. Soc.* 2000, 122(10), 2395–2396.

60. Chowdari, N. S., Ramchary, D. B., and Barbas, C. F., III. Asymmetric synthesis of quaternary α- and β-amino acids and β-lactams via proline-catalyzed mannich reactions with branched aldehyde donors. *Synlett* 2003, 1906.

61. (a) Bellis, E. and Kokotos, G. 4-Substituted prolines as organocatalysts for aldol reactions. *Tetrahedron* 2005, 61, 8669–8676. (b) Hayashi, T., Sumiya, J., Takahashi, J., Gotoh, H., Urushima, T., and Shoji, M. Highly diastereo- and enantioselective direct aldol reactions in water. *Angew. Chem. Int. Ed.* 2006, 45, 958–961. (c) Giacalone, F., Gruttadauri, M., Agrigento, P., Meo, P. L., and Noto, R. Advances towards highly active and stereoselective simple and cheap proline-based organocatalysts. *Eur J. Org. Chem.* 2010, 5696–5704. (d) Kristensen, T. E., Hansen, F. K., and Hansen, T. The selective *O*-acylation of hydroxyproline as a convenient method for the large-scale preparation of novel proline polymers and amphiphiles. *Eur. J. Org. Chem.* 2009, 387–395. (e) Hayashi, Y., Aratake, S., Okano, T., Takahashi, J., Sumiva, T., and Shoii, M. Combined proline-surfactant organocatalyst for the highly diastereo- and enantioselective aqueous direct cross-aldol reaction of aldehydes. *Angew, Chem. Int. Ed.* 2006, 45, 5527–5529. (f) List, B., Pojarleiv, P., and Castello, C. Proline-catalyzed asymmetric aldol reactions between ketones and α-unsubstituted aldehydes. *Org. Lett.* 2001, 3, 573–575. (g) Huang, J., Zhang, Z., and Armstrong, D. W. Highly efficient asymmetric direct stoichiometric aldol reactions on/in water. *Angew. Chem. Int. Ed.* 2007, 46, 9073–9077.

62. (a) Wagner, M., Contie, Y., Ferroun, C., and Revial, G. Enantioselective aldol reactions and Michael additions using proline derivatives as organocatalysts, *Int. J. Org. Chem.* 2014, 4, 55–67. (b) Hayashi, Y., Gotoh, H., Hayashi, T., and Shoji, M. Diphenylprolinol silyl ethers as efficient organocatalysts for the asymmetric Michael reaction of aldehydes and nitroalkenes, *Angew. Chem. Int. Ed.* 2005, 44, 4212–4215. (c) Marigo, M., Wabnitz, T. C., Fielenbach, D., and Jørgensen, K. A. Enantioselective organocatalyzed α sulfenylation of aldehydes, *Angew. Chem. Int. Ed.* 2005, 44, 794–797.

63. Xu, D., Wang, S., Luo, S., Yue, H., and Xu, Z. Pyrrolidine-pyridinium based organocatalysts for highly enantioselective Michael addition of cyclohexanone to nitroalkenes, *Tetrahedron: Asym.* 2007, 18, 1788–1794.

64. Montroni, E., Sanap, S. P., Lombardo, M., Quintavalla, A., Trombini, C., and Dhavale, D. D. A new robust and efficient ion-tagged proline catalyst carrying an amide spacer for the asymmetric aldol reaction, *Ad. Syn. & Cat.* 2011, 353(7), 3234–3240.

65. Bahmanyar, S., Houk, K. N., Martin, H. J., and List, B. Quantum mechanical predictions of the stereoselectivities of proline-catalyzed asymmetric intermolecular aldol reactions, *J. Am. Chem. Soc.* 2003, 125, 2475–2479.

66. Luo, S., Mi, X., Liu, S., Xu, H., and Cheng, J.-P. Surfactant-type asymmetric organocatalyst: Organocatalytic asymmetric Michael addition to nitrostyrenes in water, *Chem. Commun.* 2006, 3687–3689.

67. Lombardo, M., Pasi, F., Easwar, S., and Trombini, C. Direct asymmetric aldol catalyzed by a imidazolium-tagged trans-4-Hydroxy-L-proline under aqueous biphasic conditions. *Synlett* 2008, 2471–2474.

68. Lombardo, M., Easwar, S., Pasi, F., and Trombini, C. The ion tag strategy as a route to highly efficient organocatalysts for the direct asymmetric aldol reaction. *Adv. Synth. Catal.* 2009, 351, 276–282.

69. Lombardo, M., Easwar, S., Marco, A. D., Pasia, F., and Trombini, C. A modular approach to catalyst hydrophobicity for an asymmetric aldol reaction in a biphasic aqueous environment. *Org. Biomol. Chem.* 2008, 6, 4224–4229.
70. Omar, E. M., Dhungana, K., Headley, A. D., and Rahman, M. B. A. Ionic liquid-supported (ILS) (S)-pyrrolidine sulfonamide for asymmetric Michael addition reactions of aldehydes with nitroolefins, *Lett. Org. Chem.* 2011, 8(3), 170–175.
71. Hu, S., Jiang, T., Zhang, Z., Zhu, A., Han, B., Song, J., Xie, Y., and Li, W. Functional ionic liquid from biorenewable materials: Synthesis and application as a catalyst in direct aldol reactions. *Tetrahedron Lett.* 2007, 48, 5613–5617.
72. Xu, D.-Z., Liu, Y., Shi, S., and Wang, Y. Chiral quaternary alkylammonium ionic liquid [Pro-dabco][BF4]: As a recyclable and highly efficient organocatalyst for asymmetric Michael addition reactions, *Tetrahedron: Asym.* 2010, 21(20), 2530–2534.
73. Sarkar, D., Bhattarai, R., Headley, A. D., and Ni, B. A novel recyclable organocatalytic system for the highly asymmetric Michael addition of aldehydes to nitroolefins in water. *SynLett.*, 2011, 12, 1993–1997.
74. Qiao, Y., He, J., Ni, B., and Headley, A. D. Asymmetric Michael reaction of acetaldehyde with nitroolefins catalyzed by highly water-compatible organocatalysts in aqueous media. *Adv. Synth. Catal.* 2012, 354, 2849–2853.
75. Jiang, X., Ye, W., Song, X., Ma, W., Lao, X., and Shen, R. Novel ionic liquid with both Lewis and Bronsted acidic sites for Michael addition, *Int. J. Mol. Sci.* 2011, 12, 7438–7444.
76. Ni, B., Zhang, Q., and Headley, A. D. Pyrrolidine-based chiral pyridinium ionic liquids (ILs) as recyclable and highly efficient organocatalysts for the asymmetric Michael addition reactions, *Tetrahedron Lett.* 2008, 49, 1249.
77. Siyutkin, D. E., Kucherenko, A. S., and Zlotin, S. G. Hydroxy α-amino acids modified by ionic liquid moieties: Recoverable organocatalysts for asymmetric aldol reactions in the presence of water. *Tetrahedron* 2009, 65, 1366–1372.
78. Davis, J. H., and Forrester K. L. Thiazolium-ion based organic ionic liquids (OIL). Novel OILs which promote the benzoin condensation. *Tetrahedron Lett.* 1999, 40, 1621–1622.
79. Davis, J. H. Task-specific ionic liquids. *Chem. Lett.* 2004, 33, 1072–1077.
80. Bivona, L. A., Quertinmount, F., Beejpur, H. A., Giacalone, F., Buaki-Sogo, M., Gruttadauria, M., and Aprile, C. Thiazolium-based catalysts for the etherification of benzylic alcohols under solvent-free conditions, *Adv. Syn. & Cat.* 2015, 357, 800–810.
81. Gladyse, J. A. Experimental assay of catalyst recovery: General concepts. In *Recoverable and Recyclable Catalysts*, (Ed.) Benaglia, M. John Wiley & Sons, Chichester, UK, 2009, pp. 1–14.
82. Siyutkin, D. E., Kucherenko, A. S., Struchkova, M. I., Zlotin, S. G. A novel (S)-proline-modified task-specific chiral ionic liquid-an amphiphilic recoverable catalyst for direct asymmetric aldol reactions in water. *Tetrahedron Lett.* 2008, 49, 1212–1216.
83. Kong, Y., Tan, R., Hao, L., and Yin, D. L-proline supported on ionic liquids-modified magnetic nanoparticles as a highly efficient and reusable organocatalysts for direct asymmetric aldol reaction in water, *Green Chem.* 2013, 15, 2422–2433.
84. Deng, N., Li, M., Zhoa, L., Lu, C., de Rooy, S. L., and Warner, I. M. Highly efficient extraction of phenolic compounds by use of magnetic room temperature ionic liquids for environmental remediation, *J. Hazardous Mat.* 2011, 192(3), 1350–1357.

3 Organocatalysis Induced by the Anion of an Ionic Liquid

A New Strategy for Asymmetric Ion-Pair Catalysis

Andrea R. Schmitzer and Vincent Gauchot

CONTENTS

3.1 INTRODUCTION

The term "asymmetric ion-pair catalysis" was introduced to refer to all the catalytic processes where stereoselectivity is induced through ion-pairing between a charged intermediate and a chiral counterion. As described by Brak and Jacobsen,[1] two distinct approaches for chiral induction have emerged with respect to the definition earlier: (1) enantioselectivity can be achieved using a counterion with intrinsic chirality, or (2) through non-covalent interactions between the counterion and a chiral ligand. The concept of using ion-pairing systems for substrate activation in asymmetric catalysis was born out of the mechanistic ambiguity that exists regarding Brønsted acid catalysis. A common scenario would involve substrate activation by the Brønsted acid through hydrogen bonding. It naturally comes to mind that depending on the pKa difference between the Brønsted acid and the substrate, the proton can be completely transferred to the substrate, and that the conjugate base stays in a close environment during the critical bond-forming step.[2] Should the Brønsted acid be chiral, asymmetric catalysis becomes possible.[3]

Such a strategy for chiral induction in catalysis has been widely used for the last 30 years and led to the development of very efficient types of asymmetric ion-pair catalytic processes, ranging from phase-transfer catalysis using chiral quaternary ammonium,[4] or phosphonium salts,[5] to the emerging strategy of asymmetric counteranion-directed catalysis (ACDC).[6] This concept, as described by Benjamin List, refers to "the induction of enantioselectivity in a reaction proceeding through a cationic intermediate by means of ion-pairing with a chiral, enantiomerically pure anion provided by the catalyst."[6] The principal mechanistic uncertainty that remains in ion-pairing catalysis is whether a formal ion-pair is involved in the reaction mechanism. In organocatalysis, most examples of asymmetric ion-pair catalysis report an actual equilibrium between the real ion-pair and the corresponding Brønsted acid species interacting with the substrate. The same question arises in metal catalysis as to know whether the chiral anion can act as a ligand (Figure 3.1).

The ion-pairing system as described in ACDC can be extended to ionic liquids (ILs) and their use as catalysts, hence, belonging to the family of "task-specific ionic liquids."[7] Please note that while the term "ionic liquid" is usually used to describe solely compounds having a melting point below 100°C, it will be used here in a broader fashion to describe all kinds of molten salts. Benefitting from a permanent charged state, the cation and/or the anion cannot only be used for chiral induction, but also as catalytic species themselves. Similarly, both the cation and the anion of the ion-pair, for example the ILs, are able to play a role regarding the stereoinduction, providing chiral environments through ion-pairing. With regards to this idea, this chapter will focus on a novel and emerging concept of using the anion of ILs as catalysts for asymmetric organocatalysis. It is important to notice that this approach differs from ACDC, in the sense that the anion can now act as the actual catalyst, and that the reaction might not always proceed through a cationic intermediate.

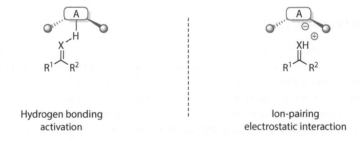

Hydrogen bonding activation   Ion-pairing electrostatic interaction

FIGURE 3.1 Equilibrium regarding asymmetric counteranion-directed catalysis.

3.2 EARLY EXAMPLES INVOLVING THE ANION OF IONIC LIQUIDS IN CATALYSIS

3.2.1 METAL CATALYSIS

While numerous reports regarding catalysis using the cation of an ionic liquid, particularly regarding the imidazolium salts family, can be found in the literature,[8,9] only a few examples of catalysis using the anion of an IL were reported. The first example of such catalysis goes back to 1986, when Wilkes et al. described the first Friedel-Crafts reaction catalyzed by the anion of an ionic liquid.[10] The authors reported the use of 1-ethyl-3-methylimidazolium chloride salt **1** that can act as Lewis acid upon addition of an excess of aluminum chloride in the reaction medium. The catalytic species was formed *in situ* and the catalytic species was identified as being the anionic Lewis acid $Al_2Cl_7^-$. The excess of aluminum chloride is necessary in this process to ensure an overall acidic medium and to allow the formation of the Lewis acid active species. This is one of the rare examples of an ionic liquid bearing an achiral anion that has been used in metal catalysis. In 2001, Cole-Hamilton et al. described the imidazolium salt **2** bearing a phosphine-based anion that can act as a ligand for a rhodium atom, and its application in the biphasic catalytic hydroformylation of alkenes, using an ionic liquid and supercritical carbon dioxide as a partitioned reaction medium (Figure 3.2).[11]

The use of such an ionic-tagged ligand allows the whole catalyst to be more soluble in the ionic liquid media, compared to its sodium salt counterpart, thus enhancing the performances of the setup of the catalytic process. Despite showing promising results in terms of catalysis, the development of such IL-based metal catalysts was left aside in favor of the design of ILs bearing catalytic functional groups on their cation.

3.2.2 TOWARD ORGANOCATALYSIS

The large availability of small chiral molecules that can act as precursors for catalytically active anions drove the focus of the research toward the development of "anionic" organocatalysts for asymmetric catalysis. Such a synthetic approach can

FIGURE 3.2 First examples of ionic liquids bearing catalytic anions.

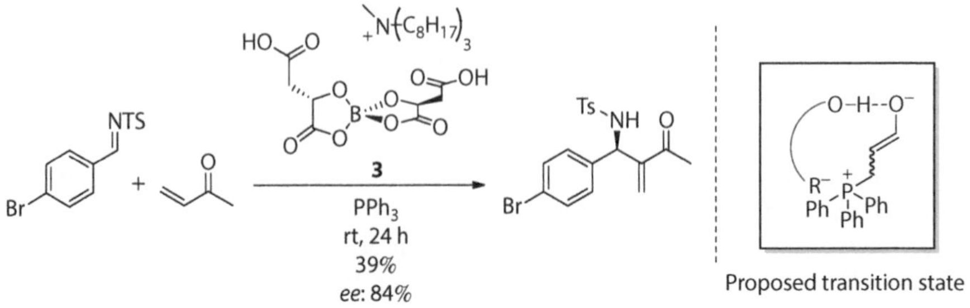

FIGURE 3.3   First example of ACDC using the anion of an ionic liquid.

efficiently yield a large variety of air-stable and catalytically active ILs through simple synthetic procedures. Gausepohl et al. were the first to report an example of organo-ACDC using the anion of an ionic liquid (Figure 3.3).[12]

The anion in this case is a chiral borate derivative prepared using the readily available L-(-) Malic acid as chiral precursor. Compound **3** acts as chiral solvent, and Leitner et al.[12] attributed the observed selectivity to the bifunctional mode of action of the anion. The high enantioselectivity was explained by a highly ordered transition state provided by the stabilization of the enolate through hydrogen-bonding and anion pairing with the phosphonium group. The results were comparable in terms of selectivity to those previously reported for the aza-Baylis-Hillman reaction in conventional organic solvents.[13]

These pioneering results opened the door for the use of chiral anions not only as chiral agents, but also as actual catalysts. In 2007, Han et al. reported the first example of IL bearing an organocatalytically active anion (**4**) and its use as catalyst in the asymmetric aldol reaction (Figure 3.4).[14]

Compound **4** formed by a cholinium cation and a proline anion ([chol]Pro) was used for the asymmetric aldol reaction of linear and cyclic ketones and aldehydes in water. Surprisingly, despite excellent yields (>99% in certain cases), enantioselectivity was only up to 10% ee in the best cases. The authors reported the recycling possibilities due to the biphasic nature of the reaction medium, but the source of the limited stereocontrol of this reaction was not discussed at that time. The aqueous solution of [chol]Pro was separated from the insoluble reaction products and reused up to three times without any decrease of activity.

The use of ILs bearing catalytic anions as catalysts for asymmetric catalysis has been proven through these reports, and at that point a deeper investigation regarding

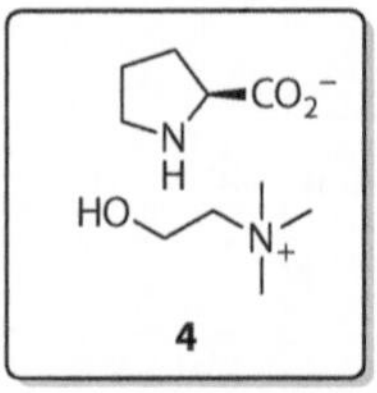

FIGURE 3.4   First L-proline-based catalytic IL described by Han in 2007.

the mechanistic aspects of such species was necessary. The goal of this chapter being to give an overview of the recent advances made in terms of anionic organocatalysis using ILs, we will focus on the mode of action of each reported example.

3.3 DEVELOPMENT OF PROLINE- AND PYRROLIDINE-BASED ANIONS

3.3.1 PREPARATION

Investigating the effect of the nature of the anion on the physicochemical properties of ionic liquids, Ohno et al. were the first to successfully report the preparation and characterization of 20 air-stable room-temperature ILs, each one containing the 1-ethyl-3-methylimidazolium (emim) cation and one naturally occurring amino acid.[15] The preparation of these species elegantly involves the use of 1-butyl-3-methylimidazolium hydroxide ([bmim]OH) as intermediate, which is readily neutralized *in situ* by the desired amino acid, yielding the corresponding IL. This method, as an alternative to the commonly used anion metathesis between halide salts and metal salts, allows the inexpensive preparation of pure and uncontaminated ILs (Figure 3.5).

Following this work, and with regards to the effectiveness of L-proline regarding organocatalysis, Wang et al.[16] focused their efforts on studying the catalytic activity of 1-ethyl-3-methylimidazolium prolinate ([emim]Pro) toward asymmetric enamine catalysis.

3.3.2 ASYMMETRIC ENAMINE CATALYSIS

Role of the imidazolium cation. Their first report represents the first successful example of organocatalysis by the anion of an ionic liquid.[16] The authors disclosed the use of [emim]Pro as a catalyst for the Michael addition reaction between the cyclohexanone and different chalcones.[16] The best yields and enantioselectivities were obtained when using 200 mol% of catalyst loading, along with more polar solvents, such as methanol, ethanol, or dimethylsulfoxide (DMSO). Interestingly, the authors also demonstrated a strong solvent effect regarding the enantioselectivity of this reaction (Figure 3.6).

Carrying out the reaction in dimethylsulfoxide yielded the (R),(S) Michael adduct, while carrying out the reaction in methanol (MeOH) yielded the other enantiomer. Performing the reaction in 1-butyl-3-methylimidazolium tetrafluoroborate ([bmim]BF$_4$) and 1-butyl-3-methylimidazolium hexafluorophosphate [bmim]PF$_6$ resulted

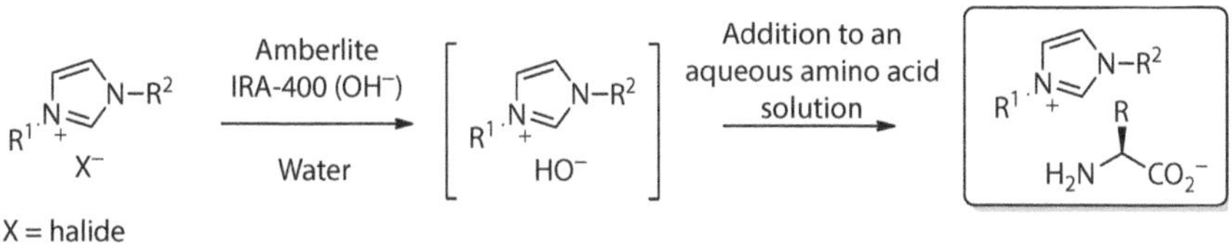

FIGURE 3.5 Wang's preparation of amino acid-based ILs[16].

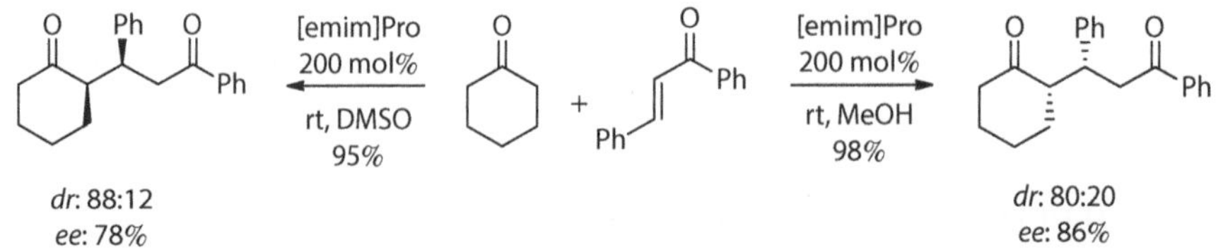

FIGURE 3.6 Solvent-dependent selectivity observed for the Michael addition.

in a consequent decrease of selectivity, yielding the major enantiomer in 31% and 46% *ee*, respectively. Wang et al.[16] rationalized the attributed enantioselectivity to the assistance of the imidazolium cation regarding the approach of the chalcone. In methanol, the electrostatic interactions between the imidazolium cation and the chalcone position the latter toward the *Re* face of the enamine (Figure 3.7a). In the case of DMSO, the reverse selectivity was explained due to the oxygen atom of the solvent occupying the place of the chalcone, hindering the *Re* face of the enamine, and thus favoring its approach toward the *Si* face (Figure 3.7b).

This example reported for the first time the involvement of the imidazolium cation in catalysis and proposed its role regarding the stereocontrol of the reaction. After this successful attempt, Wang et al.[16] reported several examples that illustrated the broad range of applications for the catalytic ionic liquid [emim]Pro. The same catalyst was successfully employed for asymmetric aza Diels-Alder reactions,[17] one-pot Mannich reactions,[18] and aldol reaction (Figure 3.8).[19]

Targeting C–C bond-forming organocatalyzed key-reactions, [emim]Pro also proved to be a powerful catalyst for the Mannich reaction between several cyclic and acyclic ketones, aromatic aldehydes, and aromatic primary amines. Performing the reaction in either dimethylformaide or DMSO at room temperature yields the best results overall, compared to those obtained in less polar solvents (hexanes, THF) or even ionic liquids ([bmim]PF_6). The influence of the temperature on the selectivity of the reaction was highlighted when performing the reaction in dimethylformaide at −20°C, where *ee* up to 97% were obtained. Interestingly, the performances of the system were improved by the addition of one equivalent of water into the reaction medium. This catalytic improvement was proposed to be the result of a transition state involving a hydrogen bond between the imine and the imidazolium moiety mediated

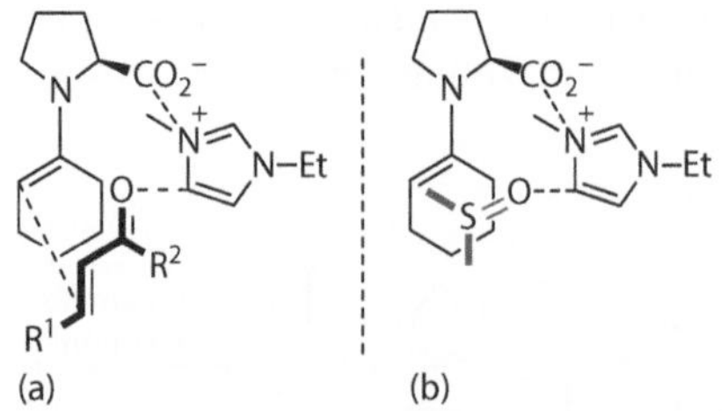

FIGURE 3.7 Influence of the imidazolium cation on enantioselectivity, as reported by Wang et al.[16] (a) Electrostatic interactions between the imidazolium cation and the chalcone and (b) DMSO intercating with the imidazolium cation hindering the *Re* face.

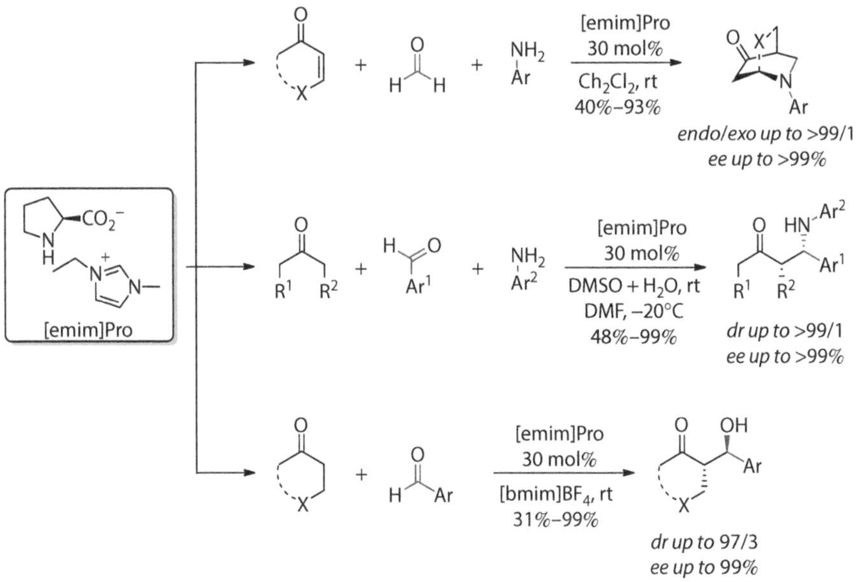

FIGURE 3.8 Reactions catalyzed by [emim]Pro, as reported by Wang et al.[17–19]

by the water molecule. The steric hindrance brought by the arylimine was identified as the discriminating factor for the facial approach of the imine toward the enamine (Figure 3.9a). The implication of the water molecule is, however, disputable, regarding the competitive results obtained when water was not present in the reaction medium (88% *ee*). This could be the case of a different mode of activation of the substrate by the catalyst. In the same way, the involvement of the imidazolium cation was once again highlighted by Wang et al.[19] when using [emim]Pro as a catalyst for the aldol reaction. Here again, the authors postulated that the preferential attack of the enamine on the *Re* face of the aldehyde was attributed to the electrostatic interactions between the imidazolium cation and the oxygen atom of the aldehyde (Figure 3.9b).

However, despite the known capacity of the most acidic H-2 as a hydrogen bond donor,[20] the authors never considered possible hydrogen bonding activation of the substrate by this H-2 proton. Interestingly, the performances of the catalyst when used in conventional organic solvents for the aldol reaction were disappointing in

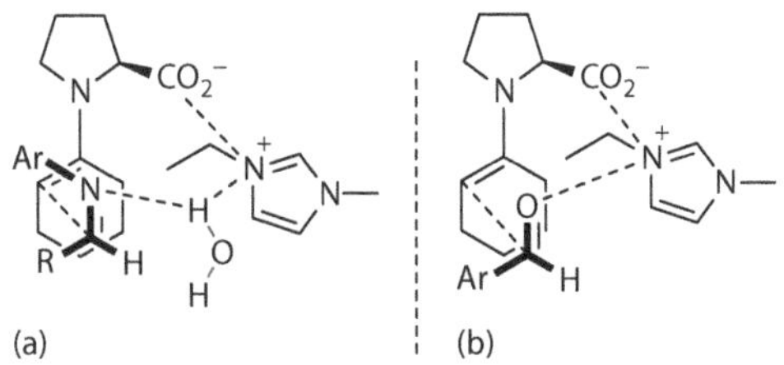

FIGURE 3.9 Proposed transition states involving the imidazolium cation for the Mannich reaction (a), and the aldol reaction (b).

terms of diastereoselectivity (*dr ca* 2/1) and enantioselectivity (*ee* up to 30% for the major isomer). However, the use of the hydrophilic ionic liquid [bmim]BF$_4$ allowed a significant improvement of both diastereo- and enantioselectivity (*dr* up to 78/22, *ee* up to 97% for the major isomer). After completion of the reaction, the aldol products were extracted with ether and the reaction medium, composed of both the catalyst and the solvent, was recycled up to four times without a significant decrease of activity.

Mechanistic insights. Our group simultaneously reported the development of catalytic ILs bearing *trans*-4-hydroxy-L-proline as catalytic anions. In an attempt to evaluate the potential activity of such catalysts for organocatalysis, we reported the preparation of two anionic catalysts and their use in the aldol reaction and the Michael addition (Figure 3.10).[21,22]

Using catalyst **5** and [bmim]NTf$_2$ as solvent, the aldol products were obtained with good yields and selectivities for reactive aldehydes such as *o*-, *m*-, and *p*-nitrobenzaldehydes. However, only traces of aldol products were observed when using bulky or less activated aldehydes, such as 2-naphtaldehyde or 2-thiophenecarboxaldehyde. Hence, in light of Wang's[19] previous reports, the authors proposed that this poor activity was the result of the imidazolium cation being unable to play its role as substrate activator due to its spatial proximity with the sulfonate group, combined with the bulkiness of the ester group.

With regards to the known dual activation of L-proline for the aldol reaction,[23] the replacement of the ester group of the catalyst by a hydrogen bond donor amide moiety (compound **6**) gave the aldol products that were not obtained previously. Remarkably, the asymmetric induction was the same for both catalysts, which led the authors to postulate that the imidazolium, being close to the sulfonate moiety, hinders the approach of the substrate toward the *Si* face of the enamine, thus leading to the preferential *Re* approach for both catalysts (Figure 3.11a). The amide group was

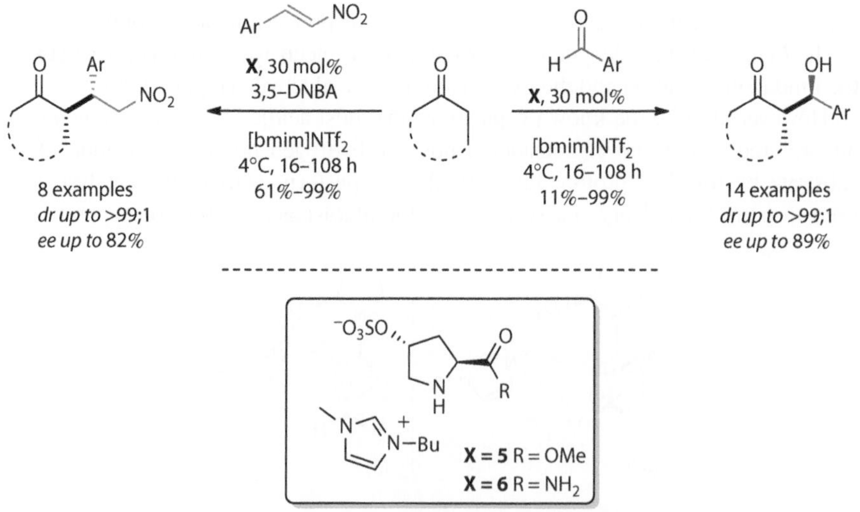

FIGURE 3.10 Aldol reaction and Michael addition catalyzed by the anion of an ionic liquid.

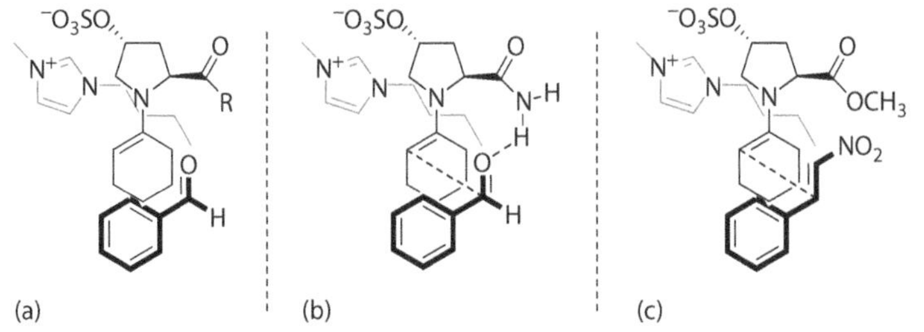

FIGURE 3.11 Proposed transition state for the aldol reaction (a and b) and the Michael addition (c) using *trans*-4-L-hydroxyproline-based counter anions.

proposed to play the role of activating agent through hydrogen bonding the aldehyde (Figure 3.11b). Gauchot and Schmitzer[22] also reported a similar transition state when using catalyst **5** for the asymmetric Michael additions in [bmim]NTf$_2$, where again, good results in terms of reactivity and selectivity were observed (Figure 3.11c).

By its close proximity with the sulfonate moiety, the imidazolium cation can promote the approach of the nitrostyrene toward the *Re* face of the enamine, despite important steric hindrance due to the ester group.

The importance of the imidazolium cation regarding the stereocontrol and the substrate activation correlates with both Wang's[19] and Gauchot et al.[21]; Gauchot and Schmitzer[22] reports. Its proximity to the catalytic center was proved to be crucial for a concrete activation of the substrate, but no clear evidence regarding its mechanism of action could be found. We recently disclosed another example of aldol reactions carried out using 3-butyl-1-methyl-1H-imidazol-3-ium (S)-pyrrolidin-2-ylmethyl sulfate **7** as the catalyst in ionic liquids as solvent (Figure 3.12).[24]

In this case, catalyst **7** slightly favors the formation of the *syn* isomer. This uncommon selectivity was explained through a transition state involving hydrogen bonding with the H-2 proton of the imidazolium cation. Being localized close to the sulfonate moiety, the imidazolium cation acts both as an activating and directing group for the substrate, favoring the approach of the aldehyde on the *Re* face of the enamine by the formation of hydrogen bonds between the imidazolium H-2 and the oxygen atom of the aldehyde (Figure 3.13a).

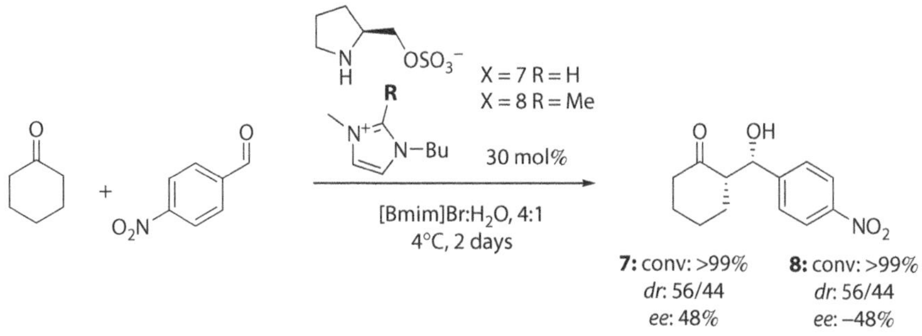

FIGURE 3.12 Aldol reaction using a pyrrolidine-based catalytic anion.

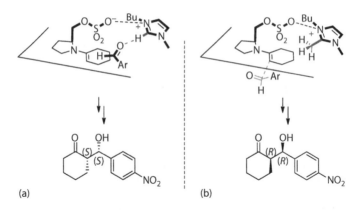

FIGURE 3.13 Role of the imidazolium H-2 in the stereocontrol of the aldol reaction. (a) Aldehyde's approach on the *Re* face of the enamine and (b) aldehydes attacks of the *Si* face of the enamine.

A control experiment was carried out by replacing the imidazolium H-2 by a methyl group (catalyst **8**) and removing the hydrogen bonding possibility. In this case, the stereocontrol of the reaction is solely ruled by steric restrictions and the aldehyde only can attack the *Si* face of the enamine (Figure 3.13b). While the structure of the chiral anion is the crucial parameter in the design of these anionic organocatalysts, it clearly appears that the cation not only acts as a "support" for the anion, but also plays the co-catalyst role in a given reaction.

Supramolecular induction of chirality using an achiral catalytic anion. While all of the examples discussed earlier target the anion as source of chirality, our group recently reported the preparation of a biotinylated imidazolium cation bearing racemic pyrrolidin-2-ylmethyl sulfate as catalytic anion.[24] In this system, the chirality is brought by the irreversible supramolecular inclusion of the biotinylated imidazolium salt inside egg-white avidin (Figure 3.14).

Benefitting from the chiral environment brought by the protein, Gauchot et al.[24] reported good yields (up to 94%) and enantioselectivity (up to 70%) for the aldol reaction between cyclohexanone and *p*-nitrobenzaldehyde in a mixed [bmim]Br/H$_2$O medium.

The pH and reaction medium composition are important factors in this case. The amount of ionic liquid in the reaction medium has an important impact on the enantioselectivity on the reaction. This was suggested to be directly related to the

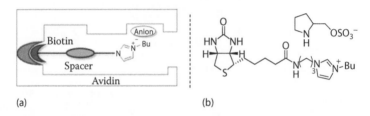

FIGURE 3.14 Schematic representation of the biotin-avidin complex (a) and biotinylated catalyst encapsulated in egg-white avidin (b).

conformation of the protein, depending on the composition of the reaction medium. Interestingly, when each enantiomer of the catalytic anions was tested apart, both yielded the exact same (S), (R) aldol product. The only plausible explanation of these results is an interaction between the avidin and the sulfonate group of the anion, diminishing the impact of the stereochemistry of the anion, and emphasizing the influence of the chiral protein architecture around the achiral catalytic anion.

To the best of our knowledge, this example is the first report of anionic IL catalysis where the chirality is brought by the supramolecular inclusion of the cation. ILs using organocatalytic achiral anions and chiral countercations have, however, yet to be developed, and with respect to the advances made regarding asymmetric ion-pair catalysis, it is reasonable to think that such systems could be extremely efficient in terms of asymmetric induction.

3.4 SUPPORTED IONIC LIQUID PHASE BEARING CATALYTIC ANIONS

Supported ionic liquid phase (SILP) catalysis involves the immobilization of ionic liquids on the surface of a highly porous support material. Over the past decade, this technique has proven to be extremely effective in solving major homogeneous catalysis issues in ionic liquids, such as catalyst efficiency,[25] environmental and cost issues,[26] and most importantly, recyclability.[27] Surprisingly enough, using supported ILs as a sorption phase for metal catalysts has been widely reported over the past 10 years,[28,29] yet much fewer reports describe the use of SILP as actual catalysts for organocatalysts.

Interestingly enough, only two examples so far report the preparation of SILP catalysts bearing potential organocatalytic counteranions. You et al. reported for the first time a polystyrene-based support for amino acid ILs.[30] Imidazolium chloride-based monomers were used in the radical preparation of the polymer, then basic treatment using aqueous KOH and subsequent anion metathesis with L-proline gave materials **9** and **10** (Figure 3.15). These polymers were used for metal scavenging and metal catalysis only, and authors did not report any possible activity regarding organocatalysis, despite the presence of the L-prolinate anion. In 2010, Salunkhe and Rachinkar reported the preparation and the complete characterization of a

X = **9** R = H
X = **10** R = CH$_3$

11

FIGURE 3.15 Examples of amino acid-based SILP.

ferrocene-labeled SILP containing L-prolinate as the organocatalytic anion and its application in catalysis (Figure 3.15, compound **11**).[31]

This material was used for the synthesis of 1-amidoalkyl-2-naphthols and Rashinkar and Salunkhe[31] reported excellent yields (up to 87% in the best cases) under solvent-free conditions, and the recycling experiments showed that simple filtration allowed the reuse of the SILP catalyst up to five times with only a slight decrease of activity. However, in accordance with the proposed mechanism, no enantioselectivity was observed and every adduct was obtained as a racemic mixture. Despite progress made toward the development of such SILP catalysts and promising performances regarding catalytic activity and recycling possibilities, work remains to be done for the preparation of efficient chiral SILP catalysts for organocatalysis and for their use as viable alternatives to common chiral ILs catalysts.

3.5 CONCLUSIONS

Our efforts here focused in listing and correlating the few examples describing the design and comprehensive studies regarding the development of ILs bearing catalytically active anions. As highlighted in this chapter, it is obvious that this concept is still in its early years and that the field has yet to be more deeply explored. However, regarding all these studies presented here, it is apparent that such "anionic ILs catalysts" can provide new methods for stereocontrol in organocatalysis, due to their unique mode of action.

Benefitting from ionic liquids credentials in terms of green chemistry, these catalysts are also intrinsically easy to prepare and their properties highly tunable, due to the virtually infinite combination of cations and anions. In light of all of these assets and their mode of action, one can easily see that rational design of such species could open the door to ultra-specific green catalysis, more compatible with the growing attention of the scientific world toward environmental and human issues.

Based upon the examples presented here, future efforts should likely be directed at broadening the scope of these "anionic" IL-based catalysts in organocatalysis and in developing the promising field of supported ionic liquid phase catalysis. At the same time, one can think of expanding this concept to asymmetric metal catalysis, using "anionic" IL-based catalysts as a basis for the design of more "greener" and easy-to-handle alternatives to existing metal catalysts.

REFERENCES

1. Brak, K. and Jacobsen, E. N., *Angew. Chem. Int. Ed.* **2013**, *52* (2), 534.
2. Phipps, R. J., Hamilton, G. L., and Toste, F. D., *Nat. Chem.* **2012**, *4* (8), 603.
3. Rueping, M., Kuenkel, A., and Atodiresei, I., *Chem. Soc. Rev.* **2011**, *40* (9), 4539.
4. Novacek, J. and Waser, M., *Eur. J. Org. Chem.* **2013**, *2013* (4), 637.
5. Enders, D. and Nguyen, T. V., *Org. Biomol. Chem.* **2012**, *10* (28), 5327.
6. Mahlau, M. and List, B., *Angew. Chem. Int. Ed. Engl.* **2013**, *52* (2), 518.
7. Lee, S. G., *Chem. Commun.* **2006**, (10), 1049.
8. Yue, C., Fang, D., Liu, L., and Yi, T.-F., *J. Mol. Liq.* **2011**, *163* (3), 99.
9. Brozinski, H. L., Delaney, J. P., and Henderson, L. C., *Aust. J. Chem.* **2013**, *66* (8), 844.
10. Boon, J. A., Levisky, J. A., Pflug, J. L., and Wilkes, J. S., *J. Org. Chem.* **1986**, *51* (4), 480.

11. Sellin, M. F., Webb, P. B., and Cole-Hamilton, D. J., *Chem. Commun.* **2001**, (8), 781.
12. Gausepohl, R., Buskens, P., Kleinen, J., Bruckmann, A., Lehmann, C. W., Klankermayer, J., and Leitner, W., *Angew. Chem. Int. Ed.* **2006**, *45* (22), 3689.
13. Shi, M. and Chen, L.-H., *Chem. Commun.* **2003**, (11), 1310.
14. Hu, S., Jiang, T., Zhang, Z., Zhu, A., Han, B., Song, J., Xie, Y., and Li, W., *Tetrahedron Lett.* **2007**, *48* (32), 5613.
15. Fukumoto, K., Yoshizawa, M., and Ohno, H., *J. Am. Chem. Soc.* **2005**, *127* (8), 2398.
16. Qian, Y., Xiao, S., Liu, L., and Wang, Y., *Tetrahedron: Asymmetry* **2008**, *19* (13), 1515.
17. Zheng, X., Qian, Y., and Wang, Y., *Catal. Commun.* **2010**, *11* (6), 567.
18. Zheng, X., Qian, Y.-B., and Wang, Y., *Eur. J. Org. Chem.* **2010**, *2010* (3), 515.
19. Qian, Y., Zheng, X., and Wang, Y., *Eur. J. Org. Chem.* **2010**, *2010* (19), 3672.
20. Noujeim, N., Leclercq, L., and Schmitzer, A. R., *Curr. Org. Chem.* **2010**, *14*, 1500.
21. Gauchot, V., Gravel, J., and Schmitzer, A. R., *Eur. J. Org. Chem.* **2012**, *2012* (31), 6280.
22. Gauchot, V. and Schmitzer, A. R., *J. Org. Chem.* **2012**, *77* (11), 4917.
23. List, B., Lerner, R. A., and Barbas, C. F., *J. Am. Chem. Soc.* **2000**, *122* (10), 2395.
24. Gauchot, V., Branca, M., and Schmitzer, A., *Chem. Eur. J.* **2014**, *20* (6), 1530.
25. Riisagera, A., Fehrmanna, R., Haumannb, M., and Wasserscheidb, P., *Top. Catal.* **2006**, *40* (1–4), 91.
26. Baudoux, J., Perrigaud, K., Madec, P.-J., Gaumont, A.-C., and Dez, I., *Green Chem.* **2007**, *9* (12), 1346.
27. Schneider, M. J., Haumann, M., and Wasserscheid, P., *J. Mol. Catal. A: Chem.* **2013**, *376*, 103.
28. Liu, Y., Wang, S.-S., Liu, W., Wan, Q.-X., Wu, H.-H., and Gao, G.-H., *Curr. Org. Chem.* **2009**, *13* (13), 1322.
29. Hagiwara, H., *Synlett* **2012**, *23* (6), 837.
30. Chen, W., Zhang, Y., Zhu, L., Lan, J., Xie, R., and You, J., *J. Am. Chem. Soc.* **2007**, *129* (45), 13879.
31. Rashinkar, G. and Salunkhe, R., *J. Mol. Catal. A: Chem.* **2010**, *316* (1–2), 146.

4 Imidazolium Hydroxides and Catalysis

Cameron C. Weber

CONTENTS

4.1 INTRODUCTION

Base catalysis is important for a wide range of organic transformations. These include, but are certainly not limited to, aldol reactions and condensations; Michael additions; Henry reactions, Dieckmann, Knoevenagel and Claisen–Schmidt condensations; and the Robinson annulation. Given the synthetic importance of base-catalyzed reactions and the use of ionic liquids (ILs) as "designer solvents," an approach to conducting base catalyzed processes has been to utilize ILs that possess a basic anion. This methodology has the advantage of simplifying the reaction medium and the subsequent purification, as the IL acts as both the solvent and as a catalyst for the reaction, reducing the need for other additives. Alternatively, this methodology can be used to enable the base to become miscible with various organic solvents and therefore provide inorganic type bases with organic solvent solubility. As one of the most studied classes of these ILs, the focus of this chapter will be on imidazolium hydroxides and their role in various catalytic transformations. The abbreviations used for the imidazolium salts will be based on the alphanumeric system used by Hallett and Welton, for example, the 1-butyl-2-methyl-3-ethylimidazolium cation would be $[C_4C_2C_1{}^2im]$.[1]

It would be remiss to introduce a chapter on imidazolium hydroxides without discussing the reactivity and general instability of these compounds.[2,3] The hydrogen in the 2-position of the imidazolium ring is mildly acidic. This hydrogen in $[C_1C_1im][I]$ was found to exchange with deuterium in borate buffer solutions at pD 8.92 with a half-life of 4.5 min.[4] This exchange occurs through the formation of an N-heterocyclic (Arduengo-type) carbene intermediate of the type depicted in Scheme 4.1.[5] Consequently, many imidazolium salts generate reactive carbene intermediates in the presence of bases, which will clearly influence their stability and reactivity. When the base is the anion of the imidazolium salt its association with the 2-position of the imidazolium ring is further enhanced due to Coulombic attraction, particularly in the absence of a diluent.[6] Consequently, the reactivity of the 2-position of the imidazolium ring increases in neat imidazolium salts, meaning that salts with anions as weakly basic as acetate behave as "proto-carbenes."[7] When more basic anions such as hydroxide are used, the reaction lies further to the right, leading to the facile formation of the carbene.[2,8] The carbene can then be hydrolyzed leading to further degradation, as illustrated in Scheme 4.1.[9–11] The hydrolysis pathway is equivalent to the nucleophilic addition of hydroxide to the imidazolium salt, which also occurs in these salts. As can be anticipated, the use of imidazolium hydroxides as solvent systems can be fraught with complications particularly if the 2-position of the imidazolium ring is not protected.

Not all imidazolium hydroxides are inherently unstable. Imidazolium salts with unsubstituted 2-positions can be stabilized by sufficiently diluting the salt in a protic solvent such as water to reduce the basicity of the hydroxide ion through hydrogen bonding interactions.[11] Importantly, this requires that these salts are used in a regime where they are not ILs under the conventional definition. Density functional theory calculations suggest that at least two water molecules are required to stabilize a single $[C_2C_1im][OH]$, and even then, many of the lowest energy structures feature a carbene.[12] These results were consistent with calculations for the hydrolysis of the imidazole-2-ylidene, which found that three water molecules were required for the formation of the imidazolium hydroxide to be thermodynamically favorable over other hydrolysis products with the stability increasing further with additional water molecules.[11] Despite these calculated values, significantly higher water concentrations were required experimentally to ensure that the imidazolium hydroxide salt did not undergo degradation under ambient conditions. For example, it has been reported that $[C_2C_1im][OH]$ is unstable at concentrations above 20 wt% in water, which corresponds to over 25 molar equivalents of water.[13] This suggests that imidazolium salts with unsubstituted 2-positions should ideally be used as dilute aqueous solutions rather than as neat liquids to prevent their degradation.

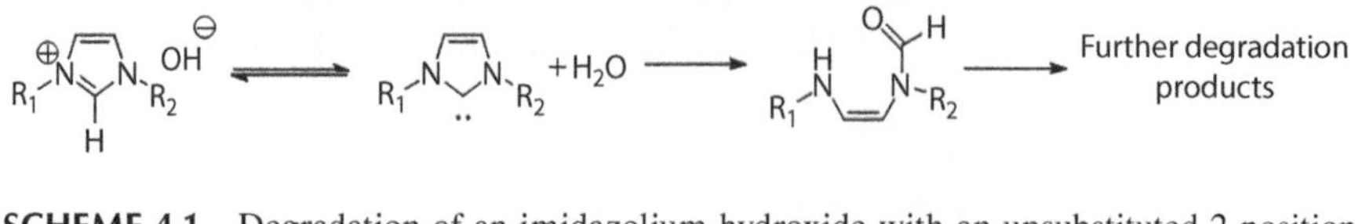

SCHEME 4.1 Degradation of an imidazolium hydroxide with an unsubstituted 2-position via hydrolysis of the intermediate carbene.

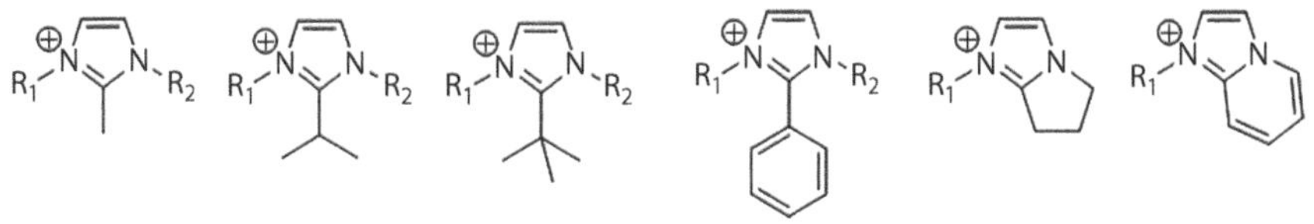

FIGURE 4.1   Some examples of 2-position substitution used to enhance the base stability of the imidazolium cation.

The other approach for the stabilization of imidazolium toward bases is the substitution of the 2-proton with alkyl or aryl groups. Some examples are depicted in Figure 4.1. The simplest substitution is the addition of a methyl group, which has been found to increase the base tolerance of the imidazolium cation.[14–16] However, the 2-methyl substituted imidazolium salts still undergo Hydrogen/Deuterium (H/D) exchange even in the presence of a mild base such as triethylamine.[14] These salts have also been found to be incompatible with strong bases such as Grignard reagents.[17] Modification of the 2-position with an isopropyl group was found to imbue sufficient base stability to the IL to allow the conduct of Grignard reactions.[18] A 2-phenyl substituent also lead to stability toward Grignard reagents for closely related imidazolinium salts.[19] The lack of conjugation around the entire ring of the imidazolinium cation may increase their base stability relative to their imidazolium counterparts, although this effect has not been studied. More exotic 2-position substitutions include the formation of fused bicyclic imidazolium salts (Figure 4.1), which did not undergo H/D substitution even under forcing basic conditions.[20,21] It is important to consider that H/D exchange reflects the acidity of the hydrogens, but not the stability of the cation. Recent publications calculated the Lowest Unoccupied Molecular Orbitals (LUMOs) for a wide range of substituted imidazolium cations which were found to correlate with experimental degradation values.[16,22] It was determined that unbranched alkyl C2 substituents actually facilitated greater stabilization of the imidazolium ring toward degradation than branched substituents due to hyperconjugation effects from the α-C-H groups, despite their increased propensity to undergo H/D exchange. This illustrates an important distinction between the acidity of the imidazolium cation and its stability, as the 2-methyl substituted imidazolium cations were found to be more acidic than the isopropyl variants despite being less prone to degradation under basic conditions.[18,21,22] Most of the stability tests were conducted in the presence of added solvent and not on the neat hydroxide salts, which means these results may not necessarily reflect their absolute stability as ILs.

It is evident from the earlier discussion that imidazolium cations can be rendered more stable toward bases by appropriate substitution of the 2-position. Unfortunately, these substitutions have not been examined for the preparation of neat hydroxide salts, so their efficacy in stabilizing the neat IL cannot be determined. However, hydroxide salts with unsubstituted 2-positions can be used in aqueous solution, provided they are sufficiently dilute to prevent the decomposition of the cation. Unfortunately, many literature reports dealing with imidazolium hydroxides do not address the issue of stability and therefore the role of carbene intermediates or degradation products cannot be clearly ascertained. It is therefore not always clear whether it is the imidazolium hydroxide itself that is responsible for the behavior observed or one of its degradation products. Given how strongly impurities are intertwined with IL chemistry generally and imidazolium hydroxide in particular, this chapter will outline approaches to the synthesis of these salts and their effect on the properties and identity of the resultant material. The application of these salts as catalysts will then be explored with special emphasis placed on the role that carbenes and IL degradation may play and how the use of imidazolium hydroxides compares to simple alternative bases.

4.2   IMIDAZOLIUM HYDROXIDE SYNTHESIS

The synthesis of imidazolium hydroxides has been conducted through conventional ion metathesis from the corresponding imidazolium halide salt in an organic solvent,[23] the use of an anion exchange resin,[24] or a combination of both methods (Scheme 4.2).[25] The metathesis process is controversial as the original report using dichloromethane as a metathesis solvent was found to result in highly colored ILs with significant halide contamination when independently repeated by several research groups.[2,25,26] Despite this, the metathesis synthesis remains widely used for those aiming to produce imidazolium hydroxide salts, while the anion exchange resin approach is more often used to produce an imidazolium hydroxide solution for immediate reaction with acids to form different ILs.[24,27] For the synthesis of imidazolium hydroxide salts free of halide contamination, the use of the anion exchange resin approach is advisable, as conventional approaches for removing residual halide following ion metathesis such as washing with water would not yield any selectivity for the halide over the hydroxide anion.[28] The preparation can also be conducted by coupling the ion metathesis with the use of an anion exchange resin. For example, Peng et al. conducted the metathesis of $[C_4C_1im][Br]$ with KOH in Tetrahydrofuran (THF) followed by passing the resultant liquid through a hydroxide loaded anion exchange resin. This methodology led to halide-free $[C_4C_1im][OH]$, although, as discussed previously, it was only stable as an aqueous solution (30 wt%–40 wt%).[25]

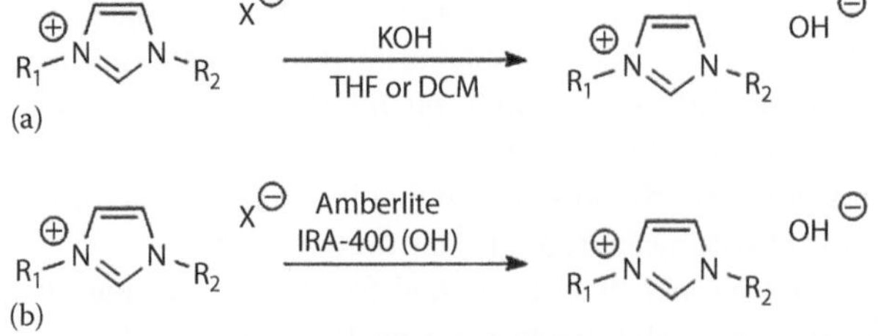

SCHEME 4.2   General synthetic approaches for the preparation of imidazolium hydroxides. X = Cl, Br. (a) Anion metathesis in organic solvent (either THF or DCM), and (b) the use of an anion exchange resin such as Amberlite IRA-400 (OH).

4.3 BASE CATALYZED REACTIONS IN IMIDAZOLIUM HYDROXIDES

4.3.1 MICHAEL ADDITIONS

Michael additions were the first base catalyzed process to be intentionally studied using an imidazolium hydroxide catalyst, in this case [C$_4$C$_1$im][OH].[23] This original study examined 1,3-dicarbonyl compounds as Michael donors with ketone, ester, and nitrile substituted acceptors (Scheme 4.3). Notably, the ester and nitrile substituted acceptors formed *bis*-addition products, while ketone acceptors formed only the anticipated monosubstituted compounds. Given the use of [C$_4$C$_1$im][OH] at 0.6 equivalents, diluted only by reagents, significant carbene concentrations would likely be present. Similar 1,3-dicarbonyl substrates have been shown to react at comparable rates using only 2.5 mol% of a closely related imidazol-2-ylidine carbene.[29] For the carbene catalyzed processes, substrates which were capable of undergoing *bis*-addition were not examined, so direct comparison with the [C$_4$C$_1$im][OH] conditions cannot be made. While it cannot be conclusively determined whether the reactivity observed is exclusively due to the presence of carbenes, it does indicate that these Michael additions can be catalyzed by the carbenes present in the [C$_4$C$_1$im][OH] solution.

Following this initial report, Michael reactions featuring a wider range of donor and acceptor groups have been reported using [C$_4$C$_1$im][OH] as a catalyst and reaction medium. These include activated methylene systems for the addition to styrylisoxazoles as well as thia- and aza-Michael reactions.[30–33] The activated methylene compounds used for the addition to styrylisoxazoles were 1,3-dicarbonyl compounds and 3,5-dimethyl-4-nitroisoxazole (Scheme 4.4).[32] The reaction conditions and yields obtained were similar to those of the original Ranu and Banerjee[23] publication, which is unsurprising given the similarity of the compounds used.

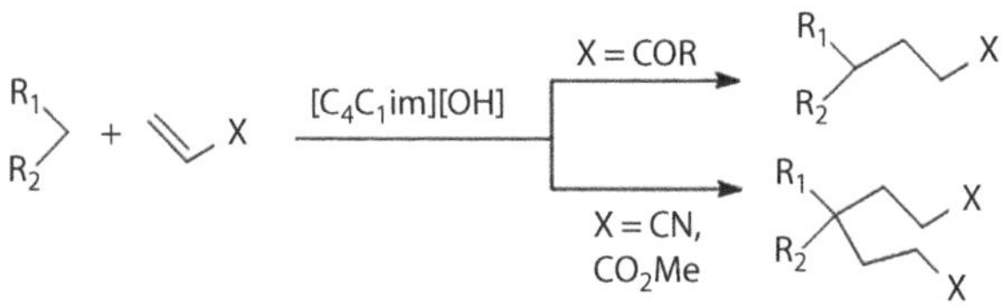

SCHEME 4.3 Michael addition of 1,3-dicarbonyl compounds in [C$_4$C$_1$im][OH]. (Reprinted with permission from Ranu, B.C. and Banerjee, S., *Org. Lett.*, 7, 3049–3052, 2005. Copyright 2005 American Chemical Society.)

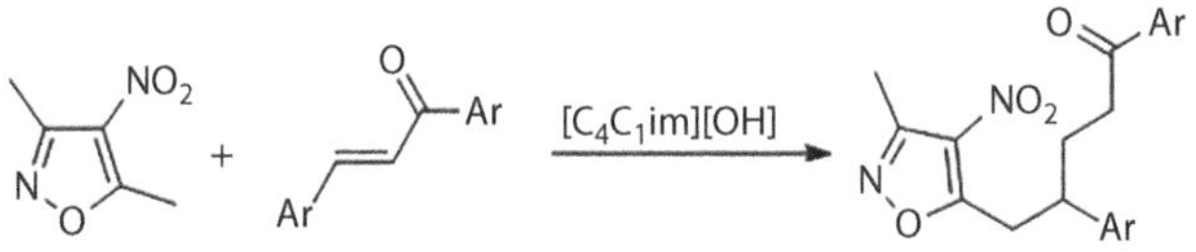

SCHEME 4.4 Reaction of 3,5-dimethyl-4-nitroisoxazole with chalcones. (From Rajanarendar, E. et al., *Indian J. Chem. B*, 50B, 587–592, 2011.)

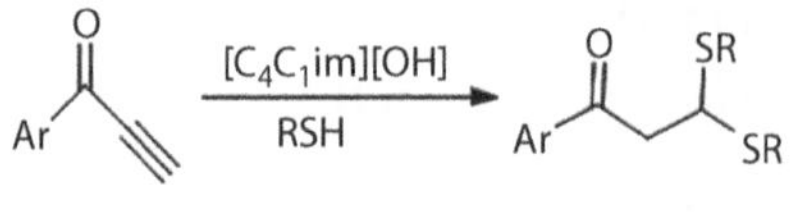

SCHEME 4.5 Thia-Michael additions as studied by Ranu et al. (Reprinted from *Tetrahedron*, 63, Ranu, B.C. et al., Ionic liquid as catalyst and solvent: The remarkable effect of a basic ionic liquid, [bmIm]OH on Michael addition and alkylation of active methylene compounds, 776–782, Copyright 2006, with permission from Elsevier.)

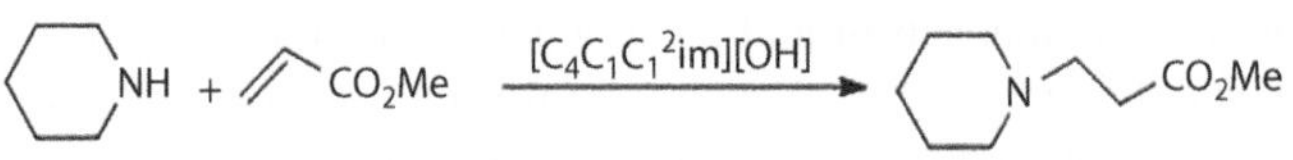

SCHEME 4.6 Aza-Michael reaction of piperidine with methyl acrylate in $[C_4C_1C_1{}^2im][OH]$. (From Xu, J.-M. et al., *Eur. J. Org. Chem.*, 2007, 1798–1802, 2007.)

These systems were found to always give faster rates than in refluxing triethylamine and no reaction was detected in the absence of a base. Unfortunately, there was no comparison with other hydroxide systems, and the use of an unstable imidazolium cation means that again the precise role of carbenes, the hydroxide anion, and other degradation products is unclear.

The thia-Michael addition reactions were studied with alkyne acceptors and generally found to form the disubstituted product (Scheme 4.5).[30] For these reactions, Ranu et al. used $[C_4C_1im][OH]$ as a 10 mol% solution in $[C_4C_1im][Br]$ to limit the reaction rate and polymerization side reactions, as the reaction proceeded too vigorously in neat $[C_4C_1im][OH]$. No control experiments were conducted with other bases or ILs, so the role of carbenes or the importance of the use of an imidazolium hydroxide rather than a simple hydroxide salt cannot be determined.

For the aza-Michael reactions (Scheme 4.6),[31] the reaction between piperidine and methyl acrylate proceeded slowly in conventional organic solvents such as THF and DMSO with an order of magnitude increase in rate in the ILs $[C_4C_1im]$ $[BF_4]$ and $[C_4C_1im][PF_6]$. However, the best results were obtained for hydroxide ILs, with yields of 98% found for $[C_4C_1im][OH]$ in 10 min compared to 60% for $[C_4C_1im]$ $[BF_4]$. In this case, it is likely the basicity of the hydroxide anion rather than *in situ* carbene formation is responsible for the activity, as similar results were obtained for $[C_4C_1C_1{}^2im][OH]$. Obviously, piperidine is inherently more basic than the 1,3-dicarbonyl compounds studied by Ranu and Banerjee,[23] which reduces the need for an added basic catalyst to enable the reaction to proceed.

Aromatic amines such as anilines, imidazoles, and pyrazoles were also studied as donors for aza-Michael additions (Scheme 4.7).[33] Due to their reduced basicity, aromatic amines had much longer reaction times than the aliphatic heterocyclic systems. Notably, no reaction for aniline was observed for a 1:1 $NaOH:[C_4C_1im][BF_4]$ solution, whereas $[C_4C_1im][OH]$ gave 90% yield over the same reaction time. This indicates that in this case, there is an inherent benefit for imidazolium hydroxides over mixtures of an imidazolium salt with a simple hydroxide salt. It is not clear whether this substantial difference in reactivity is due to chemical interactions,

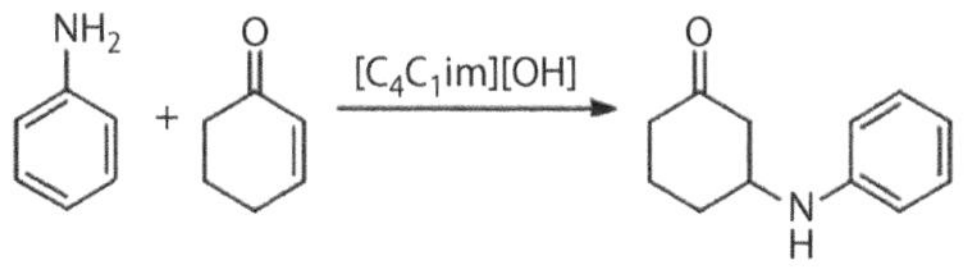

SCHEME 4.7 Reaction of aniline with cyclohexenone, typical aza-Michael addition studied by Yang et al. (Reprinted from *Tetrahedron Lett.*, 47, Yang, L. et al., Highly efficient aza-Michael reactions of aromatic amines and *N*-heterocycles catalyzed by a basic ionic liquid under solvent-free conditions, 7723–7726, Copyright 2006, with permission from Elsevier.)

physical properties of the solution such as viscosity, or the miscibility of reagents as organic bases were found to perform better than inorganic bases. Carbenes are unlikely to be implicated given the poor results for the NaOH: $[C_4C_1im][BF_4]$ system, where such species would also be present.

Overall, it appears that the role of imidazolium hydroxides in Michael additions can be complex and depend on the nature of the Michael donors used. For weakly acidic species such as the carbon acids and thiols, the importance of *in situ* carbenes cannot be ruled out, and there is literature precedent for the role of these carbenes. For the basic species used in the aza-Michael additions, it appears that carbenes are not relevant for the observed reaction behavior, and it is simply the basicity of the hydroxide anion and potentially the reduced viscosity of the imidazolium hydroxide salt relative to inorganic hydroxide/imidazolium IL mixtures that are central to the catalytic activity.

4.3.2 OTHER ADDITION REACTIONS

Given the success of imidazolium hydroxides as catalysts for Michael addition reactions, they have also been investigated for other addition processes. The Markovnikov addition of *N*-heterocycles to vinyl esters was accomplished using $[C_4C_1im][OH]$ as a solvent and catalyst (Scheme 4.8).[34] The reactivity of the imidazoles investigated increased with their acidity, and the reaction did not proceed in organic solvents in the absence of a base. Consequently, $[C_4C_1im][OH]$ likely enables this reaction to proceed through the initial deprotonation of the *N*-heterocycle to produce the more nucleophilic azolide anion followed by its nucleophilic attack on the vinyl group. No other bases were tested under these conditions, so the efficacy of $[C_4C_1im][OH]$ over other bases and the importance of carbenes cannot be determined.

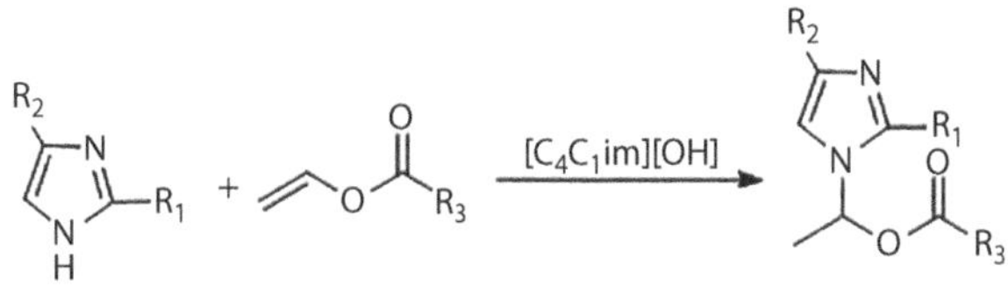

SCHEME 4.8 Markovnikov addition of substituted imidazoles to vinyl esters, a typical reaction studied by Xu et al. (Redrawn with permission from Xu, J.-M. et al., *J. Org. Chem.*, 71, 3991–3993, 2006. Copyright 2006 American Chemical Society.)

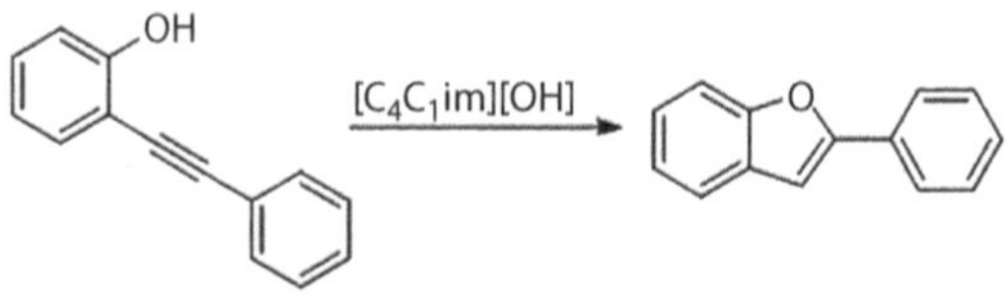

SCHEME 4.9 The heterocyclization of 2-(phenylethynyl)phenol to form 2-phenylbenzofuran, a typical heterocyclization studied by Siddiqui and coworkers. (Reprinted from *Tetrahedron Lett.*, 54, Siddiqui, I.R. et al., Recyclable [bmIm]OH promoted one-pot heterocyclization: Synthesis of substituted benzofurans, 4154–4158, Copyright 2013, with permission from Elsevier.)

The heterocyclization of various oxygen donors with alkynes and nitriles catalyzed by [C₄C₁im][OH] has been studied extensively by Siddiqui and coworkers (Scheme 4.9).[35–37] This methodology was used for the synthesis of benzofurans, isocoumarins, and benzoxazine-4-one derivatives. Ring closure was accomplished either by intramolecular Markovnikov addition, Markovnikov addition followed by a nucleophilic aromatic substitution by the resultant carbanion, or addition to the nitrile followed by nucleophilic aromatic substitution by the resultant carbamate anion. [C₄C₁im][OH] was found to be the best solvent for these reactions, in all cases performing better than a range of inorganic and organic bases in organic solvents or [C₄C₁im][Br] ILs. Part of the role of [C₄C₁im][OH] will be to activate the nucleophile by abstracting a proton in a similar fashion to the Markovnikov addition discussed. In the benzofuran reactions, it was found that inorganic bases were dramatically more active in [C₄C₁im][Br] than in polar organic solvents such as dimethylformamide which implies that carbene catalysis may also be important.

[C₈C₁im][OH] was compared with an imidazolinium hydroxide salt grafted on silica for the catalysis of the cyanosilylation of a range of carbonyl compounds (Scheme 4.10).[38] The grafted imidazolinium salt gave a 91% yield of the desired product from cyclopentanone, while [C₈C₁im][OH] achieved only 19%. The silica itself was not active even with the imidazolinium chloride salt grafted on it. Curiously, the hydroxide salts of the ILs were prepared from ammonia solution rather than the inorganic hydroxide salt or an anion exchange column, and the exchange was conducted in different solvents (THF and acetonitrile for the imidazolinium salt and [C₈C₁im][OH], respectively). This could explain the large discrepancy of activity observed, as the change in solvent as well as the presence of a modified cationic silica surface may influence the extent of ion exchange. It is important to note that *N*-heterocyclic carbenes have been shown to successfully catalyze this process with as low as 0.5 mol% loading,[39] although this wasn't explored in the IL investigation. The success of the

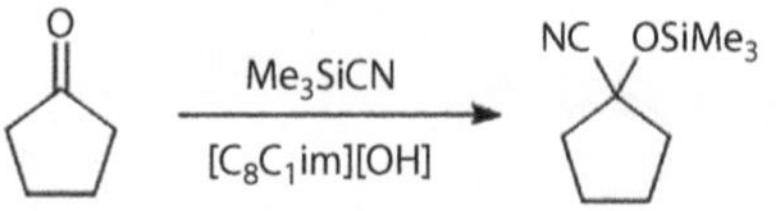

SCHEME 4.10 The cyanosilylation of cyclopentanone by trimethylsilyl cyanide catalyzed by imidazolium or imidazolinium hydroxides, as studied by Yamaguchi et al. (From Yamaguchi, K. et al., *Adv. Synth. Catal.*, 348, 1516–1520, 2006.)

imidazolinium salt for this reaction, however, does suggest that imidazolium hydroxides, with appropriately stable cations and in the presence of the pure hydroxide anion, may also be able to facilitate cyanosilylation reactions.

Collectively, these results illustrate that imidazolium hydroxides can be effective catalysts for addition reactions. The role of carbenes, however, is not clear and the relatively limited results obtained in this regard suggest they may be involved in some of the catalysis observed.

4.3.3 KNOEVENAGEL CONDENSATIONS

The Knoevenagel condensation is one of the most widely studied reactions within imidazolium hydroxides due in part to its synthetic importance. The Knoevenagel condensation of benzaldehyde and malononitrile (Scheme 4.11) was attempted in a mixture of KOH and $[C_4C_1im][PF_6]$ which would generate $[C_4C_1im][OH]$, amongst other species, *in situ*.[40] This combination was found to be favorable for the Knoevenagel reaction with 96% conversion in 6 h. Major limitations were the extraction of the product from the reaction mixture, as it was quite soluble in the IL and the recyclability of the IL system as conversion decreased over multiple uses. Titration of the reaction mixture found that up to 40% of added base was not available for reaction. This was attributed to the formation of carbenes which were identified by NMR, however, carbene formation would not reduce the basicity of the mixture. A more likely cause is the hydrolysis of the hexafluorophosphate anion which generates hydrofluoric acid.[41,42]

Building on this initial study, the Knoevenagel reaction was further explored by using 20 mol% $[C_4C_1im][OH]$ as a catalyst added to the neat reagents (Scheme 4.12).[43]

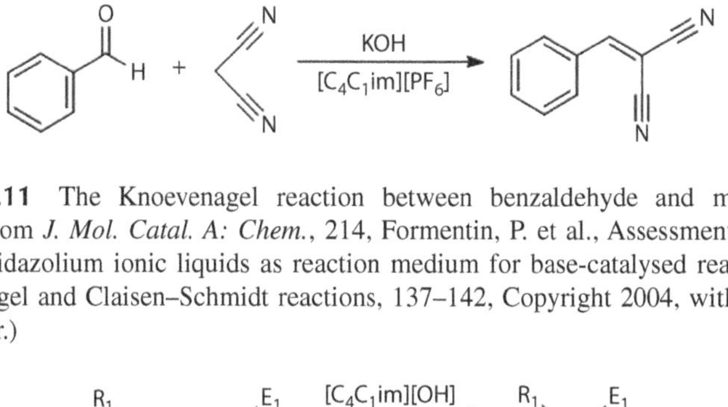

SCHEME 4.11 The Knoevenagel reaction between benzaldehyde and malononitrile. (Reprinted from *J. Mol. Catal. A: Chem.*, 214, Formentin, P. et al., Assessment of the suitability of imidazolium ionic liquids as reaction medium for base-catalysed reactions: Case of Knoevenagel and Claisen–Schmidt reactions, 137–142, Copyright 2004, with permission from Elsevier.)

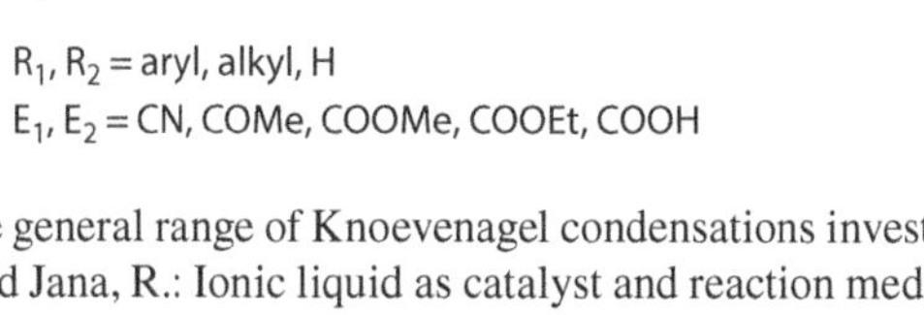

R_1, R_2 = aryl, alkyl, H
E_1, E_2 = CN, COMe, COOMe, COOEt, COOH

SCHEME 4.12 The general range of Knoevenagel condensations investigated by Ranu et al. (From Ranu, B.C. and Jana, R.: Ionic liquid as catalyst and reaction medium—A simple, efficient and green procedure for Knoevenagel condensation of aliphatic and aromatic carbonyl compounds using a task-specific basic ionic liquid. *European Journal of Organic Chemistry*. 2006. 2006. 3767–3770. Copyright Wiley-VCH Verlag GmbH & Co. KGaA. Reproduced with permission.)

This methodology was compatible with a wide range of substrates with reactions performed at room temperature, leading to good isolated yields for most products in under 30 min. This included the reaction of aliphatic aldehydes with malonic esters, which generally react much slower than their aromatic counterparts. It was noted that no side products due to carbene formation were detected, however, the importance of carbenes as catalytic agents was not explored. This methodology was extended to examine the use of grinding, microwave heating, and melting of solid reagents to enable the reaction to proceed more rapidly.[44] Unsurprisingly, given the more forcing conditions used, this approach led to faster reaction times than observed by Ranu and Jana.[43] Acidic or neutral ILs such as $[C_4C_1im][HSO_4]$ and $[C_4C_1im][BF_4]$ did not give any significant conversion even under the forcing conditions, indicating the importance of the basic anion.

Following on from the successful studies on the Knoevenagel condensation in $[C_4C_1im][OH]$, its application in the synthesis of compounds of chemical, biological, and pharmacological interest has been reported. These include 5-benzylidene rhodamine derivatives, 4-thiazolidinones (Scheme 4.13), 4-thiazolidinediones, 3-benzamidocoumarins (Scheme 4.14), and α,β-unsaturated arylsulfones (Scheme 4.15).[45–49] In the preparation of the rhodamine derivatives, $[C_4C_1im][OH]$ is used diluted in water at room temperature and gives good yields, suggesting that carbene intermediates may not be relevant for this reaction. For the syntheses of 4-thiazolidinones and thiazolidinediones, $[C_4C_1im][OH]$ is diluted only by reagents. No side reactions were reported in these cases, and yields of the desired products

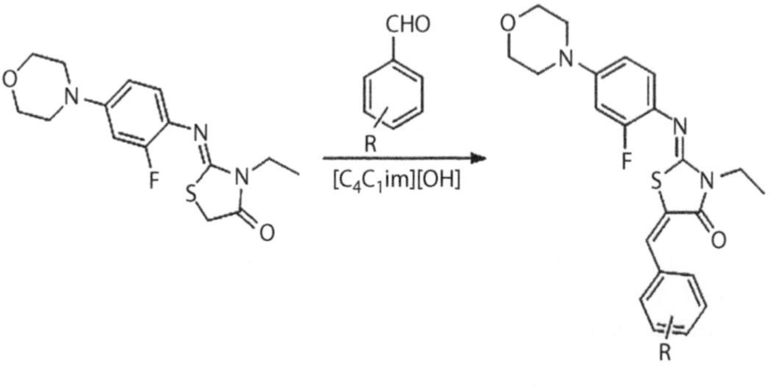

SCHEME 4.13   Knoevenagel condensation in $[C_4C_1im][OH]$ as the key step in the preparation of a variety of 4-thiazolidinones. (From Patil, S.G. et al., *Chinese Chem. Lett.*, 22, 883–886, 2011.)

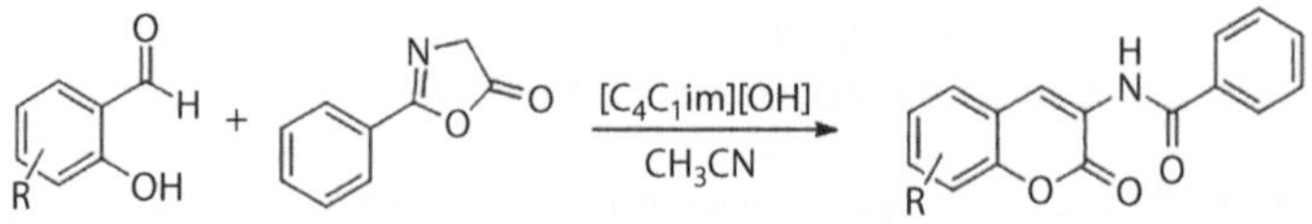

SCHEME 4.14   Preparation of 3-benzamidocoumarins from the Knoevenagel condensation. (Reprinted from *Tetrahedron Lett.*, 50, Yadav, L.D.S. et al., A one-pot [Bmim]OH-mediated synthesis of 3-benzamidocoumarins, 2208–2212, Copyright 2009, with permission from Elsevier.)

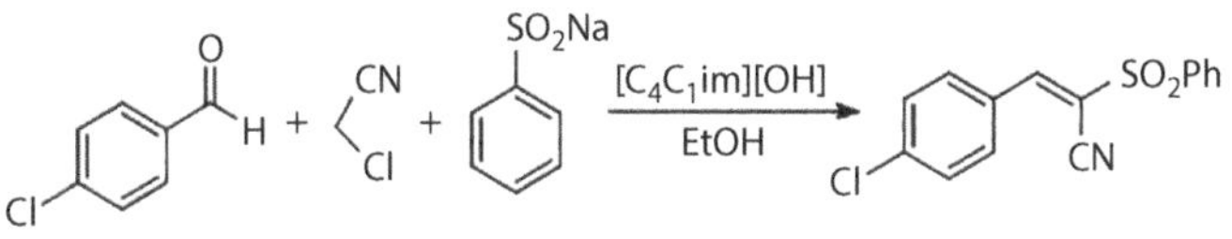

SCHEME 4.15 A typical synthesis of an α,β-unsaturated arylsulfone, as described by Zhang et al. (From Zhang, L. et al., *Chinese Chem. Lett.*, 23, 1352–1354, 2012.)

were good indicators that side reactions due to the solvent instability were not significant. Interestingly, the $[C_4C_1im][OH]$ used for the thiazolidinones was prepared using the ion metathesis method, while $[C_4C_1im][OH]$ used for the thiazolidinediones was prepared by ion exchange. The similarity of outcomes implies that, at least in these examples, the anticipated difference in solvent impurity profiles arising from the IL synthesis does not affect its reactivity.

The syntheses of 3-benzamidocoumarins (Scheme 4.14) and α,β-unsaturated arylsulfones (Scheme 4.15) both utilized $[C_4C_1im][OH]$ at low concentrations (40 mM and 50 mM, respectively) diluted by solvent. For the 3-benzamidocoumarin synthesis, it was found that yields were not improved by heating above room temperature and that acetonitrile was the best performing solvent, although methanol gave similar yields. No side products and good yields were observed with this protocol. For the arylsulfone synthesis, the reaction was conducted at reflux in a range of solvents, with alcohols giving the best yields. This was ascribed to the solubility of the reagents, although the role of the alcohol in stabilizing the IL may also be a factor. The presence of byproducts was not reported although moderate to good yields were generally observed. This demonstrates that these imidazolium hydroxides have potential utility as catalysts for synthesis once the issues regarding their stability are fully addressed.

These examples all illustrate that Knoevenagel condensations generally proceed favorably using $[C_4C_1im][OH]$ as a catalyst either diluted only by reagents or diluted by another solvent. The consistent outcomes across this range of conditions imply that degradation products and carbene formation play relatively inconsequential roles in the use of imidazolium hydroxides as catalysts for these reactions.

4.3.4 COMBINATION OF THE KNOEVENAGEL CONDENSATION AND MICHAEL ADDITION

Following on from the success of $[C_4C_1im][OH]$ as a catalyst for Knoevenagel condensations and Michael additions, their combination has been explored for the synthesis of a wide range of different compounds. These include the synthesis of functionalized pyridimidines (Scheme 4.16), furocoumarins (Scheme 4.17), tetrahydrobenzo[b]pyrans, benzo[g]chromenes (Scheme 4.18), highly substituted pyridines (Scheme 4.19), and 2-amino-2-chromenes (Scheme 4.20).[50–56]

The synthesis of the pyrimidines (Scheme 4.16) utilized microwave irradiation for a period of 2–3 min using $[C_4C_1im][OH]$ diluted only by reagents. It was found that this procedure gave improved yields over conventional heating and compared to $[C_4C_1im][OH]$ dissolved in ethanol. When compared under similar conditions,

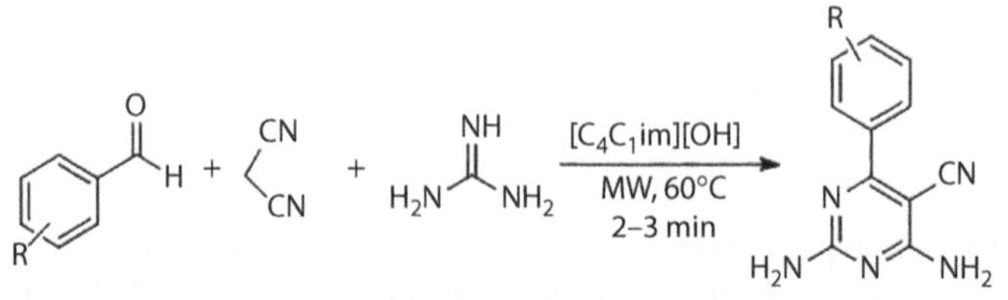

SCHEME 4.16 The synthesis of 2,4-diamino-5-pyrimidinecarbonitrile derivatives from aromatic aldehydes, malononitrile, and guanidine. (Redrawn from Raghuvanshi, D.S. and Singh, K.N., *J. Heterocyclic Chem.*, 48, 582–585, 2011. Copyright 2011, with permission from HeteroCorporation.)

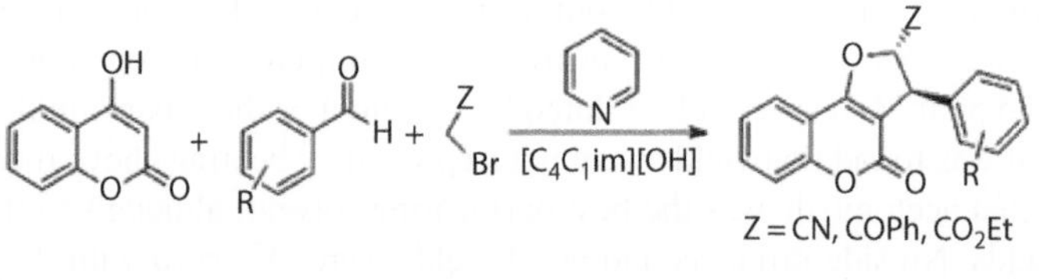

Z = CN, COPh, CO$_2$Et

SCHEME 4.17 Synthesis of furocoumarins in [C$_4$C$_1$im][OH]. (Reprinted from *Tetrahedron*, 68, Rajesh, S.M. et al., Facile ionic liquid-mediated, three-component sequential reactions for the green, regio- and diastereoselective synthesis of furocoumarins, 5631–5636, Copyright 2012, with permission from Elsevier.)

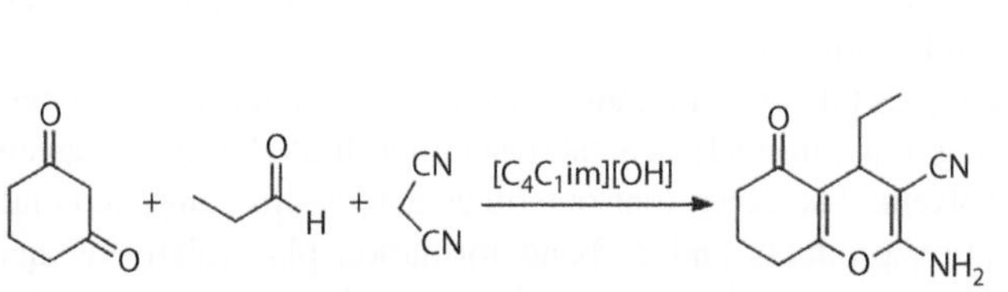

SCHEME 4.18 A typical tetrahydrobenzo[b]pyran synthesis, as described by Ranu et al. (From Ranu, B.C. et al., *Indian J. Chem. B*, 47B, 1108–1112, 2008.)

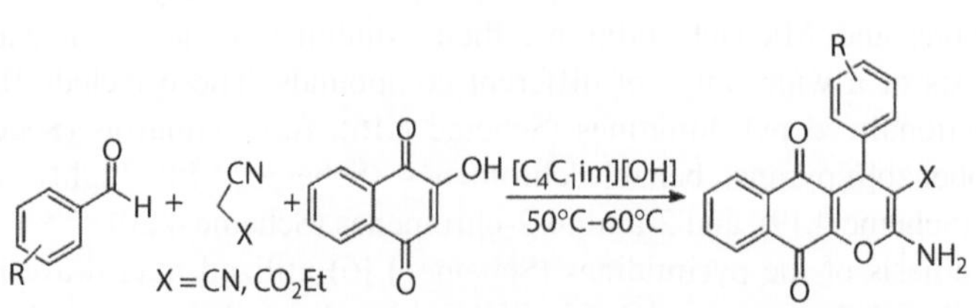

X = CN, CO$_2$Et

SCHEME 4.19 Synthesis of benzo[g]chromene derivatives. (From Khurana, J.M. et al., *Synthetic Commun.*, 42, 3211–3219, 2012.)

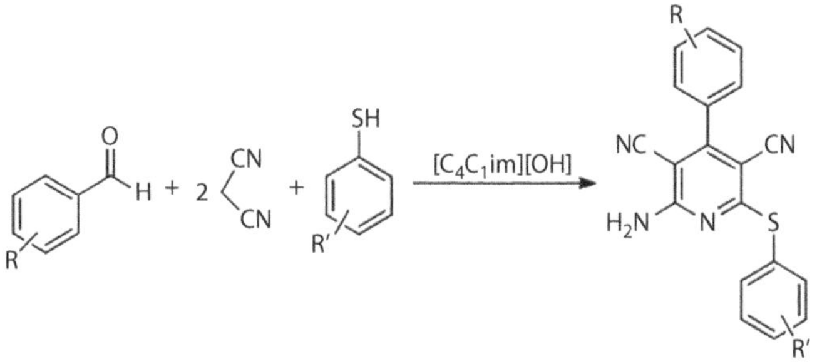

SCHEME 4.20 Synthesis of highly substituted pyridines from the condensation of malononitrile, aromatic aldehydes, and thiophenol derivatives. (Reprinted with permission from Ranu, B.C. et al., *J. Org. Chem.*, 72, 3152–3154, 2007. Copyright 2007 American Chemical Society.)

[C_4C_1im][OH] and NaOH performed equally well, which suggests that the efficacy of [C_4C_1im][OH] likely arises from its miscibility with reagents rather than any inherent synergy with the imidazolium cation.

In an archetypal example of the furocoumarin synthesis (Scheme 4.17), 1-(cyanomethyl)pyridinium bromide was formed *in situ* from pyridine and bromoacetonitrile. The 4-hydroxycoumarin then underwent a Knoevenagel condensation with *p*-tolualdehyde followed by a Michael reaction with the pyridinium-ylide formed *in situ* by deprotonation of the pyridinium salt, and finally heterocyclization occurred from the nucleophilic attack of the ketone to yield the final furocoumarins. The use of [C_4C_1im][OH] gave excellent yields, whereas organic bases such as triethylamine and 1,8-Diazabicyclo(5.4.0)undec-7-ene (DBU) gave relatively poor yields. 2 mole equivalents of [C_4C_1im][OH] were used with yields decreasing with lower loading, so its role as a catalyst is not clear. The method of [C_4C_1im][OH] synthesis and halide content was not reported, so it is possible that the actual hydroxide loading was lower than the reported 2 mole equivalents.

[C_4C_1im][OH] diluted solely by reagents was used for the synthesis of tetrahydrobenzo[*b*]pyrans (Scheme 4.18). The reactants used were malononitrile or ethylcyanoacetate, a cyclohexa-1,3-dione and an aldehyde with [C_4C_1im][OH] at 20 mol% concentration.[53] The reaction was conducted at room temperature and yields over 90% were obtained in under 20 min. While tetrahydrobenzo[*b*]pyran syntheses can proceed with excellent yields in conventional ILs such as [C_8C_1im][PF_6] for aromatic aldehydes and certain aliphatic aldehydes,[57] albeit over a slightly longer reaction time of 3 h, it was reported that the hydroxide anion is necessary for the reaction to proceed with all aliphatic substrates, although no reactions involving these substrates with other ILs were discussed.[53]

The preparation of benzo[*g*]chromene derivatives uses [C_4C_1im][OH] diluted with ethanol and good yields are obtained with slightly longer reaction times than observed for the tetrahydrobenzo[*b*]pyran derivatives.[55] Notably, the reaction does not proceed at all when water is used as a solvent, which implies that either carbene intermediates or ion-pairing may play a role in the catalysis observed. The synthesis of benzo[*g*]chromene derivatives was also explored by Khurana et al., although in this case using neat rather than diluted [C$_4$C$_1$im][OH] (Scheme 4.19).[56] Reaction times were generally comparable to those of Yu et al.,[55] where [C$_4$C$_1$im][OH] was diluted despite the use of the neat IL. Not all substrates studied were successfully able to form the desired benzochromene derivative, with 2-hydroxy-1,4-naphthoquinone not reacting with ethylcyanoacetate or its condensation product. Strangely, Yu et al.[55] report good yields for these compounds when the reaction is at room temperature in ethanol rather than diluted only by reagents at 50°C–60°C, which suggests that perhaps side reactions with the unstable ILs may be playing a role and limiting its efficacy as a reaction medium.

Highly substituted pyridines were synthesized through the Knoevenagel condensation of malononitrile with an aryl aldehyde followed by the Michael addition of a second malononitrile molecule with thiolate addition to the nitrile and cyclization to form dihydropyridine.[50] The dihydropyridine aromatizes to form the resultant pyridine (Scheme 4.20). The use of [C$_4$C$_1$im][Br] or [C$_4$C$_1$im][BF$_4$] did not allow the reaction to proceed beyond the initial Knoevenagel condensation. Whether this is due to the lack of hydroxide anions, carbenes, or a combination of the two cannot be fully established from these investigations.

2-amino-2-chromenes were synthesized in aqueous solutions of [C$_4$C$_1$im][OH], which would minimize the role of carbenes in the chemistry (Scheme 4.21).[51] This reaction was conducted at 100°C, unlike the closely related reactions of Ranu et al.[53] and Yu et al.[55] which were performed at room temperature. [C$_4$C$_1$im][OH] performed slightly better for the synthesis than NaOH, with yields of approximately 90% being obtained after 10 min for the former and 30 min for the latter. It is worth noting that the IL [C$_4$C$_1$im][BF$_4$], which is not able to act as a base, was able to achieve yields greater than 80% after 1 h, likely due to the high temperatures used and the presence of water. No control experiment in refluxing water was reported to determine the role of the added salt. Similar chromene moieties have been synthesized using neat [C$_4$C$_1$im][OH].[58] In these studies, a range of basic ILs were used, and while [C$_4$C$_1$im][OH] gave the highest yields, ILs with less basic functionalities such as acetate anions or cations functionalized with amines gave very similar results. Density functional theory calculations were conducted to give insight into the mechanism and predict that the role of the hydroxide is primarily as a base, and the imidazolium cation provides stabilization of the carbonyl and nitrile moieties.

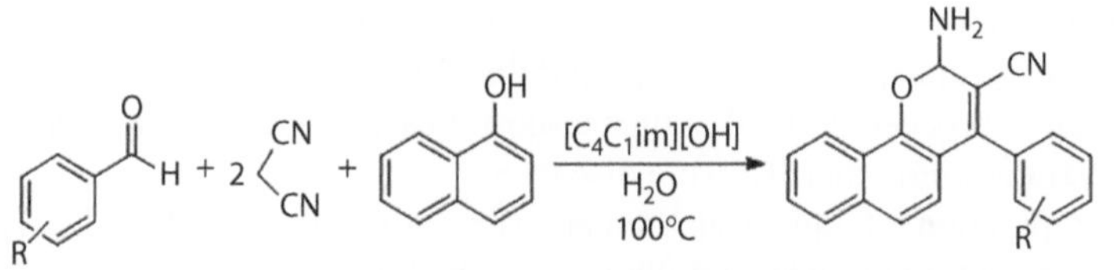

SCHEME 4.21 Typical synthesis of 2-amino-2-chromenes from an aromatic aldehyde, malononitrile and 1-naphthol. (Reprinted from *Catal. Commun.*, 9, Gong, K. et al., Basic ionic liquid as catalyst for the rapid and green synthesis of substituted 2-amino-2-chromenes in aqueous media, 650–653. Copyright 2007, with permission from Elsevier.)

These examples illustrate that imidazolium hydroxides can act as catalysts for multicomponent reactions featuring both Michael addition and Knoevenagel condensation steps, among others. Unfortunately, many of these examples do not examine the precise role or necessity of the imidazolium cation or the implication of the presence of carbenes. Studies that have examined both imidazolium hydroxides and inorganic bases suggest that the imidazolium salts may be more active for these transformations, but substantially more work needs to be conducted to confirm and understand these results.

4.3.5 TRANSESTERIFICATION

Transesterification is the main chemical conversion utilized in biodiesel production. Specifically, triglycerides in the feedstock are transesterified with a simple alcohol such as methanol or ethanol to form the desired alkyl esters (Scheme 4.22). Most commonly, a simple inorganic base such as NaOH or KOH is used as a catalyst for this process, although imidazolium hydroxides have also been investigated in this role.[59]

In one study, the transesterification of glyceryl trioleate, a representative triglyceride, with methanol was examined using a range of imidazolium hydroxide catalysts with varying alkyl chain lengths and compared to inorganic bases such as NaOH and KOH.[60] The imidazolium hydroxides generally fared poorly compared to the inorganic hydroxides with the methyl ester yields for the former in the range 50%–60%, with yields for the inorganic hydroxides above 80%. Optimization of conditions for $[C_4C_1im][OH]$ did allow for yields of up to 87.2% to be achieved. The IL was recycled six times with a slight decrease in yield to 81% after the sixth reuse.

The reaction of glyceryl trioleate with methanol has also been performed by $[(C_1=C_2)C_{12}im][OH]$ immobilized on either $CoFe_2O_4$ nanoparticles or a mesporous $SiO_2/CoFe_2O_4$ support.[61] The magnetic $CoFe_2O_4$ support was used to aid the separation of the catalyst. In this case, the supported ILs performed better than NaOH for the transesterification, although no control experiments conducted with the support were reported, so whether it is innocent in the reaction is unclear. The activity of each catalyst declined significantly upon recycling, which is likely due to decomposition of the IL and loss of hydroxide ions, as the transesterification reactions were conducted at 170°C. The mesoporous catalyst activity decreased more slowly than the other supported catalyst, which was attributed to slower degradation due to restricted diffusion within the mesoporous matrix.

In an example using a real feedstock, $[C_4C_1im][OH]$ was employed as a catalyst for the transesterification of pretreated waste oil. Optimization led to 95.7% conversion

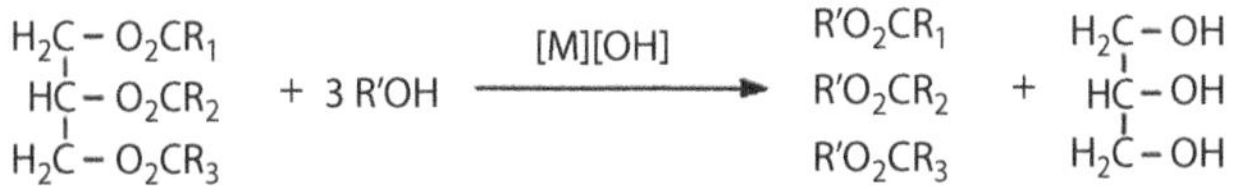

$$
\begin{array}{l}
H_2C-O_2CR_1 \\
HC-O_2CR_2 \\
H_2C-O_2CR_3
\end{array}
+ 3\ R'OH \xrightarrow{\text{[M][OH]}}
\begin{array}{l}
R'O_2CR_1 \\
R'O_2CR_2 \\
R'O_2CR_3
\end{array}
+
\begin{array}{l}
H_2C-OH \\
HC-OH \\
H_2C-OH
\end{array}
$$

SCHEME 4.22 The general transesterification process for the formation of biodiesel from triglycerides.

of the triglycerides to the desired methyl esters.[62] The optimized conditions were an 8:1 molar ratio of alcohol to oil, 70°C for 110 min and 3.0% [C_4C_1im][OH] loading. No saponification due to hydrolysis of the triglycerides was observed. Unfortunately, the results were not compared with other common methods of transesterification such as the use of simple inorganic salts with the same feedstock, so the importance of the imidazolium cation is unclear. Comparison with values obtained for other waste oils using conventional catalysts, however, suggests that these conditions are harsher and lead to lower yields than can be achieved with significantly cheaper catalysts such as KOH.[63] This difference may also be due to feedstock variation and pretreatment conditions. Collectively, these investigations demonstrate that while imidazolium hydroxides are active for the transesterification of triglycerides to form fatty acid methyl esters, they rarely display any advantage over simpler, far cheaper, and more stable inorganic hydroxide salts.

Transesterification reactions are not only used for biodiesel production. Imidazolium hydroxides have been investigated for the transesterification of dimethylcarbonate to produce various alkyl carbonates. Yi et al. utilized [C_4C_1im][OH] and [(C_1=C_2)C_1im][OH] as catalysts for the transesterification of dimethylcarbonate with glycerol to form glycerol 1,2-carbonate, which has a potential role as a renewable chemical feedstock (Scheme 4.23).[64] Both hydroxide ILs were found to give similar conversions to the desired carbonate as K_2CO_3, but with improved selectivity. These hydroxide ILs were not the best performing catalysts, however, with ILs possessing imidazolide anions giving better conversions with similar selectivity (73.4% compared to 60% for [C_4C_1im][OH]). The better performance of the imidazolide was attributed to its more effective hydrogen bonding interactions with the glycerol. A similar reaction utilizing pentanol to form dipentylcarbonate has also been studied and optimized using [C_4C_1im][OH].[65] Their approach led to 75% yield of the desired dipentylcarbonate, although no direct comparison was made with inorganic bases, so the importance of the cation is not clear.

The production of dimethylcarbonate through a transesterification route has also been examined.[66] In this case, the transesterification of ethylene carbonate with methanol was investigated using 1-(triethoxy)silylpropyl-3-methylimidazolium hydroxide ([SiC_1im][OH]) that was condensed onto a mesocellular silica foam, a 3D structured mesoporous material with large mesopores, as a heterogeneous catalyst. This material was able to achieve yields of 82.9%, which is comparable to the neat IL (86.1%). Both yields were substantially higher than could be achieved using either the

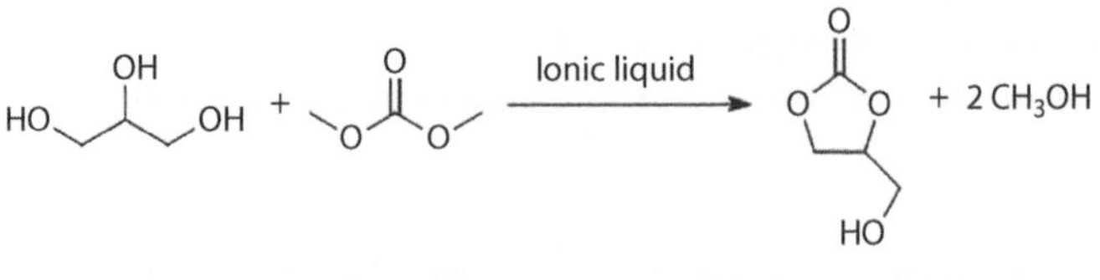

SCHEME 4.23 Transesterification of glycerol with dimethylcarbonate to form glycerol 1,2-carbonate. (Redrawn from Yi, Y. et al., *Chinese J. Catal.*, 35, 757–762, 2014. Copyright 2014, with permission from Dalian Institute of Chemical Physics, the Chinese Academy of Sciences.)

[SC$_1$im][Cl] salt or the support alone. Catalytic performance decreased slightly upon recycling with an approximately 10% decrease in conversion after five cycles. The yields obtained by this catalyst exceeded those reported for this reaction by other catalysts despite being conducted at a lower temperature. These reports indicate that imidazolium hydroxides can be used for the conduct of transesterification reactions, although at present it has not been established whether they offer substantial benefits over inorganic bases or the role played by *in situ* carbenes.

4.3.6 ACTIVATION OF CO$_2$

CO$_2$ is an attractive reagent for chemical synthesis due to its low cost, abundance, and desire to prevent its escape to the atmosphere due to its role as a greenhouse gas. CO$_2$ can also act as a replacement single carbon building block over toxic compounds such as phosgene. One of the initial studies on the utilization of CO$_2$ within ILs found that the use of CsOH within [C$_4$C$_1$im]Cl could give good yields of symmetric *N,N'*-disubstituted ureas (Scheme 4.24) without the addition of a drying agent.[67] No explicit rationale was given for these results. The most likely explanation is that the strong coordination of water to the [Cl]$^-$ and [OH]$^-$ anions drives the dehydration reaction. The [OH]$^-$ anion would also increase the effective solution concentration of CO$_2$ through the *in situ* formation of [HCO$_3$]$^-$. The use of KOH rather than CsOH led to significantly reduced yields (53.5% versus 98%), which was ascribed to the different basicity of the salts, although given the strong dissociation power of ILs on ionic compounds, serendipitous water may play a larger role than the cation of the base. If the strong coordination of water to the anion drives this reactivity, then imidazolium hydroxides, notwithstanding their stability issues, should enable similar reactions to proceed in the absence of added alkali hydroxides.

The use of [C$_4$C$_1$im][OH] as a catalyst both in the presence and absence of added solvent for the earlier reaction has been explored.[68] The maximum yield obtained with [C$_4$C$_1$im][OH] was 47.9% for the *N,N'*-dicyclohexylurea described previously, compared with 98% in the CsOH/[C$_4$C$_1$im][Cl] system. However, [C$_4$C$_1$im][OH] was used at 15 mol% compared to the much higher [C$_4$C$_1$im][Cl] concentrations (~100 mol%), so if the equilibrium was affected by the coordination of water to the anion, the low anion concentration would account for the reduced yields. Under these conditions, it would be expected that the [HCO$_3$]$^-$ anion would form from the reaction of [OH]$^-$ with CO$_2$, however, according to Jiang et al.,[68] [C$_4$C$_1$im][OH] exposed to 5.5 MPa at 170°C for 19 h did not show evidence of undergoing any reaction. This may also indicate that the reaction was reversible upon removal of CO$_2$. That the [C$_4$C$_1$im][OH] salt did not undergo significant degradation under such conditions is itself remarkable, given reports[2,3,8–11] of its instability under

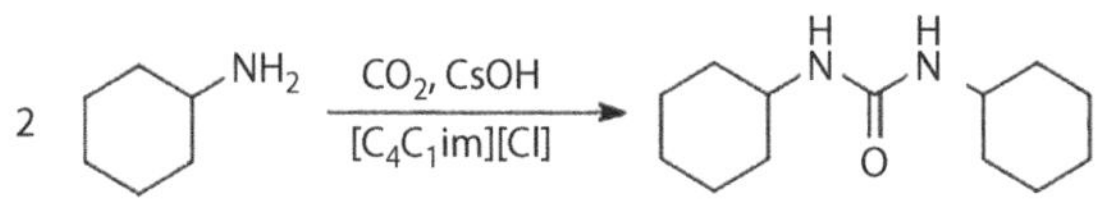

SCHEME 4.24 Typical synthesis of substituted ureas from amines and CO$_2$. (From Shi, F. et al., *Angew. Chem. Int. Ed.*, 42, 3257–3260, 2003.)

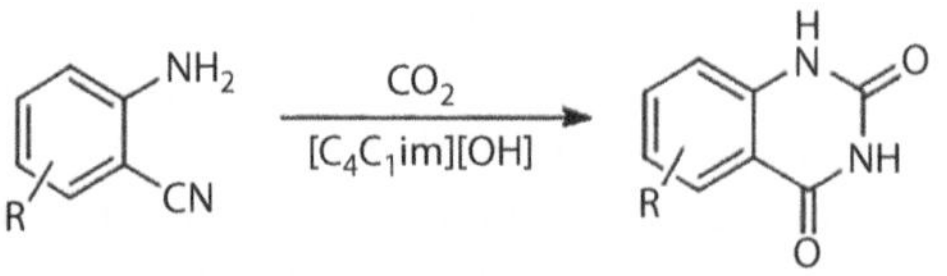

SCHEME 4.25 Synthesis of quinazoline-2,4(1*H*,3*H*)-diones from 2-aminobenzonitriles. (Reprinted from *Catal. Today*, 148, Patil, Y.P. et al., Synthesis of quinazoline-2,4(1*H*,3*H*)-diones from carbon dioxide and 2-aminobenzonitriles using [Bmim]OH as a homogeneous recyclable catalyst, 355–360, Copyright 2009, with permission from Elsevier.)

much milder conditions by a number of authors, and may be due to the *in situ* formation of the less basic $[HCO_3]^-$ anion.

The use of imidazolium hydroxides for the carbonylation of amines has been extended to the synthesis of quinazoline-2,4(1*H*,3*H*)-diones, which are used as pharmaceutical intermediates, from 2-aminobenzonitriles (Scheme 4.25).[69–71] The initial study on this synthesis found that 5 mol% $[C_4C_1im][OH]$ under 3 MPa of CO_2 at 120°C for 18 h with no added solvent was capable of producing the desired product in 90% yield. Further increases in catalyst loading did not improve the yield, and the use of relatively strong inorganic bases such as potassium *tert*-butoxide gave no product or very low yields. Meanwhile, triethylamine, a weaker base, but stronger nucleophile, gave yields of 28%. A computational study on this reaction proposed that the reaction is aided by nucleophilic catalysis of the carbene rather than the base catalysis as was proposed.[71] Preparation of a silica-supported $[C_6C_1im][OH]$ enabled this reaction to be conducted heterogeneously, but led to lower yields of around 40%–50% except when DMF was used as a solvent.[70] Given the harsh reaction conditions and the tendency of DMF to decompose into CO, it is likely the results in DMF are an artifact of the experimental conditions. This reiterates the inherent complexities of catalysis using imidazolium hydroxides due to the potential for carbene formation.

The other major synthetic application of CO_2 explored using imidazolium hydroxide catalysis is the formation of dialkylcarbonates or cyclic carbonates, namely, dimethylcarbonate and propylenecarbonate.[72–74] The formation of dimethylcarbonate directly from the reaction of methanol with CO_2 was attempted using the imidazolium hydroxide salts $[C_2C_1im][OH]$, $[C_4C_1im][OH]$, and $[(HOC_2)C_1im][OH]$ among others.[72] The imidazolium salts performed poorly compared to choline hydroxide, which was able to produce over three times the amount of dimethylcarbonate as $[(HOC_2)C_1im][OH]$, the best performing imidazolium salt. The use of electrolysis to accomplish this reaction has also been attempted with a 4.0 V potential applied for 60 h to a solution of IL electrolyte in methanol.[73] While $[C_4C_1im][OH]$ was active for this reaction, it led to lower yields of dimethylcarbonate than either $[C_4C_1im][Cl]$ or $[BzC_1im][Cl]$, which when combined with its instability indicates that imidazolium hydroxides do not appear to be useful catalysts for this process. Attempts to use a two-step synthesis of dimethylcarbonate from the cycloaddition of CO_2 to ethylene oxide followed by its transesterification with methanol catalyzed by imidazolium ILs in the presence of basic anions were also unsuccessful.[75] This result was attributed to the imidazolium cation reducing the basicity of the added base.

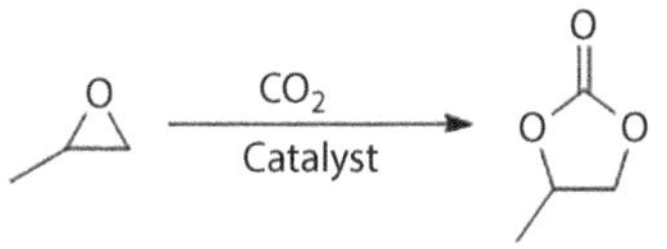

SCHEME 4.26 Cycloaddition of CO_2 with propylene oxide to form propylene carbonate. (From Zhang, X. et al., *Catal. Commun.*, 11, 43–46, 2009.)

The cycloaddition of CO_2 to propylene oxide to form propylene carbonate (Scheme 4.26) was studied with ILs that were grafted onto a silica support through the use of silyl side chains.[74] The grafted imidazolium hydroxide ILs were found to exhibit excellent activity, achieving yields of 95% compared to 83% yield obtained by the industrially used homogeneous KI catalyst. The role of the hydroxide anion is not clear given the reaction generally proceeds via nucleophilic rather than basic catalysis and indicates that this may be another example where carbenes, much better nucleophiles than the $[OH]^-$ anion, may be implicated in the activity.

In conclusion, the carbonylation of amines is favored in imidazolium hydroxides due to the strong affinity of water for the anion and potentially their reactivity with CO_2. Carbenes may also participate through nucleophilic catalysis for appropriate substrates, an effect that may also be responsible for their role in the cycloaddition of CO_2 to epoxides. The synthesis of carbonates from simple alcohols is ineffective using imidazolium hydroxide catalysts due to their instability. Revisiting the participation of imidazolium hydroxides in all of these reactions using more base stable cations may be worthwhile to fully elucidate their role.

4.4 CONCLUSIONS

Imidazolium hydroxides have clearly been shown to be useful for a range of catalytic transformations, particularly Michael additions and Knoevenagel condensations, both as a solvent and diluted by another solvent. Nonetheless, the inherent instability of these compounds is currently underreported and casts doubt on the active catalytic species, particularly the role of carbenes, and the actual identity of the material used for many of these investigations. Despite the discovery of more base stable cations, most studies have focused on the use of $[C_4C_1im][OH]$ and closely related dialkylimidazolium cations. If more stable and pure imidazolium hydroxides can be used for the transformations highlighted, then the utility of this class of compounds can be fully realized, and their role in basic catalysis delineated. Without this advance, imidazolium hydroxides may remain underutilized due to reproducibility and stability concerns.

REFERENCES

1. J. P. Hallett and T. Welton. Room-temperature ionic liquids. solvents for synthesis and catalysis. 2. *Chem. Rev.* 111 (2011): 3508–3576.
2. A. K. L. Yuen, A. F. Masters, and T. Maschmeyer. 1,3-Disubstituted imidazolium hydroxides: Dry salts or wet carbenes? *Catal. Today* 200 (2013): 9–16.

3. S. Chowdhury, R. S. Mohan, and J. L. Scott. Reactivity of ionic liquids. *Tetrahedron* 63 (2007): 2363–2389.

4. R. A. Olofson, W. R. Thompson, and J. S. Michelman. Heterocyclic nitrogen ylides. *J. Am. Chem. Soc.* 86 (1964): 1865–1866.

5. A. J. Arduengo. Looking for stable carbenes: The difficulty in starting anew. *Acc. Chem. Res.* 32 (1999): 913–921.

6. T. Cremer, C. Kolbeck, K. R. J. Lovelock et al. Towards a molecular understanding of cation-anion interaction—Probing the electronic structure of imidazolium ionic liquids by NMR spectroscopy, X-ray photoelectron spectroscopy and theoretical calculations. *Chem. Eur. J.* 16 (2010): 9018–9033.

7. H. Rodriguez, G. Gurau, J. D. Holbrey, and R. D. Rogers. Reaction of elemental chalcogens with imidazolium acetates to yield imidazole-2-chalcogenones: Direct evidence for ionic liquids as proto-carbenes. *Chem. Commun.* 47 (2011): 3222–3224.

8. S. Sowmiah, V. Srinivasadesikan, M. C. Tseng, and Y. H. Chu. On the chemical stabilities of ionic liquids. *Molecules* 14 (2009): 3780–3813.

9. H. Long and B. Pivovar. Hydroxide degradation pathways for imidazolium cations: A DFT Study. *J. Phys. Chem. C* 118 (2014): 9880–9888.

10. Y. Ye and Y. A. Elabd. Relative chemical stability of imidazolium-based alkaline anion exchange polymerized ionic liquids. *Macromolecules* 44 (2011): 8494–8503.

11. O. Hollóczki, P. Terleczky, D. Szieberth, G. Mourgas, D. Gudat, and L. Nyulászi. Hydrolysis of imidazole-2-ylidenes. *J. Am. Chem. Soc.* 133 (2011): 780–789.

12. Z. Song, H. Wang, and L. Xing. Density functional theory study of the ionic liquid [emim] OH and complexes [emim]OH(H$_2$O)$_n$ (n=1,2). *J. Solution Chem.* 38 (2009): 1139–1154.

13. S. Himmler, A. König, and P. Wasserscheid. Synthesis of [EMIM]OH *via* bipolar membrane electrodialysis—Precursor production for the combinatorial synthesis of [EMIM]-based ionic liquids. *Green Chem.* 9 (2007): 935–942.

14. S. T. Handy and M. Okello. The 2-position of imidazolium ionic liquids: Substitution and exchange. *J. Org. Chem.* 70 (2005): 1915–1918.

15. B. Lin, H. Dong, Y. Li, Z. Si, F. Gu, and F. Yan. Alkaline stable C2-substituted imidazolium-based anion-exchange membranes. *Chem. Mater.* 25 (2013): 1858–1867.

16. W. Wang, S. Wang, X. Xie, Y. Iv, and V. Ramani. Density functional theory study of hydroxide-ion induced degradation of imidazolium cations. *Int. J. Hydrogen Energ.* 39 (2014): 14355–14361.

17. E. Ennis and S. T. Handy. The chemistry of the C2 position of imidazolium room temperature ionic liquids. *Curr. Org. Synth.* 4 (2007): 381–389.

18. S. T. Handy. Grignard reactions in imidazolium ionic liquids. *J. Org. Chem.* 71 (2006): 4659–4662.

19. V. Jurčik and R. Wilhelm. An imidazolinium salt as ionic liquid for medium and strong bases. *Green Chem.* 7 (2005): 844–848.

20. J. Y. Cheng and Y. H. Chu. 1-Butyl-2,3-trimethyleneimidazolium bis(trifluoromethylsulfonyl) imide ([b-3C-im][NTf$_2$]): A new, stable ionic liquid. *Tetrahedron Lett.* 47 (2006): 1575–1579.

21. S. C. Price, K. S. Williams, and F. L. Beyer. Relationships between structure and alkaline stability of imidazolium cations for fuel cell membrane applications. *ACS Macro Lett.* 3 (2014): 160–165.

22. H. Dong, F. Gu, M. Li et al. Improving the alkaline stability of imidazolium cations by substitution. *ChemPhysChem* 15 (2014): 3006–3014.

23. B. C. Ranu and S. Banerjee. Ionic liquid as catalyst and reaction medium. The dramatic influence of a task-specific ionic liquid, [bmIm]OH, in Michael addition of active methylene compounds to conjugated ketones, carboxylic esters, and nitriles. *Org. Lett.* 7 (2005): 3049–3052.

24. J. Golding, S. Forsyth, D. R. MacFarlane, M. Forsyth, and G. B. Deacon. Methanesulfonate and *p*-toluenesulfonate salts of the *N*-methyl-*N*-alkylpyrrolidinium and quaternary ammonium cations: Novel low cost ionic liquids. *Green Chem.* 4 (2002): 223–229.

25. Y. Peng, G. Li, J. Li, and S. Yu. Convenient synthesis of various ionic liquids from onium hydroxides and ammonium salts. *Tetrahedron Lett.* 50 (2009): 4286–4288.

26. K. Guo, M. J. Thompson, and B. Chen. Exploring catalyst and solvent effects in the multicomponent synthesis of pyridine-3,5-dicarbonitriles. *J. Org. Chem.* 74 (2009): 6999–7006.

27. K. Fukumoto, M. Yoshizawa, and H. Ohno. Room temperature ionic liquids from 20 natural amino acids. *J. Am. Chem. Soc.* 127 (2005): 2398–2399.

28. A. K. Burrell, R. E. Del Sesto, S. N. Baker, T. M. McCleskey, and G. A. Baker. The large scale synthesis of pure imidazolium and pyrrolidinium ionic liquids. *Green Chem.* 9 (2007): 449–454.

29. T. Boddaert, Y. Coquerel, and J. Rodriguez. N-Heterocyclic carbene-catalyzed Michael additions of 1,3-dicarbonyl compounds. *Chem. Eur. J.* 17 (2011): 2266–2271.

30. B. C. Ranu, S. Banerjee, and R. Jana. Ionic liquid as catalyst and solvent: The remarkable effect of a basic ionic liquid, [bmIm]OH on Michael addition and alkylation of active methylene compounds. *Tetrahedron* 63 (2007): 776–782.

31. J. M. Xu, Q. Wu, Q. Y. Zhang, F. Zhang, and X. F. Lin. A basic ionic liquid as catalyst and reaction medium: A rapid and simple procedure for aza-Michael addition reactions. *Eur. J. Org. Chem.* 2007 (2007): 1798–1802.

32. E. Rajanarendar, K. R. Murthy, F. P. Shaik, and M. N. Reddy. A fast, highly efficient and green protocol for Michael addition of active methylene compounds to styrylisoxazoles using task-specific basic ionic liquid [bmIm]OH as catalyst and green solvent. *Indian J. Chem. B* 50B (2011): 587–592.

33. L. Yang, L. W. Xu, W. Zhou, L. Li, and C. G. Xia. Highly efficient aza-Michael reactions of aromatic amines and *N*-heterocycles catalyzed by a basic ionic liquid under solvent-free conditions. *Tetrahedron Lett.* 47 (2006): 7723–7726.

34. J. M. Xu, B. K. Liu, W. B. Wu, C. Qian, Q. Wu, and X. F. Lin. Basic ionic liquid as catalysis and reaction medium: A novel and green protocol for the Markovnikov addition of *n*-heterocycles to vinyl esters, using a task-specific ionic liquid, [bmIm]OH. *J. Org. Chem.* 71 (2006): 3991–3993.

35. I. R. Siddiqui, M. A. Waseem, S. Shamim, A. Srivastava, and A. Srivastava. Recyclable [bmIm]OH promoted one-pot heterocyclization: Synthesis of substituted benzofurans. *Tetrahedron Lett.* 54 (2013): 4154–4158.

36. M. A. Waseem, A. A. Abumahdi et al. [bmIm]OH catalyzed hetero-cyclisation of o-halobenzoic acid and alkyne: A green approach to synthesize isocoumarins. *Catal. Commun.* 55 (2014): 70–73.

37. M. A. Waseem, S. Shireen, A. Srivastava, and I. R. Siddiqui. [bmIm]OH: An efficient basic catalyst for the synthesis of 4*H*-benzo[*d*][1,3-]oxazin-4-one derivatives in solvent-free conditions. *Tetrahedron Lett.* 55 (2014): 6072–6076.

38. K. Yamaguchi, T. Imago, Y. Ogasawara, J. Kasai, M. Kotani, and N. Mizuno. An immobilized organocatalyst for cyanosilylation and epoxidation. *Adv. Synth. Catal.* 348 (2006): 1516–1520.

39. J. J. Song, F. Gallou, J. T. Reeves, Z. Tan, N. K. Yee, and C. H. Senanayake. Activation of TMSCN by N-heterocyclic carbenes for facile cyanosilylation of carbonyl compounds. *J. Org. Chem.* 71 (2006): 1273–1276.

40. P. Formentin, H. Garcia, and A. Leyva. Assessment of the suitability of imidazolium ionic liquids as reaction medium for base-catalysed reactions: Case of Knoevenagel and Claisen-Schmidt reactions. *J. Mol. Catal. A: Chem.* 214 (2004): 137–142.

41. M. G. Freire, C. M. S. S. Neves, I. M. Marrucho, J. A. P. Coutinho, and A. M. Fernandes. Hydrolysis of tetrafluoroborate and hexafluorophosphate counter ions in imidazolium-based ionic liquids. *J. Phys. Chem. A* 114 (2010): 3744–3749.
42. R. P. Swatloski, J. D. Holbrey, and R. D. Rogers. Ionic liquids are not always green: Hydrolysis of 1-butyl-3-methylimidazolium hexafluorophosphate. *Green Chem.* 5 (2003): 361–363.
43. B. C. Ranu and R. Jana. Ionic liquid as catalyst and reaction medium—A simple, efficient and green procedure for Knoevenagel condensation of aliphatic and aromatic carbonyl compounds using a task-specific basic ionic liquid. *Eur. J. Org. Chem.* 2006 (2006): 3767–3770.
44. D. Y. Wang, G. H. Xi, J. J. Ma, C. Wang, X. C. Zhang, and Q. Q. Wang. Condensation reactions of aromatic aldehydes with active methylene compounds catalyzed by alkaline ionic liquid. *Synthetic Commun.* 41 (2011): 3060–3065.
45. L. Zhang, M. H. Ding, and H. Y. Guo. One-step synthesis of α,β-unsaturated arylsulfones by a novel multicomponent reaction of aromatic aldehydes, chloroacetonitrile, benzenesulfinic acid sodium salt. *Chinese Chem. Lett.* 23 (2012): 1352–1354.
46. S. G. Patil, R. R. Bagul, M. S. Swami, N. Kotharkar, and K. Darade. Synthesis of 5-benzylidene-3-(3-fluoro-4-yl-morpholin-4-yl-phenylimino)-thiazolidin-4-one derivatives catalyzed by [BmIm]OH and their anti-microbial activity. *Chinese Chem. Lett.* 22 (2011): 883–886.
47. K. Gong, Z. W. He, Y. Xu, D. Fang, and Z. Liu. Green synthesis of 5-benzylidene rhodanine derivatives catalyzed by 1-butyl-3-methyl imidazolium hydroxide in water. *Monatsh. Chem.* 139 (2008): 913–915.
48. Y. Hu, T. Xie, K. M. Fu, H. Kang, P. Wei, and H. Huang. A convenient synthesis of 5-Arylidenethiazolidine-2,4-diones catalyzed by alkaline ionic liquid. *Heterocycles* 78 (2009): 757–761.
49. L. D. S. Yadav, S. Singh, and V. K. Rai. A one-pot [Bmim]OH-mediated synthesis of 3-benzamidocoumarins. *Tetrahedron Lett.* 50 (2009): 2208–2212.
50. B. C. Ranu, R. Jana, and S. Sowmiah. An improved procedure for the three-component synthesis of highly substituted pyridines using ionic liquids. *J. Org. Chem.* 72 (2007): 3152–3154.
51. K. Gong, H. L. Wang, D. Fang, and Z. L. Liu. Basic ionic liquid as catalyst for the rapid and green synthesis of substituted 2-amino-2-chromenes in aqueous media. *Catal. Commun.* 9 (2008): 650–653.
52. S. M. Rajesh, S. Perumal, J. C. Menéndez, S. Pandian, and R. Murugesan. Facile ionic liquid-mediated, three-component sequential reactions for the green, regio- and diastereoselective synthesis of furocoumarins. *Tetrahedron* 68 (2012): 5631–5636.
53. B. C. Ranu, S. Banerjee, and S. Roy. A task specific basic ionic liquid, [bmIm] OH-promoted efficient, green and one-pot synthesis of tetrahydrobenzo[*b*]pyran derivatives. *Indian J. Chem. B* 47B (2008): 1108–1112.
54. D. S. Raghuvanshi and K. N. Singh. Microwave-assisted one-pot synthesis of functionalized pyrimidines using ionic liquid. *J. Heterocyclic Chem.* 48 (2011): 582–585.
55. Y. Yu, H. Guo, and X. Li. An improved procedure for the three-component synthesis of benzo[*g*]chromene derivatives using basic ionic liquid. *J. Heterocyclic Chem.* 48 (2011): 1264–1268.
56. J. M. Khurana, D. Magoo, and A. Chaudhary. Efficient and green approaches for the synthesis of 4*H*-Benzo[*g*]chromenes in water, under neat conditions, and using task-specific ionic liquid. *Synthetic Commun.* 42 (2012): 3211–3219.
57. J. Zhao-Qin, J. Shun-Jun, L. Jun, and Y. Jin-Ming. A mild and efficient synthesis of 5-Oxo-5,6,7,8-tetrahydro-4*H*-benzo-[*b*]-pyran derivatives in room temperature ionic liquids. *Chinese J. Chem.* 23 (2005): 1085–1089.

58. Y. Wang, Y. Wu, Y. Wang, and L. Dai. Experimental and theoretical investigation of one-pot synthesis of 2-amino-4H-chromenes catalyzed by basic-functionalized ionic liquids. *Chinese J. Chem.* 30 (2012): 1709–1714.

59. F. Ma and M. A. Hanna. Biodiesel production: A review. *Bioresource Technol.* 70 (1999): 1–15.

60. S. Zhou, L. Liu, B. Wang, F. Xu, and R. C. Sun. Biodiesel preparation from transesterification of glycerol trioleate catalyzed by basic ionic liquids. *Chinese Chem. Lett.* 23 (2012): 379–382.

61. Y. Zhang, Q. Jiao, B. Zhen, Q. Wu, and H. Li. Transesterification of glycerol trioleate catalyzed by basic ionic liquids immobilized on magnetic nanoparticles: Influence of pore diffusion effect. *Appl. Catal. A: Gen.* 453 (2013): 327–333.

62. Y.-J. Zhang, Y.-D. Xu, A.-H. Zhang, and Z.-H. Xiao. Research on producing biodiesel from waste oil with 1-Methyl-3-butyl imidazole hydroxide ionic liquid as a catalyst. *Energ. Source Part A* 35 (2013): 1691–1697.

63. Z. Al-Hamamre and J. Yamin. Parametric study of the alkali catalyzed transesterification of waste frying oil for Biodiesel production. *Energ. Convers. Manage.* 79 (2014): 246–254.

64. Y. Yi, Y. Shen, J. Sun, B. Wang, F. Xu, and R. Sun. Basic ionic liquids promoted the synthesis of glycerol 1,2-carbonate from glycerol. *Chinese J. Catal.* 35 (2014): 757–762.

65. S. Han, M. Luo, X. Zhou, Z. He, and L. Xiong. Synthesis of dipentyl carbonate by transesterification using basic ionic liquid [bmIm]OH catalyst. *Ind. Eng. Chem. Res.* 51 (2012): 5433–5437.

66. J. Xu, H. T. Wu, C. M. Ma, B. Xue, Y. X. Li, and Y. Cao. Ionic liquid immobilized on mesocellular silica foam as an efficient heterogeneous catalyst for the synthesis of dimethyl carbonate via transesterification. *Appl. Catal. A: Gen.* 464–465 (2013): 357–363.

67. F. Shi, Y. Deng, T. SiMa, J. Peng, Y. Gu, and B. Qiao. Alternatives to phosgene and carbon monoxide: Synthesis of symmetric urea derivatives with carbon dioxide in ionic liquids. *Angew. Chem. Int. Ed.* 42 (2003): 3257–3260.

68. T. Jiang, X. Ma, Y. Zhou, S. Liang, J. Zhang, and B. Han. Solvent-free synthesis of substituted ureas from CO_2 and amines with a functional ionic liquid as the catalyst. *Green Chem.* 10 (2008): 465–469.

69. Y. P. Patil, P. J. Tambade, K. M. Deshmukh, and B. M. Bhanage. Synthesis of quinazoline-2,4($1H$,$3H$)-diones from carbon dioxide and 2-aminobenzonitriles using [Bmim]OH as a homogeneous recyclable catalyst. *Catal. Today* 148 (2009): 355–360.

70. D. B. Nale, S. D. Saigaonkar, and B. M. Bhanage. An efficient synthesis of quinazoline-2,4($1H$,$3H$)-dione from CO_2 and 2-aminobenzonitrile using [Hmim]OH/SiO_2 as a base functionalized supported ionic liquid phase catalyst. *J. CO_2 Util.* 8 (2014): 67–73.

71. Y. Ren, T. T. Meng, J. Jia, and H. S. Wu. A computational study on the chemical fixation of carbon dioxide with 2-aminobenzonitrile catalyzed by 1-butyl-3-methyl imidazolium hydroxide ionic liquids. *Comput. Theor. Chem.* 978 (2011): 47–56.

72. J. Sun, B. Lu, X. Wang, X. Li, J. Zhao, and Q. Cai. A functionalized basic ionic liquid for synthesis of dimethyl carbonate from methanol and CO_2. *Fuel Process. Technol.* 115 (2013): 233–237.

73. X. Yuan, B. Lu, J. Liu, X. You, J. Zhao, and Q. Cai. Electrochemical conversion of methanol and carbon dioxide to dimethyl carbonate at Graphite-Pt electrode system. *J. Electrochem. Soc.* 159 (2012): E183–E186.

74. X. Zhang, D. Wang, N. Zhao et al. Grafted ionic liquid: Catalyst for solventless cycloaddition of carbon dioxide and propylene oxide. *Catal. Commun.* 11 (2009): 43–46.

75. J. Q. Wang, J. Sun, C. Y. Shi, W. G. Cheng, X. P. Zhang, and S. J. Zhang. Synthesis of dimethyl carbonate from CO_2 and ethylene oxide catalyzed by K_2CO_3-based binary salts in the presence of H_2O. *Green Chem.* 13 (2011): 3213–3217.

5 Organocatalysis of S$_N$2 Reactions by Multifunctional Promotors
Ionic Liquids and Derivatives

Sungyul Lee and Dong Wook Kim

CONTENTS

5.1 INTRODUCTION

Ionic liquids have found a wide range of applications in chemistry because of their unique and useful physicochemical properties. In addition to being ideal solvents due to high polarity, low volatility, and easy recovery, ionic liquids are also gaining importance as catalysts in various chemical reactions.[1–3] Many reactions that proceeded *via* reactive intermediates (alkyl, vinyl, arenium carbocations, and oxygen radical anion) showed pronounced positive ionic liquid effects such as rate acceleration and increased regio- and stereoselectivity originating from the stabilization of the reactive intermediates.[4,5] Significantly enhanced reactivity and selectivity have also been observed in concerted reactions such as nucleophilic substitution reactions in ionic liquids.[6–8] As shown in Scheme 5.1, the use of an ionic liquid such as [bmim][OMs] for S$_N$2 reactions with CsF proceeds with remarkable rate acceleration and high chemoselectivity, in contrast to the corresponding reactions in common organic solvents.[6,9]

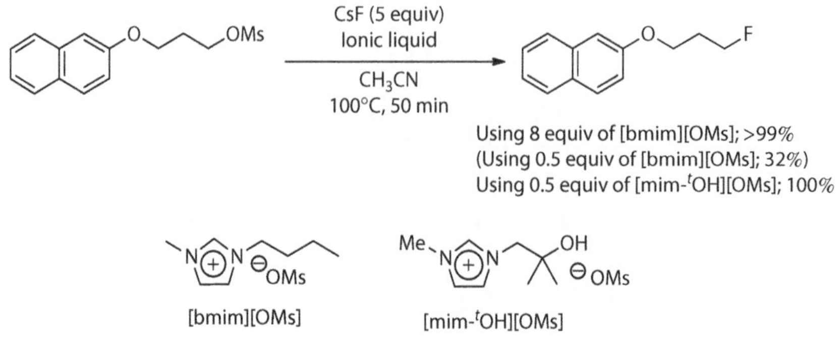

SCHEME 5.1 S_N2 fluorination with CsF in ionic liquids.

Recent progress in the bimolecular nucleophilic substitution (S_N2) reaction has demonstrated that this fundamental and very useful type of chemical reaction must still be studied systematically, especially in terms of the role of counterion and solvent in the efficiency of the reaction.[10,11] The use of metal salts and ionic liquids seems to have opened a breakthrough for this purpose.

Herein, we present a brief review of the catalytic activity of ionic liquids and their derivatives for S_N2 reactions, focusing on the mechanistic features. The roles of ionic liquid cations and anions and those of the functional groups of side chains are discussed. Two mechanisms, the contact ion-pair and the naked nucleophile mechanism, are discussed, demonstrating that different mechanistic features are exhibited depending on the structures of the modified ionic liquid. Systematic analysis is also described for the origin of the experimentally observed trend of reactivity ($F^- > Cl^- > Br^-$) of the halides under the influence of the promotor [dihexaEGim][OMs], which is in direct contrast with the reactivity ($F^- < Cl^- < Br^-$) usually observed for S_N2 reactions in polar aprotic solvents.[11]

5.2   ACCELERATION OF S_N2 REACTIONS BY IONIC LIQUID [bmim][OMs]

As shown from the examples in Scheme 5.1, the use of an ionic liquid, such as [bmim][OMs] for the S_N2 reaction with CsF, proceeds with remarkable rate acceleration, high yield (usually >90%) and high chemoselectivity, in contrast to the corresponding reactions in common organic solvents.[6–9] This scheme of S_N2 fluorination is very desirable because the very stable and inexpensive metal salt CsF (and KX, X = F, Cl, Br, I), which has been known to be very inefficient for S_N2 reactions in organic solvents, is employed. At first sight, however, this catalysis by ionic liquids may look puzzling. The ionic liquid cation may only retard the reactivity of the nucleophile F^- due to its strong Coulombic influence on F^-, whereas the ionic liquid anion may choose to be as far away as possible from the nucleophile also because of the repelling electrostatic interactions with F^-. Another initially puzzling feature is the use of Cs^+, which is known to be avoided for S_N2 reactions as the counterion because of strong Coulombic force influence on F^-.

The key to understanding this new type of S$_N$2 reactions is the ion-pair mechanism proposed for the rate acceleration of S$_N$2 reactions by bulky protic solvents (*t*-butyl alcohol, *t*-amyl alcohol)[12,13] and oligo-ethylene glycols.[14,15] In this mechanism, the oxygen atoms of the solvent promotors bind as Lewis bases to the counterion Cs$^+$ or K$^+$, dramatically reducing its Coulombic influence on the nucleophile. Thus, although the metal salts still exist as contact ion-pair, the nucleophile is almost "freed" from the retarding influence of the counterion.

Figure 5.1 presents the calculated transition state of S$_N$2 fluorination reaction [Cs$^+$F$^-$ + C$_3$H$_7$OMs → C$_3$H$_7$F + Cs$^+$OMs$^-$] in ionic liquid [bmim][OMs]. The cation [bmim], anion [OMs], nucleophile F$^-$, counterion Cs$^+$, and the leaving group of the substrate form a very stable and compact cyclic structure. The counterion Cs$^+$ binds both to F$^-$ and the leaving group, allowing an ideal configuration for the nucleophilic attack. The role of the ionic liquid cation [bmim] and anion [OMs] for accelerating the reaction is clearly seen here: [OMs] (and the leaving group) interacts with the counterion Cs$^+$, reducing the retarding Coulomblic influence of Cs$^+$ on F$^-$ (thus "freeing" F$^-$), whereas [bmim] "collects" [OMs], Cs$^+$, and F$^-$ in this ideal arrangement. The role of Cs$^+$ promoting the approach of F$^-$ to the leaving group (the substrate) and that of [OMs] (acting as a Lewis base) to neutralize the Coulombic influence of Cs$^+$ on F$^-$ seems to be the key factor in this mechanism, suggesting that choice and "design" of the ionic liquid anion and the counterion will be critical. Binding of [bmim] to F$^-$ may partially decrease the nucleophilicity of F$^-$, but the stronger influence of [OMs] on Cs$^+$ seems to overcome this, giving the reaction barrier ($E^{\ddagger} = 20.3$, $G^{\ddagger}_{100°C} = 20.8$ kcal/mol) that is quite similar to those in the very efficient S$_N$2 reactions in protic solvents *t*-butanol (20.4 kcal/mol) and ethylene glycol (20.0 kcal/mol) presented in earlier works. Based on these results, we demonstrate that the S$_N$2 rate constants in *t*-butanol[16] and ethylene glycol[17] are quite comparable, in good agreement with the observed excellent S$_N$2 rates in ionic liquids.[6–8] According to the S$_N$2 mechanism depicted in Figure 5.1, it seems that an ionic liquid with strong coordinating ability (via stronger influence on Cs$^+$) would be favorable.

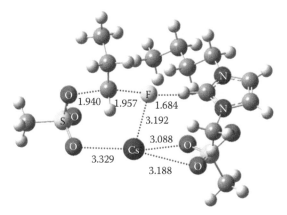

FIGURE 5.1　Calculated transition state in S$_N$2 reaction using [bmim][OMs] as a promotor. Bond lengths in Å [MPW1K/6-311++G**; ECP for Cs, Hay-Wadt VDZ(n+1)].

5.3   USE OF THE MODIFIED IONIC LIQUID [mim-^{t}OH][OMs] AS A PROMOTOR FOR S$_N$2 FLUORINATION

Derivatives of oligoethylene glycols were also found to be an efficient promotor of S$_N$2 reactions. We first describe the use of [mim-^{t}OH][OMs], in which t-butyl alcohol and [mim][OMs] are fused into a molecule by covalent bonding. Scheme 5.1 also depicts this case. The efficiency of nucleophilic fluorination was found to be profoundly improved (S$_N$2 fluorination proceeded to completion in less than 1 h) by employing this modified ionic liquid (Table 5.1), which can complete the reaction within 1 h with catalytic amounts.[9] It is quite obvious that some synergistic collaboration of the two moieties contributes to the enhanced catalysis. The roles of the t-butyl alcohol and [mim][OMs] moieties for lowering the activation barrier $G^{\ddagger}_{100°C}$ by 1.7 kcal/mol relative to that for [bmim][OMs] were studied by quantum chemical methods (MPWPW1K/6-311++G**[18,19], ECP for Cs, Hay-Wadt VDZ(n+1))[20], revealing that the t-BuOH moiety acts as an "anchor" to the leaving group for facile nucleophilic attack by F$^-$, rather than as a Lewis base to Cs$^+$, also helping to decrease the retarding effects of the H-F$^-$ interaction (Figure 5.2).

TABLE 5.1

Experimentally Observed Relative Efficiency and Calculated Barriers of S$_N$2 Reactions [Cs$^+$F$^-$ + C$_3$H$_7$OMs → C$_3$H$_7$F + Cs$^+$OMs$^-$] (Energy and Gibbs Free Energy in kcal/mol) in Ionic Liquids

Ionic Liquids	$E^{\ddagger}$	$G^{\ddagger}_{100°C}$	Observed S$_N$2 Yield (%)
[bmim][BF$_4$]	19.8	21.9	24[a]
[bmim][OMs]	20.3	20.8	32[a]
[mim-^{t}OH][OMs]	16.0	19.1	100[a]

Notes:

[a] Reactions were carried out on a 1.0 mmol scale of substrate 3-(naphthalen-2-yloxy)propyl methanesulfonate with 5 equivalent of CsF using 0.5 equivalent of ionic liquid in CH$_3$CN (3.0 mL) for 50 min at 100°C.

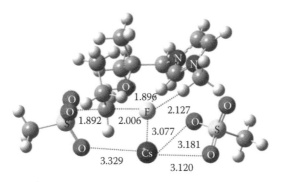

FIGURE 5.2   Calculated transition state in S_N2 reaction in [mim-^{t}OH][OMs]. Bond lengths in Å [MPW1K/6-311++G**; ECP for Cs, Hay-Wadt VDZ(n+1)].

5.4   USE OF THE OLIGOETHYLENE GLYCOLIC IMIDAZOLIUM SALTS AS A PROMOTOR FOR S_N2 REACTION

Oligoethylene glycols (OligoEGs) were the first organocatalyst of S_N2 reactions using KF.[14] The former metal salt is clearly favorable over CsF in terms of cost and availability. Because of smaller size and more concentrated positive charge in K^+ than in Cs^+, K^+ exerts much larger Coulombic influence on the nucleophile than Cs^+. The use of K^+ as a counterion would therefore necessitate a much more efficient promotor than bulky alcohols, such as OligoEGs. An improvement of catalytic efficiency was thus devised by synthesizing a complex in which OligoEGs were covalently bonded as a side chain to the ionic liquid cations. Scheme 5.2a depicts the synthetic route to 1-hexaethyleneglycolic-3-methylimidazolium [hexaEGmim] salts with various anions such as mesylate [OMs], tosylate [OTs], triflate [OTf], tetrafluoroborate [BF_4], hexafluorophosphate [PF_6], and triflate [OTf]. They could be prepared by a simple treatment of 1-methylimidazole with hexaethylene glycolic (hexaEG) mesylate **1a** or hexaEG tosylate **1b** in acetonitrile for 24 h to afford [hexaEGmim][OMs] and [hexaEGmim][OTs] in good yields, respectively.[21,22] The ionic liquids [hexaEGmim][OTf], [hexaEGmim][BF_4], and [hexaEGmim][PF_6] were obtained by further treatment of [hexaEGmim][OMs] with KOTf, $NaBF_4$, or $NaPF_6$ in acetone for two days in 79%–83% yield, respectively.[22] As shown in Scheme 5.2b, the N-alkylation of imidazole with hexaEG bromide **2** in the presence of K_2CO_3, followed by treatment of **3** with hexaEG mesylate **1a** for two days, produced a 1,3-dihexaethylene glycolic imidazolium mesylate [dihexaEGim][OMs], which was found to exhibit the best promoting activity after carefully monitoring the efficiency as a function of the side chain length and the structure of the anion (Figure 5.3).[21,22]

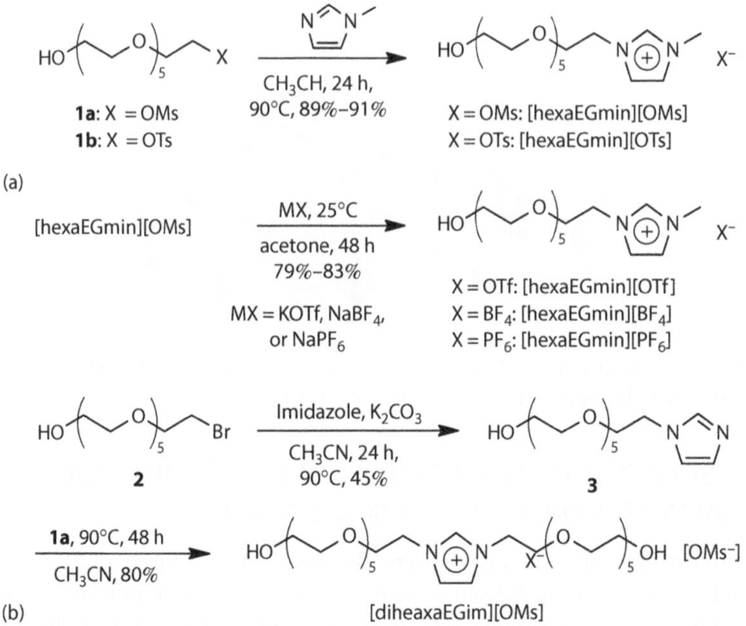

(a)

(b)

SCHEME 5.2 (a) Preparation of 1-hexaethyleneglycolic-3-methylimidazolium salts [hexaEGmim][X] (X = OMs, OTs, OTf, BF$_4$, PF$_6$), and (b) 1,3-dihexaethylene glycolic imidazolium mesylate salts: [dioligoEGim][OMs].

The promoting activity of these hexaEGILs was investigated by the nucleophilic fluorination using 3 equiv. of CsF in the presence of the hexaEGILs (0.5 equiv) in acetonitrile under a uniform reaction condition (at 90°C for 2 h), and these results were compared with the same reaction in the presence of a conventional ionic liquids (IL) [bmim][OMs] (bmim = 1-*n*-butyl-3-methylimidazolium) or in the absence of any catalyst, as shown in Figure 5.3a. From these results, it is found that hexaEGILs ([hexaEGmim][OMs] and [hexaEGmim][OTs]), whose anions do not contain F atoms, exhibited significantly better promoting activity than those ([hexaEGmim] [OTf], [hexaEGmim][BF$_4$], and [hexaEGmim][PF$_6$]) containing F atoms in the anions because the fluorine-containing anions (particularly [PF$_6$]) can inhibit the initial interaction by H-bonding between the hexaEGIL and fluoride nucleophile[9] and decrease the reaction rates. Therefore, the mesylate [OMs] was found to be the best among these oligoEG substituted imidazolium salts. Furthermore, Figure 5.3b shows that [dihexaEGim][OMs] containing two hexaEG components demonstrated high catalytic activity compared with mono-substituted [hexaEGmim][OMs], as well as hexaethylene glycol (hexaEG) and [bmim][OMs]. Consequently, [dihexaEGim] [OMs] exhibited the best promoting activity in the nucleophilic fluorination.[21,22]

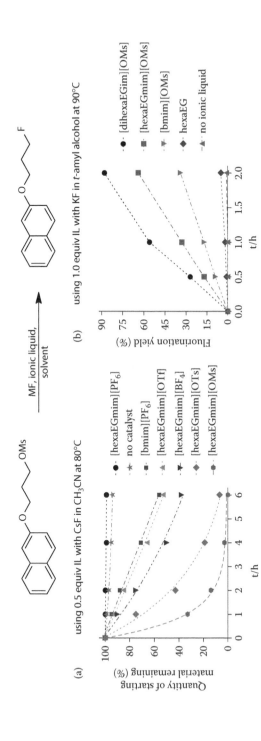

FIGURE 5.3 (a) Influence of ionic liquid anions in nucleophilic fluorination by CsF in the presence of various [hexaEGmim][X] (X = OMs, OTs, OTf, BF$_4$, and PF$_6$). (b) Catalytic activity of [hexaEGmim][OMs] and [dihexaEGmim][OMs] with KF. The quantity of product was determined by ^{1}H NMR. R = naphthyl.

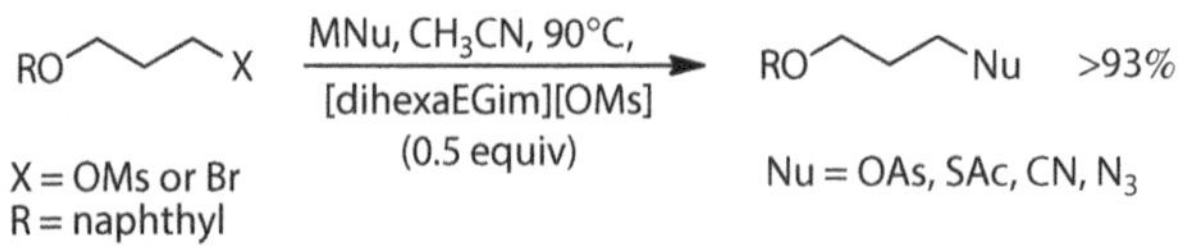

RO⌒⌒X → $\xrightarrow[\text{[dihexaEGim][OMs]}]{\text{MNu, CH}_3\text{CN, 90°C,}}$ → RO⌒⌒Nu   >93%

(0.5 equiv)

X = OMs or Br   Nu = OAs, SAc, CN, N$_3$
R = naphthyl

SCHEME 5.3 Diverse S$_N$2 reactions using various alkali metal salts (MNu) with [dihexaEGim][OMs].

As shown in Scheme 5.3, the [dihexaEGim][OMs] as an organic promoter showed good performance in diverse nucleophilic substitution reactions such as acetoxylation, thioacetoxylation, iodination, azidation, and nitrilation. The various transformation reactions using various alkali metal salts (such as KOAc, KSAc, KCN, or NaN$_3$) as nucleophile sources at 90°C in acetonitrile proceeded almost quantitatively, affording the corresponding product in very high yields (more than 93%).[22]

Table 5.2 shows that this type of promotor may be applied to a variety of S$_N$2 reaction substrates. The results confirm that the acetoxyl-, thioacetoxyl-, and azido-groups were incorporated into the benzylic or phenyl ethyl site in almost quantitative

TABLE 5.2

Nucleophilic Transformations of the Various Substrates in the Presence of [dihexaEGim][OMs]

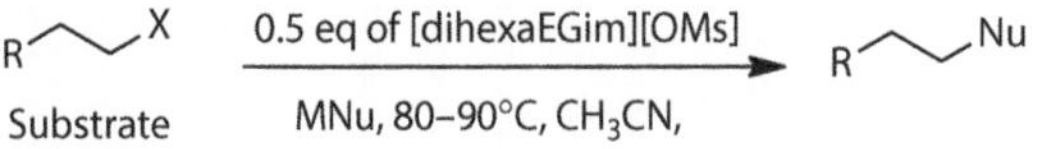

R⌒⌒X $\xrightarrow[\text{MNu, 80–90°C, CH}_3\text{CN,}]{\text{0.5 eq of [dihexaEGim][OMs]}}$ R⌒⌒Nu

Substrate

Substrate	MNu	Time (h)	Yield (%)
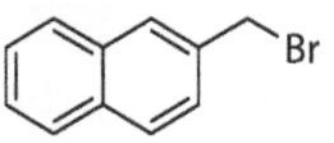 (naphthyl-CH$_2$-Br)	KOAc	0.5	99
	KSAc	0.5	98
	NaN$_3$	0.5	98
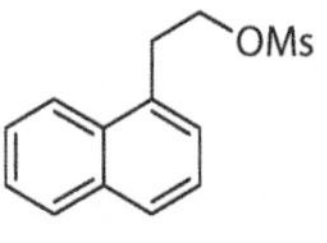 (naphthyl-CH$_2$CH$_2$-OMs)	KOAc	0.5	99
	KSAc	0.5	98
	NaN$_3$	0.5	99
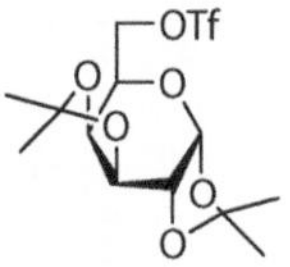 (OTf sugar derivative)	KOAc	1	98
	KSAc	1	97
	NaN$_3$	1	98
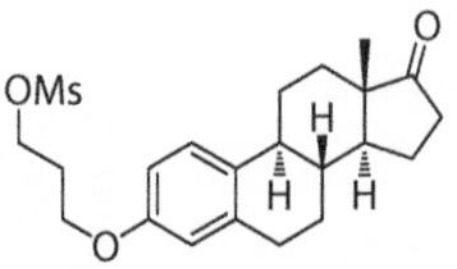 (steroid OMs derivative)	KOAc	3	96
	KSAc	4	93
	NaN$_3$	3	94

yield (98%–99%). In addition, [dihexaEGim][OMs] can also act as an efficient promoter for the nucleophilic substitution of various bioactive substrates (such as a sugar-triflate or an estrone mesylate) using KOAc, KSAc, and NaN_3.[22]

The mechanism of organocatalysis by [hexaEGmin][OMs] was systematically examined by quantum chemical calculations (MPWPW91/G-31*[18,19] for substrate and promotor and IEFPCM for solvent continuum[23–25]). Figure 5.4 shows the calculated S_N2 reaction mechanisms of KCl promoted by [hexaEGmim][OMs] in CH_3CN. For this reaction, two mechanisms proceeding from the pre-reaction complexes (Figure 5.4a and b) are feasible. It would be useful to note that the distance R_{K-Cl} in pre-reaction complex (A) is 3.356 A, a bit larger than the gas phase value (2.667 Å), whereas in (B), R_{K-Cl} is = 5.085 A, indicating that K^+ and Cl^- are essentially separated by the influence of the promoter [hexaEGmim][OMs]. These two complexes have similar thermodynamic stability (the difference in Gibbs energy is only 0.1 kcal/mol), and therefore, the reaction mechanism would be determined by the relative activation barriers. The difference between the barriers from complexes (A) and (B) is, however, also quite small (~0.7 kcal/mol). Therefore, it seems that the two mechanisms are competing, and thus it is not clear whether the metal salt KCl is in contact ion-pair form or in separated form.

The role of contact ion-pair (CIP) versus naked nucleophile for S_N2 reactions is a very important issue. Although our previous works[6–9,12,13,21,22] demonstrated that metal halides as CIP can be very effective reagents for S_N2 reactions in the presence of Lewis base solvents, "separation" of the metal cation (Cs^+ and K^+) from the nucleophile may further enhance the rate constants of S_N2 reactions. Considering that the electrostatic attraction between the metal cation and the nucleophile is so strong, only extremely powerful interactions between the promotor and the nucleophile may be enough to overcome it. Even strong cryptands such as the crown ethers were shown to work for S_N2 reactions according to the CIP mechanism, in which the

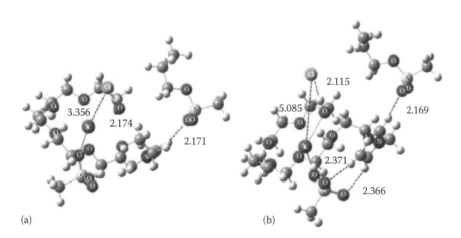

(a)   (b)

FIGURE 5.4 Pre-reaction complexes in the S_N2 reaction of KBr (a) and KCL (b) in [hexaEGmim][OMs]/CH_3CN.

metal salts are in CIP form and the oxygen atoms coordinate on the counterion to neutralize its Coulombic influence on the nucleophile.

The indication that the CIP and the separated forms of metal salts are more or less equally feasible in S_N2 reactions using [hexaEGmin][OMs] is very important, because a little further modification of the promotor may achieve the separation of metal salts into separated counterion and the "naked" nucleophile that has been the goal of S_N2 community for long. To the best of our knowledge, the promotor [dihexaEGmin] [OMs] was the first to exhibit such capacity to produce such a naked nucleophile that is essentially detached from the counterion. In the presence of this newly developed organocatalyst, the S_N2 reaction using KF was observed to show the fastest reaction rate, providing the corresponding fluoro-product **2** in 98% yield within 2 h. This certainly is a significant advance in S_N2 catalysis, because the observed rate increase is larger than that (~83% yield in 2 h) achieved by the promotor [monohexaEGim][OMs] in our previous work[21,22] using CsF, in which Cs$^+$ of much larger polarizability (and thus, much smaller retarding Coulombic influence on the nucleophile) was employed. Figure 5.5 shows the rates of halogenation using this powerful promotor. Depending on their strengths of interaction between potassium cation and halide nucleophiles, the generalized reactivity-order of KX is known to be KBr > KCl > KF. In particular, KF alone shows feeble reactivity for S_N2 reactions in polar aprotic medium compared with other KXs.[11] Interestingly, the use of [dihexaEGim][OMs] in the bulky protic solvent could allow the S_N2 reaction using KF to have the fastest reaction rate. Moreover, it was observed that the order of KX reactivity was KF > KCl > KBr in this reaction condition. These interesting observations suggest that very intricate interactions between the potassium cation, the halide nucleophile, the promotor, solvent molecules, and the leaving group may cause the nonconventional relative S_N2 reactivity of F$^-$ > Cl$^-$ > Br$^-$ in these S_N2 halogenation reactions.[26]

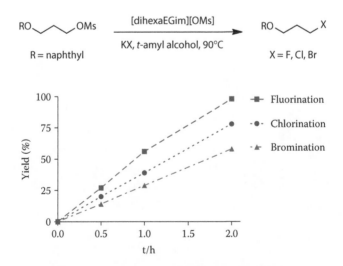

FIGURE 5.5 Comparison of relative rates of nucleophilic halogenation reactions using KX (X = F, Cl, Br) in the presence of [dihexaEGim][OMs] in *t*-amyl alcohol.

Figure 5.6 presents the calculated lowest energy pre-reaction complexes for S_N2 halogenation reaction using KX (X = F, Cl, Br). Several general features are to be noted: First, the counterion K$^+$ and the nucleophile X$^-$ are far apart from each other with the distance of 4.3 ~ 6.6 Å in all pre-reaction complexes. Because the bond lengths of gas phase KF, KCl, and KBr are 2.171, 2.667, 2.821 Å, respectively, this indicates that the metal salt KX is essentially separated to form a "naked" nucleophile that is almost free from the Coulombic influence of the counterion K$^+$. The six O atoms of a hexaEG moiety coordinate K$^+$, whereas the other hexaEG interacts with

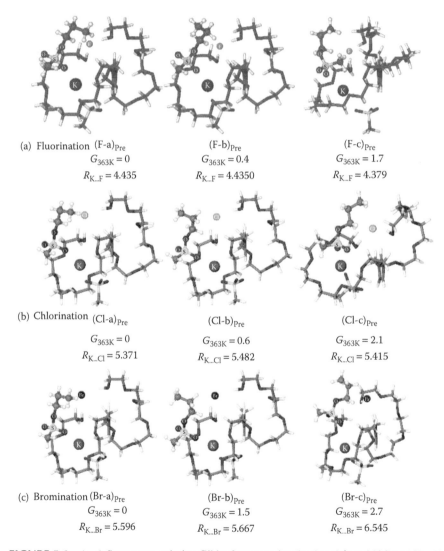

(a) Fluorination (F-a)$_{Pre}$
$G_{363K} = 0$
$R_{K...F} = 4.435$

(F-b)$_{Pre}$
$G_{363K} = 0.4$
$R_{K...F} = 4.4350$

(F-c)$_{Pre}$
$G_{363K} = 1.7$
$R_{K...F} = 4.379$

(b) Chlorination (Cl-a)$_{Pre}$
$G_{363K} = 0$
$R_{K...Cl} = 5.371$

(Cl-b)$_{Pre}$
$G_{363K} = 0.6$
$R_{K...Cl} = 5.482$

(Cl-c)$_{Pre}$
$G_{363K} = 2.1$
$R_{K...Cl} = 5.415$

(c) Bromination (Br-a)$_{Pre}$
$G_{363K} = 0$
$R_{K...Br} = 5.596$

(Br-b)$_{Pre}$
$G_{363K} = 1.5$
$R_{K...Br} = 5.667$

(Br-c)$_{Pre}$
$G_{363K} = 2.7$
$R_{K...Br} = 6.545$

FIGURE 5.6 (a–c) Structures, relative Gibbs free energies (kcal mol^{-1}) at 90°C and K... X (X = F, Cl, Br) distances (Å) of pre-reaction complexes with K$^+$ and X$^-$ separated.

the imidazolium cation, with a hydroxyl O atom binding to the acidic 2-H. To the best of our knowledge, this is the first observation of such separation of metal salts by any agents for S_N2 reactions. Second, the two –OH groups, each from the two hexaEGs, act as "anchor" to the nucleophile X^- in a configuration such that the backside attack of X^- on the substrate is most favorable. Third, the ionic liquid anion [OMs]$^-$ lying between the imidazolium and K^+ helps to mitigate the positive charge of K^+. The acidic 2-H in imidazolium is surrounded by the O atoms of hexaEG not interacting with K^+. Overall, the electronegative oxygen atoms, the acidic –OH, the imidazolium cation, [OMs]$^-$ anion, K^+, X^-, and the leaving group OMs$^-$ in the substrate exert very intricate interactions with one another in a compact configuration that is extremely favorable for S_N2 reaction. The counterion K^+ is coordinated by nine O atoms in all. The structures of pre-reaction complexes (Cl-a)$_{Pre}$ and (Cl-b)$_{Pre}$ for chlorination, and those [(Br-a)$_{Pre}$, (Br-b)$_{Pre}$, (Br-c)$_{Pre}$] for bromination are similar to those described for fluorination, in that the metal salts KCl and KBr are in promotor-separated form.

Figure 5.7 illustrates the transition states and energy profiles of the most feasible reaction pathways of S_N2 reactions, including the Gibbs free energy of activation $G^{\ddagger}_{363K}$. We find that all S_N2 reactions proceed from the lowest energy pre-reaction complexes (F-a)$_{Pre}$, (Cl-a)$_{Pre}$, and (Br-a)$_{Pre}$ given in Figure 5.6. The distances $\tilde{R}_{K\cdots O}$, $R_{K\cdots -OMs}$, $R_{K\cdots [OMs]}$ in the transition states slightly decrease as compared with the corresponding pre-reaction complexes. In (F-a)$_{TS}$ and (Cl-a)$_{TS}$, K^+ is coordinated by eight O atoms, whereas in (Br-a)$_{TS}$, it is surrounded by nine O atoms. The experimentally observed relative rates of fluorination, chlorination, and bromination presented in Figure 5.5 are in the order of $F^- > Cl^- > Br^-$, which is in direct contrast with the generalized order of potassium halide reactivity of KBr > KCl > KF. This is very intriguing because the S_N2 reaction using KF in polar aprotic medium proceed

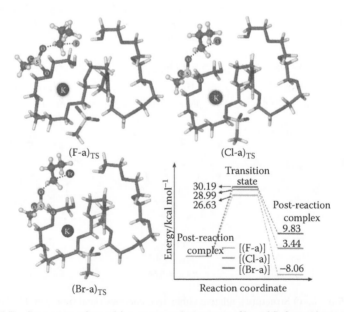

FIGURE 5.7   Structures of transition states and energy profiles of S_N2 reaction pathways.

hardly compared with other KXs. The calculated $G^{\ddagger}_{363K}$ for fluorination, chlorination, and bromination are 26.6 kcal mol^{-1}, 29.0 kcal mol^{-1}, and 30.2 kcal mol^{-1}, respectively, indicating that the rates of reaction are in the order F$^-$ > Cl$^-$ > Br$^-$, in good agreement with the experimental observations in Figure 5.5. It seems that the order of nucleophilicity of F$^-$, Cl$^-$, and Br$^-$ is the same as the order of basicity F$^-$ > Cl$^-$ > Br$^-$, because the nucleophiles are essentially free from the influence of the counterion and solvent molecules. The usually observed side reactions such as E$_2$ elimination for strongly basic nucleophiles such as F$^-$ don't seem to be problematic here due to the very compact arrangement of the participants (nucleophile, counterion, promotor, solvent, and the substrate) that are ideal configurations for the S_N2 attack by the nucleophile on the substrate in the pre-reaction complexes.[26]

The role of protic solvent *t*-amyl alcohol is also of considerable interest because S_N2 reactions were previously observed to be promoted by bulky alcohols alone, which acts as Lewis base on the counterion.[24–26] Figure 5.8 presents the structures of pre-reaction complexes for S_N2 reactions in the presence of a *t*-amyl alcohol molecule. The *t*-amyl alcohol molecule may either interact with the nucleophile X$^-$ or with the ionic liquid anion [OMs]$^-$. In the former case, [(F-a-1)$_{Pre}$, (Cl-a-1)$_{Pre}$, (Br-a-1)$_{Pre}$], the *t*-amyl alcohol molecule acts as a Lewis acid on X$^-$, exerting a direct and retarding effect on the nucleophilicity of X$^-$. In the latter situation, [(F-a-2)$_{Pre}$,

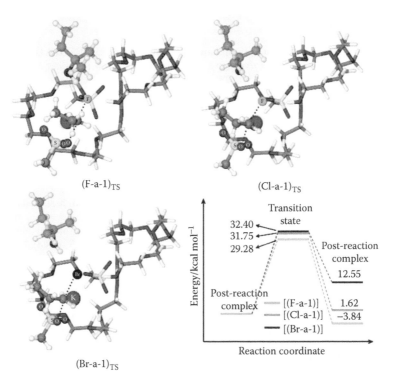

(F-a-1)$_{TS}$ (Cl-a-1)$_{TS}$

(Br-a-1)$_{TS}$

FIGURE 5.8 Structures of transition states and energy profiles of S_N2 reaction pathways under the influence of a solvent (*t*-amyl alcohol in circle) molecule. Effective barriers (relative energy from the global minimum energy structure plus the activation barrier) in kcal mol^{-1}.

(Cl-a-2a)$_{Pre}$, (Cl-a-2b)$_{Pre}$, (Br-a-2)$_{Pre}$], the solvent molecule interacts with [OMs]$^-$ as a hydrogen bond donor, affecting the S$_N$2 rates indirectly. The thermodynamic stabilities of these two types of pre-reaction complexes are quite similar, the differences being only 0.3 kcal mol^{-1}–1.2 kcal mol^{-1}. Therefore, the relative feasibility of the reactions proceeding from each type of pre-reaction complexes would mostly be determined by the magnitudes of the activation barriers.

5.5   CONCLUSION

In this brief review, we demonstrated that structural modification of ionic liquids may bring about substantial improvement in their capacity as organocatalysts. This is in fact the merit of using ionic liquids, because of their tremendous variety of structures, the possibility of tailor-making them for accelerating specific chemical reactions. We gave examples of a bulky alcohol (*t*-butanol) and oligoethylene glycols as side chains of imidazolium, but other structural modifications would, of course, be possible. Systematic study by quantum chemical techniques will be very useful to elucidate the reaction mechanism, and also help to design the tailor-made ionic liquids for specific purposes in the spirit of solvent engineering. Further developments, both experimental and theoretical, in this interesting subject will be highly desirable.

ACKNOWLEDGMENTS

We thank the National Research Foundation of Korea (NRF-2016R1D1A1B03931188, -2017R1A2A2A10001451, -2015M2A2A6A01045378) for financial support and the KISTI Supercomputing Center (2016).

REFERENCES

1. Zhao, H. and Malhotra, S.V. (2002). Applications of ionic liquids in organic synthesis. *Aldrichimica Acta* 35(3): 75–83.
2. Wasserscheid, P. and Keim, W. (2000). Ionic liquids - new "solutions" for transition metal catalysis. *Angew. Chem., Int. Ed.* 39(21): 3772–3789.
3. Welton, T. (1999). Room-temperature ionic liquids. Solvents for synthesis and catalysis. *Chem. Rev.* 99(8): 2071–2083.
4. Lei, Z., Chen, B., and Li, C. et al. (2008). Predictive molecular thermodynamic models for liquid solvents, solid salts, polymers, and ionic liquids. *Chem. Rev.* 108(4): 1419–1455.
5. Lee, J.W., Shin, J.Y., and Chun, Y.S. et al. (2010). Toward understanding the origin of positive effects of ionic liquids on catalysis: Formation of more reactive catalysts and stabilization of reactive intermediates and transition states in ionic liquids. *Acc. Chem. Res.* 43(7): 985–994.
6. Kim, D.W., Song, C.E., and Chi, D.Y. (2002). New method of fluorination using potassium fluoride in ionic liquid: Significantly enhanced reactivity of fluoride and improved selectivity. *J. Am. Chem. Soc.* 124(35): 10278–10279.
7. Kim, D.W., Song, C.E., and Chi, D.Y. (2003). Significantly enhanced reactivities of the nucleophilic substitution reactions in ionic liquid. *J. Org. Chem.* 68(11): 4281–4285.
8. Kim, D.W. and Chi, D.Y. (2004). Polymer-supported ionic liquids: Imidazolium salts as catalysts for nucleophilic substitution reactions including fluorinations. *Angew. Chem., Int. Ed.* 43(4): 483–485.

9. Shinde, S.S., Lee, B.S., and Chi, D.Y. (2008). Synergistic effect of two solvents, tert-alcohol and ionic liquid, in one molecule in nucleophilic fluorination. *Org. Lett.* 10(5): 733–735.

10. Reichaedt, C. (1998). *Solvents and Solvent Effects in Organic Chemistry*, 3rd ed. Wiley VCH: Cambridge, UK.

11. Smith, M.B. and March, J. (2001). *Advanced Organic Chemistry*, 5th ed. Wiley-Interscience: New York, pp. 462–467.

12. Kim, D.W., Ahn, D.S., and Oh, Y.H. et al. (2006). A new class of S$_N$2 reactions catalyzed by protic solvents: Facile fluorination for isotopic labeling of diagnostic molecules. *J. Am. Chem. Soc.* 128(50): 16394–16397.

13. Kim, D.W., Jeong, H.J., and Lim, S.T. et al. (2008). Tetrabutylammonium tetra (tert-Butyl alcohol)-coordinated fluoride as a facile fluoride source. *Angew. Chem., Int. Ed.* 47(44): 8404–8406.

14. Lee, J.W., Yan, H., Jang, H.B. et al. (2009). Bis-terminal hydroxy polyethers as all-purpose, multifunctional organic promoters: A mechanistic investigation and applications. *Angew. Chem., Int. Ed.* 48(41): 7683–7686.

15. Jadhav, V.H., Jang, S.H., and Jeong, H.J. et al. (2012). Oligoethylene glycols as highly efficient mutifunctional promoters for nucleophilic-substitution reactions. *Chem.-Eur. J.* 18(13): 3918–3924.

16. Oh, Y.H., Ahn, D.S., and Chung, S.Y. et al. (2007). Facile S$_N$2 reaction in protic solvent: Quantum chemical analysis. *J. Phys. Chem. A* 111(40): 10152–10161.

17. Kim, J.Y., Kim, D.W., and Song, C.E. et al. (2013). Nucleophilic substitution reactions promoted by oligoethylene glycols: A mechanistic study of ion-pair S$_N$2 processes facilitated by Lewis base. *J. Phys. Org. Chem.* 26(1): 9–14.

18. Lynch, B., Fast, P.L., Harris, M. et al. (2000). Adiabatic connection for kinetics. *J. Phys. Chem. A* 104(21): 4811–4815.

19. Adamo, C. and Barone, V. (1998). Exchange functionals with improved long-range behavior and adiabatic connection methods without adjustable parameters: The mPW and mPW1PW models. *J. Chem. Phys.* 108(2): 664–675.

20. Wadt, W.R. and Jeffrey, H.P. (1985). Ab initio effective core potentials for molecular calculations. Potentials for K to Au including the outermost core orbitals. *J. Chem. Phys.* 82(1): 299–310.

21. Jadhav, V.H., Jeong, H.J., Lim, S.T. et al. (2011). Tailor-made hexaethylene glycolic ionic liquids as organic catalysts for specific chemical reactions. *Org. Lett.* 13(5): 2502–2505.

22. Jadhav, V.H., Kim, J.Y., Chi, D.Y. et al. (2014). Organocatalysis of nucleophilic substitution reactions by the combined effects of two promoters fused in a molecule: Oligoethylene glycol substituted imidazolium salts. *Tetrahedron* 70(2): 533–542.

23. Cances, E., Mennucci, B., and Tomasi, J. (1997). A new integral equation formalism for the polarizable continuum model: Theoretical background and applications to isotropic and anisotropic dielectrics. *J. Chem. Phys.* 107(8): 3032–3041.

24. Cossi, M., Barone, V., Mennucci, B. et al. (1998). Ab initio study of ionic solutions by a polarizable continuum dielectric model. *Chem. Phys. Lett.* 286(3–4): 253–260.

25. Cances, E., Mennucci, B., and Tomasi, J. (1998). Analytical derivatives for geometry optimization in solvation continuum models. II. Numerical applications. *Chem. Phys.* 109(1): 260–266.

26. Lee, S.-S., Jadhav, V.H., Kim, J.-Y., Chun, J.-H., Lee, A., Kim, S.-Y., Lee, S., and Kim, D.W. (2015). Quantum chemical investigation of the origin of activation of SN2 type halogenation by oligo-ethylene glycol-ionic liquids. *Tetrahedron* 71(19): 2863–2871.

6 Sustainable Organic Synthesis Using Ionic Liquids

Toshiyuki Itoh and Toshiki Nokami

CONTENTS

6.1 INTRODUCTION

Ionic liquids (ILs) [1] have been extensively investigated in recent years, and the chemistry of ILs has rapidly progressed during these past two decades [2,3]. The liquids have attractive properties as follows: (1) ILs have a very low vapor pressure (this means that ILs are less flammable and cause no air pollution). (2) ILs have a low melting point and theoretically no boiling point (the means that we can use ILs as the solvent in a wide temperature range). (3) ILs dissolve various types of organic and inorganic materials and their solubility could be controlled by the design of the ILs. (4) We can design the properties of ILs and add functionality to the liquids. (5) And it could be possible to realize reaction systems that allow recycling of the catalysts anchored by the ILs [1,4–9]. These properties of the ILs are quite attractive for the aim of realizing sustainable chemical reactions using ILs as the reaction media.

The first example of an organic reaction using an IL as a reaction medium was the Friedel-Crafts alkylation or acylation of benzene using 1-butyl-3-methylimidazolium aluminum tetrachloride ($[C_4mim][AlCl_4]$) that was reported by Wilkes and co-workers in 1986 [10]. However, since $[C_4mim][AlCl_4]$ was a moisture sensitive liquid, the results gained interest by a few chemists and only several papers, for application of the ILs for organic syntheses have been published during the 25 years from 1975 from 1999. An explosion of the research projects about the ILs began in 2000, and the total number of papers regarding organic synthesis in ILs reached 4,523 among 82,992 papers in all the fields of ILs this year [2].

ILs offer a high potential to replace classic flammable and toxic organic solvents [1,4–10]. We can also use ILs as a solvent even for enzymatic reactions, and the recycling use of an enzyme by anchoring it in the ILs like transition metal-catalyzed reactions has been established [7]. In this chapter, we report the typical application of ILs for organic synthesis from the standpoint of realizing sustainable organic syntheses using ILs.

6.2 TREND IN THE USE OF ILs FOR SUSTAINABLE ORGANIC SYNTHESES

6.2.1 How to Use ILs as Solvents for Organic Reactions

6.2.1.1 Transition Metal-Catalyzed Reactions

Dupont et al. reported the Rh-catalyzed hydration in a mix solvent of an IL with an organic molecular liquid in 1996 [11]. This the first example of a transition metal-catalyzed reaction using ionic liquids as the reaction media. The same group reported the enantioselective version using Ru-BINAP in 1997 (Figure 6.1) [12]. However, since a mixed solvent system of imidazolium IL with a conventional organic solvent was used in their reaction and the results were modest, unfortunately, these papers did not provide a strong impact on the synthetic organic chemistry community.

The milestone example of the transition metal-catalyzed reaction using an IL solvent was the Mizoroki-Heck reaction [13]. Carmichael et al. reported that iodobenzene and ethyl acrylate reacted in $[C_4mim][PF_6]$ in the presence of 2 mol% $Pd(OAc)_2$, 4 mol% triphenylphosphine, and 1.2 eq. of trimethylamine as a base at 100°C to afford ethyl cinnamate [14]. After the reaction, the addition of cyclohexane gave a two-phasic layer of an organic solvent layer and the IL. The desired product, ethyl cinnamate, was quantitatively extracted from the cyclohexane layer, then the ionic layer was washed with water to successfully remove the byproduct, triethylammonium hydroioide. Since the Pd catalyst remained in the IL layer, after drying of the IL solution under reduced pressure, the addition of the second set of

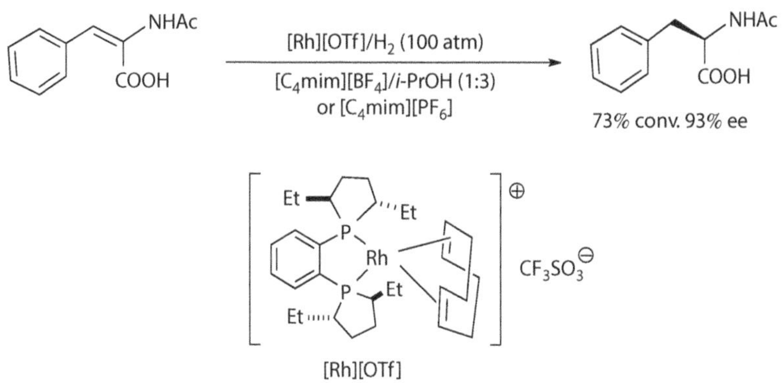

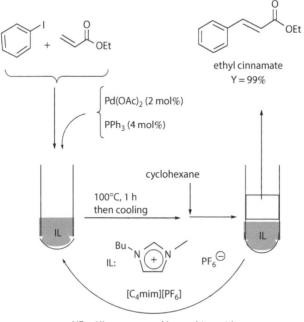

FIGURE 6.1 Chiral Rh-complex mediated enantioselective hydration using IL-*i*-PrOH mixed solvent system.

the substrate (iodebenzene and ethyl acrylate) caused the next reaction. The authors thus succeeded in demonstrating the recycling use of the catalyst, as illustrated in Figure 6.2 [14]. The recycling use of an expensive Pd catalyst is a very important benefit of the use of ILs as the solvent from the standpoint of sustainable syntheses. Hermann and Bölm also reported a similar reaction in the same year [15]. A highly polar solvent like *N,N*-dimethylformamide or acetonitrile (CH$_3$CN) is generally

FIGURE 6.2 Mizoroki-Heck reaction using the [C$_4$mim][PF$_6$] solvent system.

necessary for the Mizoroki-Heck reaction. However, these solvents require the complex process of the extraction of the products and removing the solvent. Therefore, the results had a strong impact on the synthetic organic chemistry community, and this year became the turning point year in this field.

Dupont and co-workers reported the first successful preparation of iridium nanoparticles when an iridium complex (Ir(cod)Cl)$_2$) was reduced with hydrogen gas in [C$_4$mim][PF$_6$] in 2002 [16]. As expected, the produced Ir nanoparticles showed a unique reactivity toward several reactions [17]. In 2006, Torimoto et al. accomplished the preparation of Au-Ag hybrid nanoparticles by the electro-reduction procedure in an IL [18]. It is now recognized that the IL engineering is a key to preparing novel metal nanoparticles.

Sodeoka et al. reported an excellent example of the chiral Pd-catalyzed enantioselective fluorination using an IL solvent system in 2003 [19]. The 2-methyl-3-oxo-3-phenylpropionic acid t-butyl ester was reacted with bis(phenylsulfonyl)ammoium fluoride in the presence of the 2.5 mol% chiral Pd-aqua complex in [C$_4$mim][BF$_4$] to afford the fluorinated product in 93% yield with 92% ee (Figure 6.3). They demonstrated the recycling use of the catalyst and succeeded in obtaining the product in 82% (91% ee) even after repetition of the reaction 11 times. Although it has been 15 years after publication of this report, the results still maintain a certain impact in the field of synthetic organic chemistry.

Ryu and Fukuyama reported the copper-free Sonogashira coupling reaction using an IL solvent system in 2002 (Figure 6.4) [20]. Although the Sonogashira coupling reaction is well respected as the useful means for preparing alkynylated aromatic compounds, the method has a serious weak point from the standpoint of sustainable chemistry that a large excess of copper powder is necessary to achieve the reaction. The authors solved this problem using ILs, they obtained the desired alkynylated

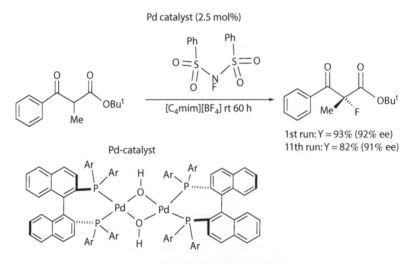

FIGURE 6.3 Chiral Pd-aqua complex-catalyzed enantioselective fluorination using [C$_4$mim][BF$_4$] solvent system.

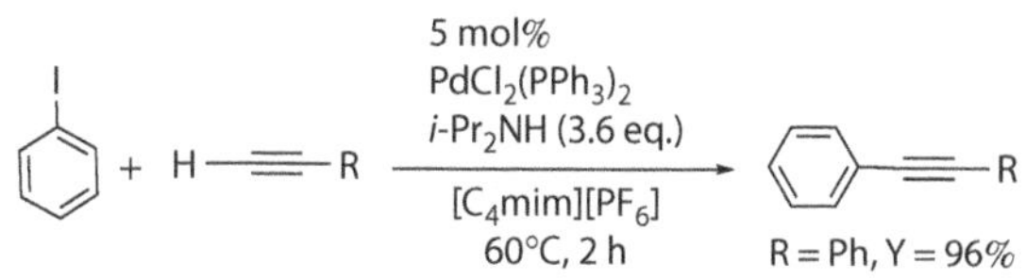

FIGURE 6.4 Cu-free Sonogashira-coupling reaction using an IL solvent system.

benzene derivatives in an excellent yield in the absence of copper metal using the $[C_4mim][PF_6]$ solvent system (Figure 6.4) [20].

Ryu and co-workers group recently reported interesting results, they discovered that use of an imidazolium IL as the solvent inhibited the Sonogashira coupling reaction when the reaction was conducted in the presence of a strong base [20]. To investigate the details of the reaction course, they found that the 2-arylated imidazolium IL was produced under the reaction conditions. Using this, the authors accomplished the synthesis of 2-aryl-substituted imidazolium ILs after optimization of the reaction conditions (Figure 6.5) [21].

Arylation was also reported by Punzi et al. using the deep eutectic solvent system [22]. The authors found that duplicated arylation took place when a thiophene derivative was treated with 5 mol% $Pd_2(dba)_3$ in the presence of tris(2-methoxyphenyl) phosphine (10 mol%), 2 eq. of Cs_2CO_2 as a base and PivOH to give the arylated compound in good yield (Figure 6.6). They postulated that the hydrogen atom at the neighbor of the sulfur atom in the thiophene ring might be activated in a highly-polar deep eutectic solvent.

Recycling of the catalyst using ILs is applicable for the one-pot multiple reactions. We found that the Nazarov cyclization of the 3-(pyrrole-2-yl or thiophene-2-yl)-3-oxo-2-alkenlyl ester smoothly proceeded in the presence of 0.5 mol%–5 mol%

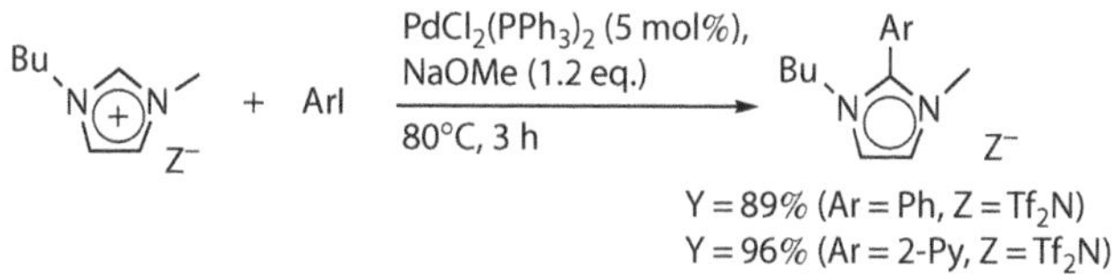

FIGURE 6.5 Pd-catalyzed arylation of imidazolium ILs through C-H activation at the 2-position.

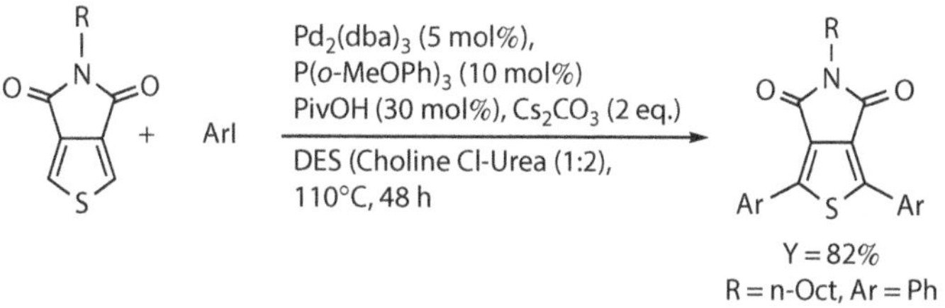

FIGURE 6.6 Synthesis of 1,3-diophenyl-4H-thieno[3,4-c]pyrrole-4,6-(5H)-dione derivative using Pd-catalyzed C-H activation.

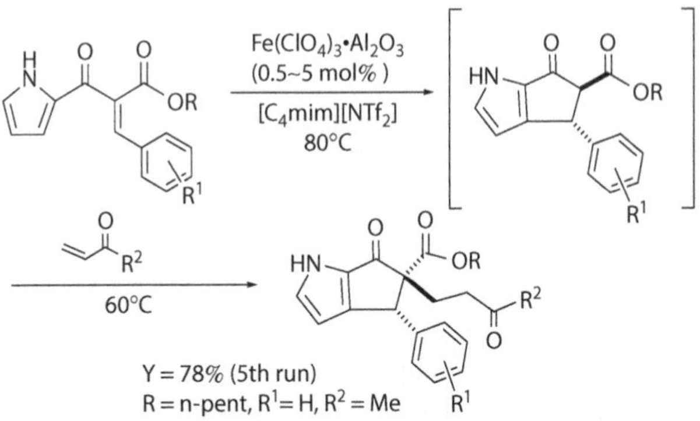

FIGURE 6.7   Iron(III) salt-catalyzed Nazarov cyclization using an IL solvent system.

alumina-supported iron(III) perchlorate ($Fe(ClO_4)_3 \cdot Al_2O_3$) in ILs, and the subsequent Michael reaction took place when a vinyl ketone was added to the reaction mixture to afford the Nazarov-Michael products in good yield (Figure 6.7) [23]. During testing of recyclable use of catalyst, however, we encountered unexpected difficulty that isolation of the final product, 4,5-dihydrocyclopenta[*b*]pyrrol-6(1*H*)-one derivative (R = Me, R^2 = Me) was unsuccessful from the reaction mixture because the product was very soluble in $[C_4mim][Tf_2N]$. After evaluation several esters, we finally succeeded in isolating the desired products from the IL reaction mixture in acceptable yields when n-pentyl ester or benzyl ester was used as a starting material. We thus demonstrated the catalyst recyclable use system employing n-pentyl ester in $[C_4mim][Tf_2N]$ as solvent (Figure 6.7) [23c].

6.2.1.2   Chemical Reactions Using IL-Tag

As noted, it is now well recognized that the recycling use of the transition metal catalyst is possible using IL solvent systems because many types of transition metal complexes easily dissolve in the ILs. However, the recyclable use of several catalysts failed due to the poor solubility of the catalyst in the ILs. To solve this difficulty, IL-tag-substituted catalysts have been developed, such as Ru@IL for the olefin methathesis reaction [24], Cu(OTf)2-bisoxazoline@IL for the enantioselective Diels-Alder reaction [25], VOsalen@IL for the asymmetric oxidation [26], and Salen-Mn@Il for the oxidative kinetic resolution of racemic alcohols [27]. The concept of IL-tag is, of course, applicable to organocatalyst and IBX@IL and has been developed for the oxidation of alcohols to ketones [28]. Typical examples are shown in Figure 6.8 [24–28]. The use of these IL-tag-type catalysts not only enabled recycling of the catalyst, but also a significant reduction of the catalyst leaching from the reaction media.

We accomplished the electrolyte-free electrochemical glycosylation using glycosyl acceptors containing an IL-tag as an intramolecular electrolyte and for realizing easy isolation of the resulting glycosides from the reaction mixture. A one-pot electrochemical glycosylation-Fmoc deprotection sequence has thus been performed under an electrolyte-free condition (Figure 6.9) [29]. The IL-tag was proved to play

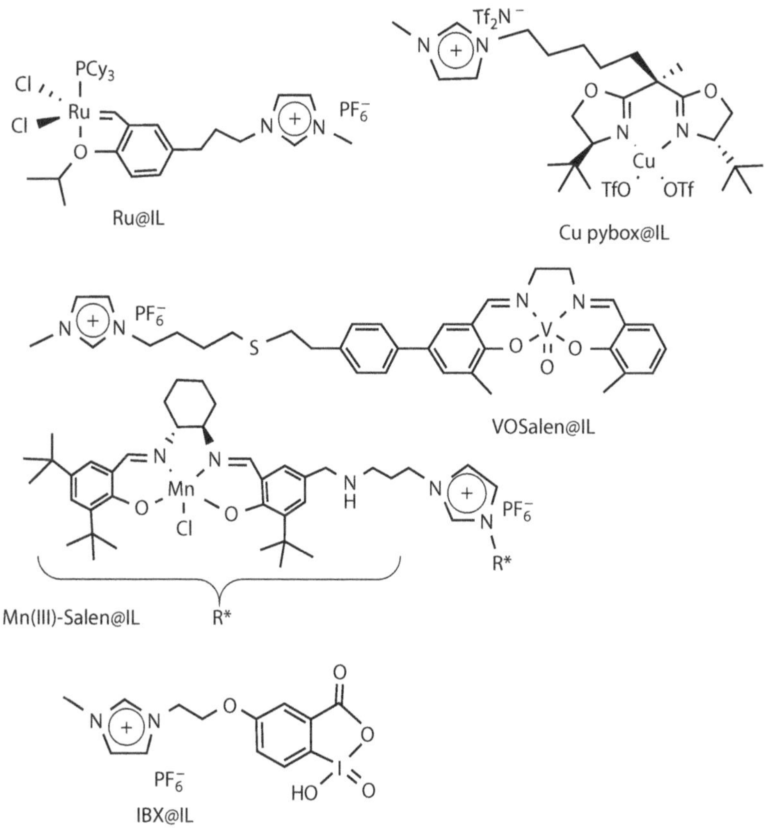

FIGURE 6.8 List of typical IL-tag-type catalysts.

multifunctional roles in the glycosylation, the tag in the thioglycosides worked as an intramolecular electrolyte for the anodic oxidation, worked as a stereo-controlling group, and made possible easy extraction of the products after the reaction. We successfully applied this to the synthesis of a tetrasaccharide which is a precursor of TMG-chitotriomycin [29b]. Li and co-workers also succeeded in synthesizing origosugars using glycosyl acceptors containing an ionic-liquid tag (Figure 6.10) [30].

6.2.1.3 Design of Appropriate ILs for Chemical Reactions

It is known that the acidity of the 2-position of the imidazolium ring of ILs is higher than the usual proton of which the pKa values are estimated to be ca. 20 [31]. Therefore, carbene complexes were easily formed when imidazolium ILs were treated with transition metal complexes in the presence of a strong base. Although this generally provided favorable results in the transition metal catalyst-mediated reactions by improving the stability of the complex [31a,b], poor results were also sometimes obtained due to the acidic proton at 2-position of the imizadolium ring. For example, we found that no product was obtained when the Pd-catalyzed allylic alkylation was carried out in the presence of sodium bis(trimethylsilyl)amide

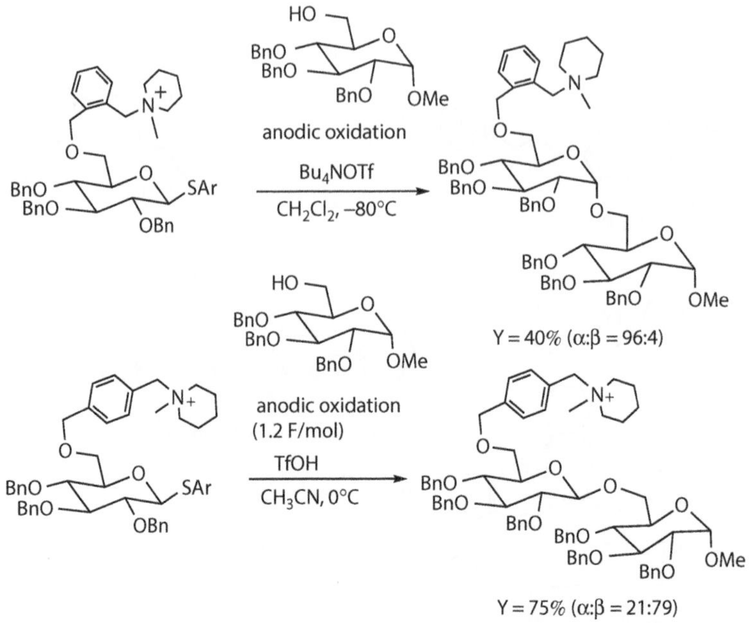

FIGURE 6.9 IL tag effect on the stereochemistry of electrochemical glycosylation.

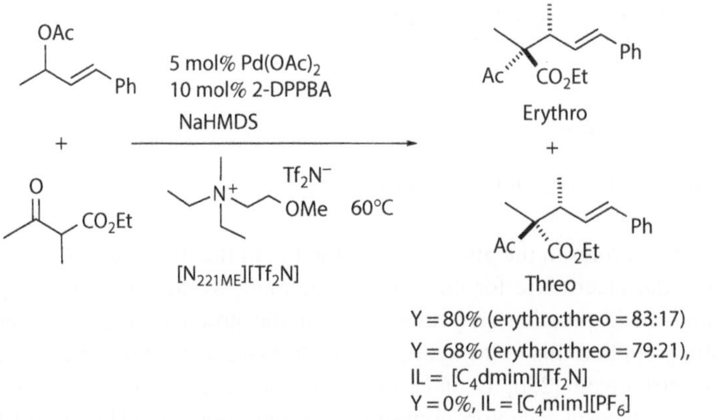

FIGURE 6.10 Pd-catalyzed allylic alkylation using an IL solvent system.

(NaHDMS) in [C$_4$mim][Tf$_2$N] because NaHDMS was decomposed by the acidic proton of the [C$_4$mim] cation, and the resulting imidazolium carbene complexes exhibited no allylation activity. We solved this problem by using the 1-butyl-2,3-dimethylimkidazolium salt IL ([C$_4$dmim][PF$_6$]), and the best results were obtained when the reaction was carried out in quaternary ammonium IL, [N$_{221ME}$][Tf$_2$N], as the solvent (Figure 6.10) [32].

ILs have been considered inappropriate for strong base-mediated reactions [1,4]. Clyburne and co-workers solved this problem by demonstrating Grignard reactions in a mixture of THF and a phosphonium ionic liquid, [P$_{666(14)}$][Tf$_2$N] (Figure 6.11) [33].

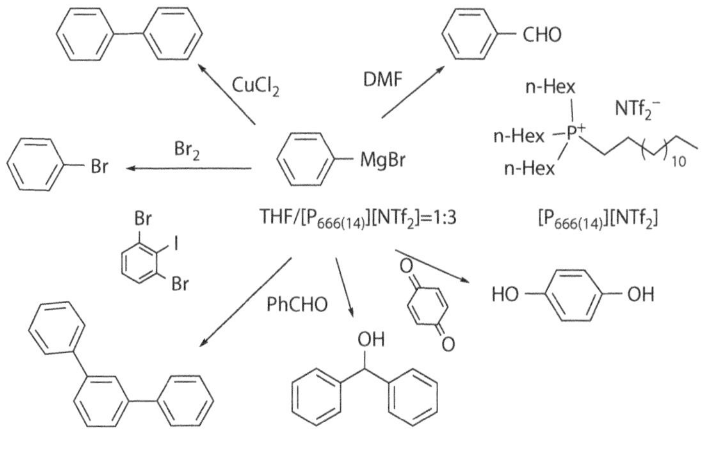

FIGURE 6.11 Grignard reactions in a phosphonium IL, $[P_{666(14)}][Tf_2N]$ as the solvent.

The authors showed that six types of reactions were possible in $[P_{666(14)}][Tf_2N]$, including the addition of carbonyl, benzyne reaction, halogenation, and coupling reactions because the long alkyl chain in the cationic part inhibits the Grignard reagent to remove the highly acidic proton. Wilhelm and Jurcik [34a] and Handy [34b] also reported that Grignard reactions were accomplished in the 2-substituted imidazolium-type ionic liquid solvent systems. Furthermore, Chan et al. reported the preparation of ethyl magnesium iodide in the pure ionic liquid, N-butylpyridinium tetrafluoroborate ([bpy][BF_4]) [35].

We designed a phosphonium IL by introduction of an alkyl ether moiety on the side arm of the phosphonium cation, methoxyethyl(tri-n-butyl)phosphonium bis(trifluoromethanesulfonyl)imide ($[P_{444ME}][NTf_2]$), was developed for this aim [36]. We then applied this IL as a solvent for the homo-coupling reaction of the aryl Grignard reagent in the presence of 1 mol% $FeCl_3$ using 1,2-dichloroethane or 1,2-diiodoethane as an oxidant. The desired coupling reaction proceeded at amazing speed, biaryl derivatives were obtained in less than 5 min at 0°C in excellent yields except for the bulky 2,4,6-trimethylphenylMgBr [37]. The IL solvent system was also useful for the homo-coupling reaction of the alkynyl Grignard reagent [37]. In this case, the use of 1,2-diiodoethene was essential to afford the coupling product, and we obtained the desired homo-coupling product in good yield (Figure 6.12) [37].

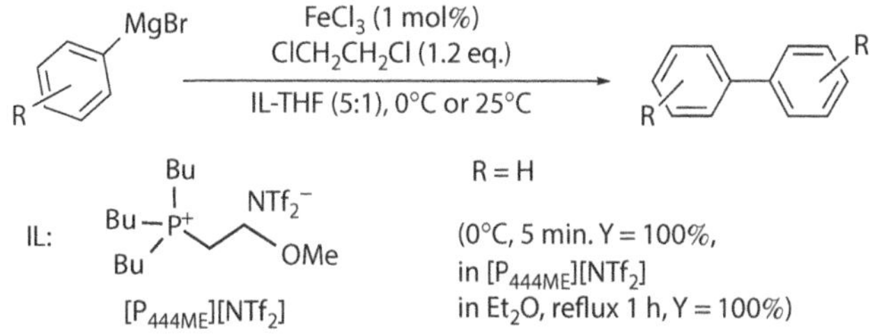

FIGURE 6.12 Iron(III)-catalyzed rapid homo-coupling reaction of aryl Grignard reagents in the IL solvent.

These results clearly indicated that we should evaluate (design) ILs by considering the reaction type. Fortunately, many types of pure ILs are now commercially
available. We believe that the solvent engineering using ILs would become more
important in future from the standpoint of realizing sustainable synthesis.

6.2.2 ACTIVATION OF CHEMICAL REACTIONS USING ILs
FOR REALIZING SUSTAINABLE ORGANIC SYNTHESIS

6.2.2.1 IL-Type Catalyst-Mediated Reactions

ILs offer a high potential to replace classic flammable and toxic organic solvents and
almost all types of reactions are now possible in IL solvent systems except for strong
basic conditions using butyl lithium. However, replacement of the reaction media
from classical organic solvents or water to ILs in industry has still not yet been realized. One serious barrier is the high viscosity of the ILs because viscous solvents
require complete replacement of the reaction process. Therefore, it is required to
develop the means or reactions that are only possible in ILs. In order for ILs to
become more popular, the key is how to reduce the amount of the ILs used, while
obtaining a certain benefit. To this aim, investigations using ILs as not just solvents,
but catalysts or activating agents, have been attempted. A typical example is the
task-specific ILs [38,39] by introduction of the sulfonyl group on the side arm of the
imidazolium cation that was first developed by Davis, Jr. et al. [38].

Cao et al. recently reported unique task-specific IL-catalyzed reaction;
one equivalent of 1,3-bis(4-sulfonylbutyl)imidazolium hydrogen sulfate worked as
a catalyst to achieve the diazenylation of the indole derivatives (Figure 6.13) [40].
The authors postulated that the terminal sulfonyl group substituted on the imidazolium salt reacted with aryltriazene to give tetrahydropyrrolyl (THP) cation, which
was transformed into azo cation and releasing tetrahydropyrrole. The azo cation
then reacted with indole to afford the product *via* electrophilic substitution reaction.
Although numerous examples have been reported in this field, since good reviews
for acidic ILs have already been published [41], we focused on recent different
methodogies in this chapter.

Y = 99%, R = H
Y = 88%, R = 4-OMe
Y = 79%, R = 4-Cl

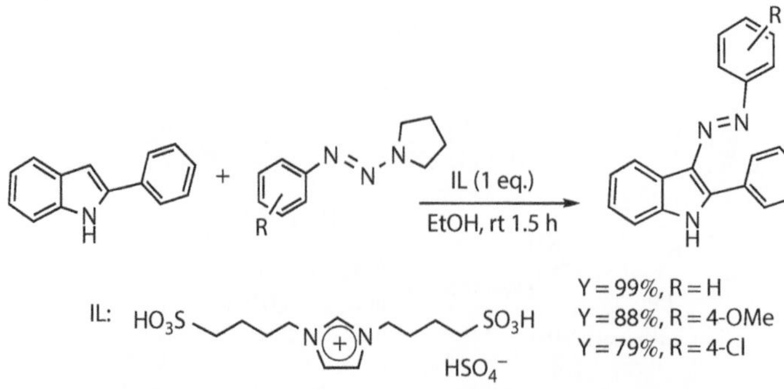

FIGURE 6.13 IL-catalyzed diazenylation of N-heterocyclic compounds.

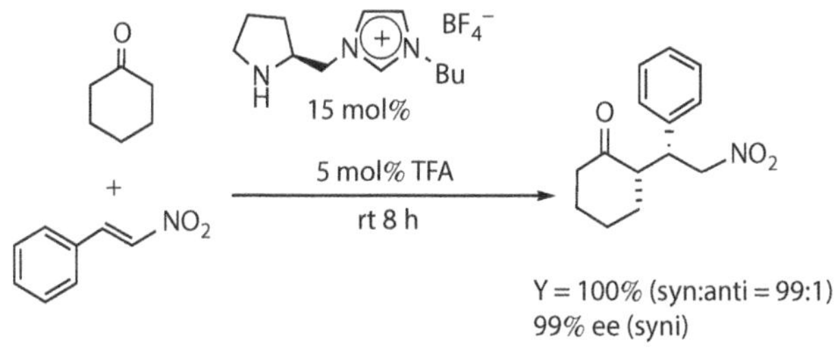

FIGURE 6.14 Chiral imidazolium salt-catalyzed asymmetric Michael reaction.

Luo et al. reported very successful results of asymmetric Michael reaction using IL-type organocatalyst in 2006 (Figure 6.14) [42]. Cyclohexanone was reacted with 1-nitro-2-phenyethene in the presence of 15 mol% of chiral imidazolium BF$_4$, which was derived from L-proline and 5 mol% trifluoroacetic acid as a co-catalyst to afford the Michael product in quantitative yield with excellent enantio- and *syn*-selective manner.

Since then, numerous examples have been reported for similar types of chiral IL-type organocatalyst-mediated reactions [43]. Obregón-Zúniga and co-workers recently reported the preparation of chiral solvate IL that was derived from triglyme and the lithium salt of L-proline, and the corresponding IL acted as an efficient catalyst for the enantioselective aldol condensation, and the authors demonstrated the recyclable use of the catalyst (Figure 6.15) [44].

Varyani and co-workers prepared a hybrid of the amino acid IL with the iron-salen complex and accomplished oxidative conversion of benzylamine into cyanobenzene using 7.5 mol% iron-salen complex as a catalyst under aerobic conditions in excellent yield (Figure 6.16) [45].

As noted before, since the 2-position of the imidazolium ring is highly acidic, an imidazolium IL sometimes acted as a Brønsted catalyst for acylation [46]. Dupont et al. recently reported an hydrogen-deuterium (H-D) exchange reaction using imizadolium ILs as a catalyst (Figure 6.17) [47]. The authors found that partial deuteration at the 2-methyl group of the proline salt of 1-butyl-2,3-dimethylimidazolium ([C$_4$dmim][Pro]) took place when the salt was dissolved in

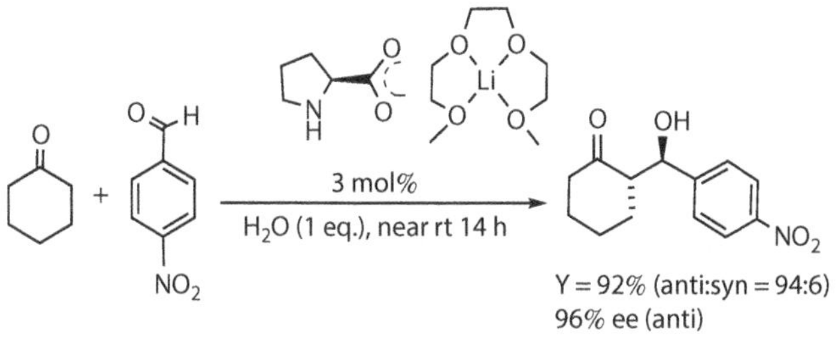

FIGURE 6.15 Enantioselective aldol reaction in the presence of solvate IL derived from (*S*)-proline.

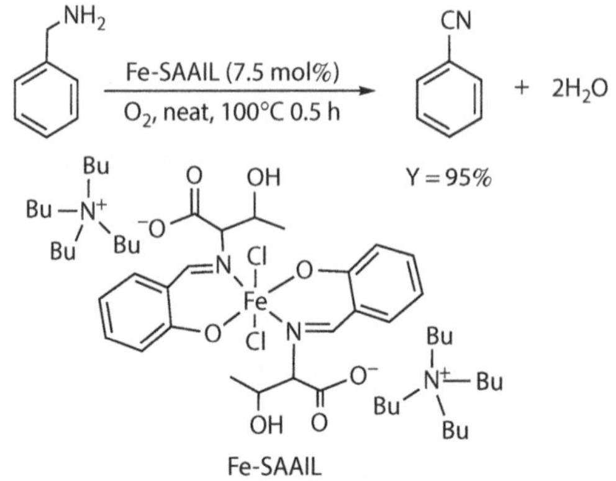

FIGURE 6.16 Aerobic oxidation of phenylmethanamine to cyanobenzene using iron-salen complex.

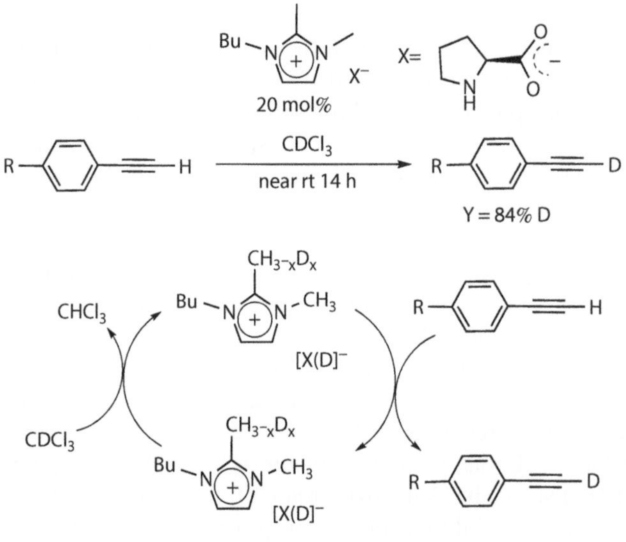

FIGURE 6.17 Imidazolium salt-mediated deuteration of alkynes in $CDCl_3$.

$CDCl_3$, and this salt acted as deuteration agent, the H-D exchange reaction took place when the arylacetylene was treated with $CDCl_3$ in the presence of 20 mol% [C_4dmim][Pro] to provide the deuteriated arylacetylene, as illustrated in Figure 6.17.

6.2.2.2 IL-Coating Catalyst (SILC or SILP)

Hagiwara et al. reported the preparation of a silica gel-supported IL-Pd-catalyst (SILC), the silica gel and Pd/C was mixed with [C_4mim][PF_6], and the SILC displayed an excellent reactivity for the Mizoroki-Herck reaction in 2001 [48]. It is well known that the stability of the transition metal was improved by being supported on

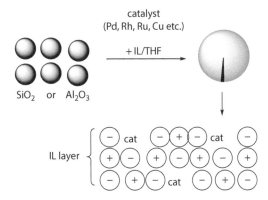

FIGURE 6.18 Model structure of the IL-coated silica gel-supported catalyst (SILC or SILP).

a highly porous solid such as zeolite or mesoporous silica, while the reaction rate is generally inferior compared to that of a molecular catalyst. On the contrary, the SILC-catalyzed reactions rapidly proceeded almost the same as a corresponding molecular catalyst. The authors hypothesized that the IL on the surface on the silica works as carbene ligand and the substrates easily absorbed into the IL phase, thus the reaction proceeded smoothly. Washing with the reaction mixture after completion of the reaction and dried under reduced pressure, the SILC was repeatedly used. The authors prepared various types of SILC catalysts, as illustrated in Figure 6.18 [48].

Mehnert and co-workers reported the preparation of the silica gel-IL-supported Rh complex in 2003 [49]. The authors named their catalyst supported ionic liquid-phase (SILP), then many excellent applications have been reported by Fehrmann and Wasserscheid et al. [50]. These types of silica gel-IL-supported transition metal catalysts are now widely known as SILP. Wasserscheid and co-workers reported the preparation of Pt-SILP, which was prepared by Al_2O_3 and 30 wt% Li/K/Cs[OAc] and revealed that this catalyst exhibited not only an increased reaction rate, but also improved stability under high reaction temperature conditions [51].

Since viscous solvents require complete replacement of the reaction process, replacement of the reaction media from classical organic solvents or water to ILs in industry has still not yet been realized. However, there is no barrier using the SILP in industry. Therefore, SILP is now used as a catalyst for removing mercury from natural gas in industry [52].

6.2.2.3 Activation of S_N2-Type Halogenation by the IL

Numerous examples of halogenation reactions using ILs have been reported [53]. Fuchigami et al. developed the electro-halogenation in $[C_2mim][OTf]$, which contains $Et_3N \cdot 5HF$ and reported many good examples [54]. Chi and co-workers reported a very successful example of halogenation using IL engineering, a mesylate compound reacted with potassium fluoride (KF) in $[C_4mim][BF_4]$ to afford the corresponding fluorinated compound, while this type of S_N2-type reaction was generally unsuccessful due to the poor nucleophilic property of the fluorine anion (Figure 6.19) [55].

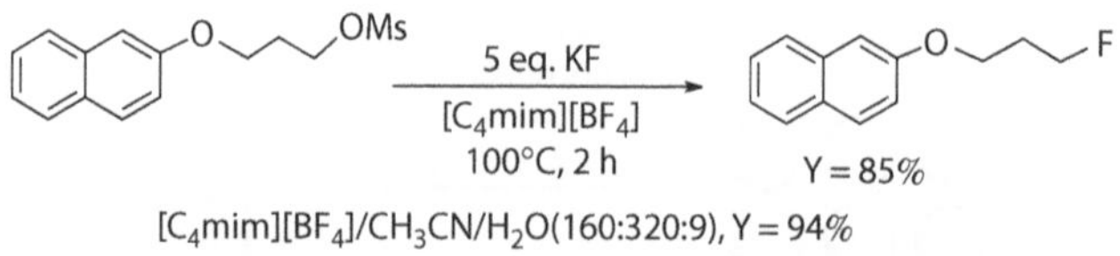

FIGURE 6.19 S_N2-type fluorination using KF in the presence of an IL.

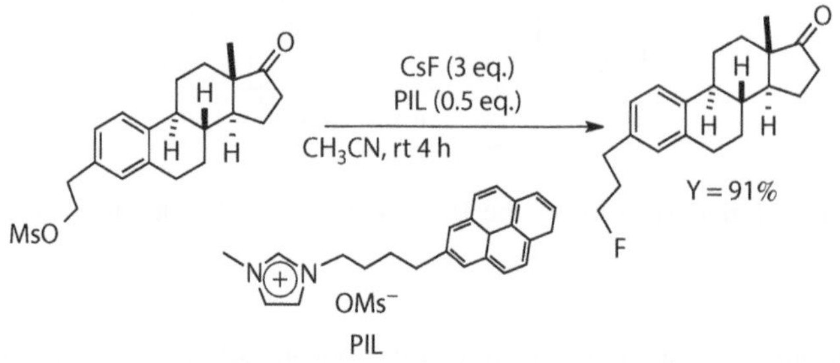

FIGURE 6.20 IL-catalyst-mediated fluorination of a mesylate using KF.

Kim et al. recently reported the preparation of the pyrene-substituted imidazolium salt, and this acted as a catalyst for the CsF-mediated fluorination of mesylate (Figure 6.20) [56]. The authors hypothesized that the Cs cation could be trapped by the pyrene moiety, then produced the naked fluoride anion, which contributed to activating the S_N2 substitution reaction. The authors also demonstrated recovery of the catalyst by absorbing it on graphite.

6.2.2.4   Activation of Aromatic Halogenation

Sutherland et al. found that $Fe(Tf_2N)_3$ was produced by the anion exchange reaction of $FeCl_3$ with $[C_4mim][Tf_2N]$, then $Fe(Tf_2N)_3$ activated N-chlorosuccinimide (NCS) or N-bromosuccinimide (NBS) to catalyze the chlorination or bromination of aromatic compounds [57]. The authors discovered that the addition of 7.5 mol% $[C_4mim][Tf_2N]$ in THF activated NCS and NBS, and the desired halogenation reaction proceeded smoothly. Since the halogenation took place regioselectively, they succeeded in preparing 2-chloro-4-bromoanisole or 2-bromo-4-chloroanisole by a simple exchange of the order of NCS and NBS in a 1-pot reaction system, as illustrated in Figure 6.21 [57].

6.2.2.5   Activation of Lewis Acid-Mediated Reactions

We found that the [2+3]-type cycloaddition reaction of styrene derivatives with 1,4-benzoquinone proceeded very rapidly in the presence of 3 mol% $Fe(ClO_4)_3 \cdot Al_2O_3$ in an IL as solvent (Figure 6.22) [58]. The reaction was completed in less than 10 min in 1-butyl-3-methylimidazolium hexafluorophosphate ([bmim][PF$_6$]), while it took 2 h in acetonitrile at the same concentration. Furthermore, the reaction was completed when 3 mol% $Fe(BF_4)_2 \cdot 6H_2O$ reacted under ambient conditions in an IL solvent, while no reaction took place in CH_3CN. A very rapid reaction was recorded for $[C_4mim][PF_6]$ and $[C_4mim][Tf_2N]$, while the reaction rate decreased for $[C_4mim]$

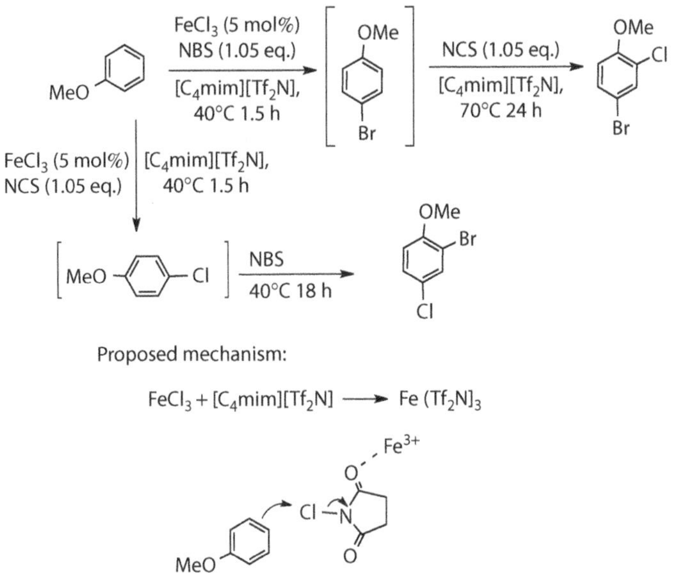

FIGURE 6.21 Iron(III) triflimide-catalyzed chlorination in the presence of an IL.

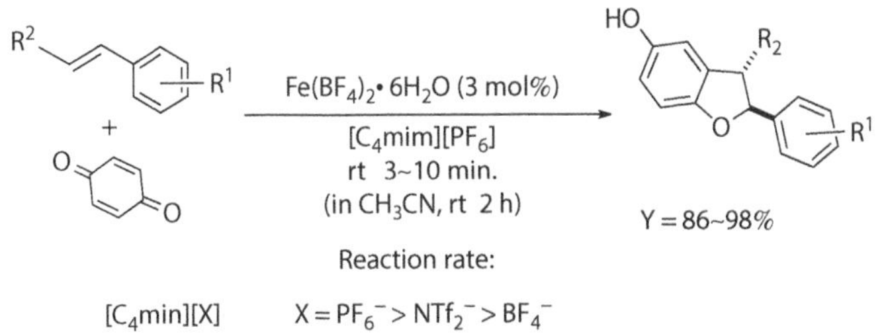

FIGURE 6.22 Iron salt-catalyzed rapid [2 + 3]-cycloaddition of styrene derivatives with bernzoquinone.

[BF$_4$], and no reaction occurred when [C$_4$mim][OTf] or [C$_4$mim][n-C$_5$H$_{11}$SO$_4$] was used as the solvent. It was thus found that the reaction rate depended on the anion of the imidazolium salts, and the order of the reaction rate was PF$_6^-$ > Tf$_2$N$^-$ > BF$_4^-$ [58b].

Lee and co-workers reported the Friedel-Crafts type alkynylation in the presence of 10 mol% Sc(OTf)$_3$ in an IL solvent system (Figure 6.23) [59,60]. The reaction was strongly activated using an IL solvent, and the reaction rate depended on the anionic part of the solvent IL. The order of the reaction rate was reported as SbF$_6^-$ > PF$_6^-$ > Tf$_2$N$^-$ > BF$_4^-$ >>OTf$^-$ [61].

The same group also reported the interesting activation of the Pd/C mediated hydration of aromatic compounds using IL as the solvent [62]. It is well known that harsh reaction conditions are necessary to reduce the aromatic ring to the corresponding cycloalkanes. For example, the hydration of C$_{60}$ fullerene by a Pd/C

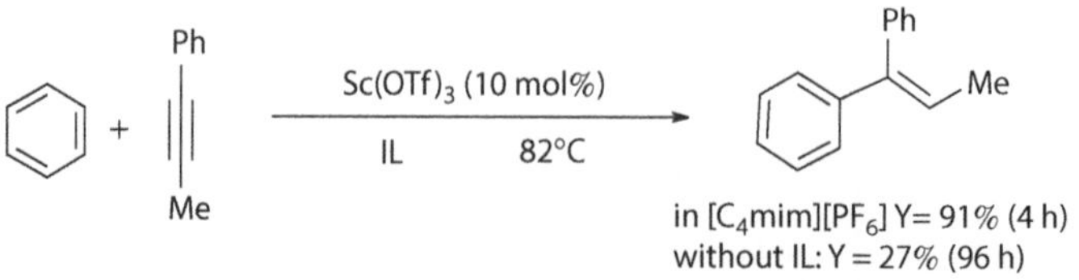

FIGURE 6.23 Lewis acid-catalyzed Friedel-Crafts-type alkenylation in an IL.

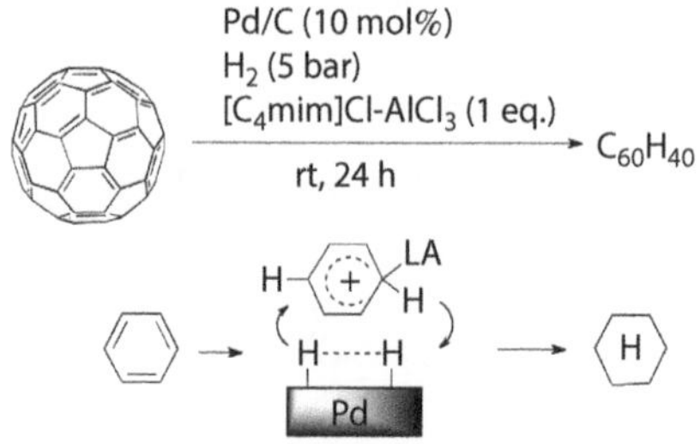

FIGURE 6.24 Lewis acid-mediated hydrogenation of C_{60} fullerene.

catalyst should be conducted at 120 bar at 400°C, while the reaction was completed at ambient pressure and room temperature (rt) for 48 hrs in the presence of 1 eq. of $[C_4mim]Cl-(AlCl_3)_{0.67}$ (Figures 6.22 and 6.24) [62].

6.2.3 LIPASE-MEDIATED REACTION FOR SUSTAINABLE ORGANIC SYNTHESIS USING ILs

The value of the enzymatic reaction in organic synthesis is highly respected from the standpoint of green chemistry and has now reached an industrially proven level in the pharmaceutical and food industries [63]. I focus on topics concerning the improved performance of the lipase-catalyzed reactions using ionic liquid engineering because lipase (triglycerol acylhydrolases EC 3.1.1.3) is the most popular biocatalyst for organic chemists, and the enzyme can catalyze a number of reactions including hydrolysis, transesterification, alcoholysis, acidolysis, esterification, and aminolysis [64].

We reported the enantioselective lipase-catalyzed reactions in ILs in early 2001. Typically, the lipase-catalyzed enantioselective transesterification using an IL solvent system has been carried out, as illustrated in Figure 6.25 [65]. To a mixture of lipase in the ionic liquid were added a racemic alcohol [(±)-1] and vinyl acetate as the acyl donor, and the resulting mixture was stirred at 35°C. After the reaction, ether or a mixed solvent of hexane and ether was added to the reaction mixture to form a biphasic layer, and the acetate product and unreacted alcohol were quantitatively extracted with the organic solvent layer. The enzyme remained into the ionic liquid phase, the addition of the next set of substrate (±)-1 and acyl donor 3 caused the next cycle of the reaction to afford the chiral ester (R)-2 and (S)-1. Therefore,

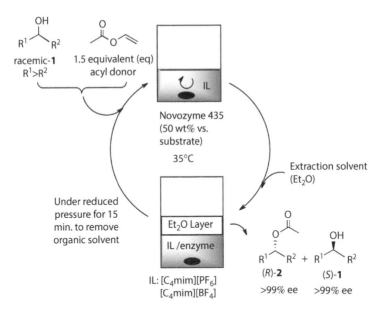

FIGURE 6.25 Typical lipase-catalyzed transesterification using the IL solvent system.

the lipase could be used repeatedly anchored in the reaction media. This is a typical example indicating a certain benefit of using ILs as the reaction media of enzymatic reactions. Selection of the appropriate ILs as a reaction medium is the key to realizing the desired reaction [65,66]. The hydrophobic IL, [C₄mim][PF₆], was selected as a favorable solvent for the reaction because the system allowed a simple work-up process due to the insolubility of this IL in both water and extraction of the organic solvent. On the contrary, a hydrophilic IL, such as [C₄mim][BF₄], was soluble in water and, therefore, it was difficult to remove a water soluble by-product by a simple work-up process, though the enzyme exhibited a high activity in the solvent.

6.2.3.1 Activation of Lipase-Catalyzed Reaction by IL-Coating Immobilization

Lipases are generally tolerable to a wide number of substrates, and ILs allow various types of reactions that are impossible to realize using conventional molecular solvents or water as the reaction medium [7]. However, very slow reactions or poor enantioselective reactions are also sometimes obtained [64].

We have established that the ionic liquid engineering was very successful to solve this problem. We synthesized the 3-methyl-1-butylimidazolium cetyl-PEG(10)-sulfate ionic liquid (IL1) and prepared the "IL1 coated *Burkholderia cepacia* lipase (IL1-PS)" through lyophilization process [67], the activation effect depended on the substrates, and the IL1-PS exhibited an excellent reactivity for many substrates [67]. We further discovered that coating the lipase PS with an amino acid with IL1 further activated the transesterification [68]. Selection of the appropriate ILs as a reaction medium is, of course, the key to realizing the desired reaction [69–71].

We further found amazingly successful results of the recycling use of the IL-coated enzyme in the [N$_{221MEM}$][Tf$_2$N] solvent system. We prepared triazolium cetylPEG10 sulfate ILs, **Tz1** (Figure 6.23), and found that **Tz1-PS** exhibited excellent results when the enzyme was used as a catalyst for the transesterification of 1-phenylethanol in [N$_{221MEM}$][Tf$_2$N] [72]. The lipase-catalyzed reaction in an IL solvent system is carried out as follows. To a mixture of **Tz1-PS** in the IL were added a racemic 1-phenyethanol and vinyl acetate as the acyl donor, and the resulting mixture was stirred at 35°C. After the reaction, ether or a mixed solvent of ether and hexane (1:4) was added to the reaction mixture to form a biphasic layer, and the acetate product and unreacted alcohol were quantitatively extracted with the organic solvent layer. The enzyme remained in the ionic liquid phase, the addition of the next set of substrate alcohol and acyl donor caused the next reaction cycle to afford the chiral (R)-ester and (S)-alcohol. Therefore, the lipase could be used repeatedly anchored in the reaction media [66,67]. Almost the same activity was obtained as that of the freshly prepared one even after storage for two years in the IL (Figure 6.23) [72]. These results clearly indicated the possibility of the IL engineering for enzyme stabilization (Figure 6.26).

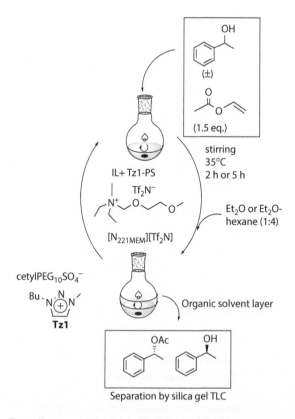

FIGURE 6.26   Recycling use of Tz1-PS in the [N$_{221MEM}$][Tf$_2$N] solvent system.

6.2.3.2 Biodiesel Oil Production Using Lipase-Catalyzed Reaction

The enzymatic production of biodiesel oil has recently gained strong interest from the standpoint of sustainable energy production. Diesel oils are hydrophobic compounds, and organic solvent-free separation from the IL reaction mixture has been easily realized (Figure 6.27) [73]. ILs could thus be used as the appropriate solvents for biodiesel oil production from a bioresource, such as vegetable oil etc., through the lipase-catalyzed transesterification. Lozano et al. prepared [C$_{16}$tam][NTf$_2$], which was termed a "sponge-like IL(SLIL)" [74]. SLIL exhibited a unique property as a reaction medium of the lipase-catalyzed reaction. The authors thus established an efficient protocol for biodiesel oil production employing the SLIL (Figure 6.27) [74].

In this chapter, we focused on only the lipase-catalyzed reactions because the reaction can be carried out in pure ILs, and it has now reached an industrially proven level in the pharmaceutical and food industries. However, numerous examples of

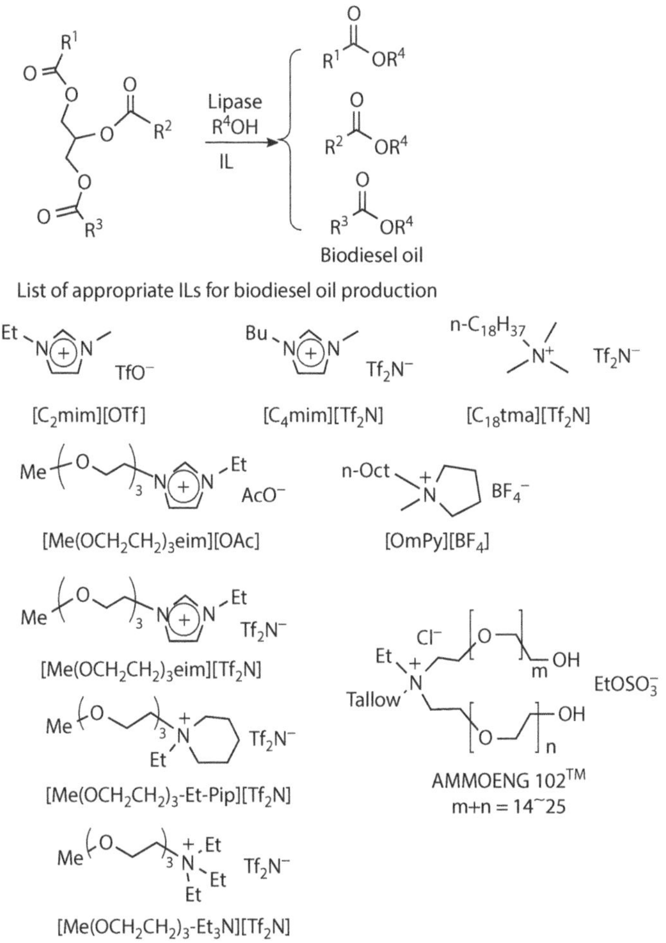

List of appropriate ILs for biodiesel oil production

[C$_2$mim][OTf]   [C$_4$mim][Tf$_2$N]   [C$_{18}$tma][Tf$_2$N]

[Me(OCH$_2$CH$_2$)$_3$eim][OAc]   [OmPy][BF$_4$]

[Me(OCH$_2$CH$_2$)$_3$eim][Tf$_2$N]

[Me(OCH$_2$CH$_2$)$_3$-Et-Pip][Tf$_2$N]   AMMOENG 102^{TM}
m+n = 14~25

[Me(OCH$_2$CH$_2$)$_3$-Et$_3$N][Tf$_2$N]

FIGURE 6.27 Lipase-catalyzed transesterification for practical biodiesel oil production.

various biocatalysts-mediated reactions have been reported. For such examples, please see my recent review [7].

6.3   CONCLUSIONS AND PROSPECTS

Innovation of the reaction medium used in a chemical reaction can provide a breakthrough in sustainable organic synthesis. In this chapter, we describe how to use ILs from the standpoint of suitainable organic stynthesis. The use of ILs as reaction media changes the kinetics of the chemical reactions compared to those in aqueous or classical molecular solvents. Unfortunately, replacement of the reaction media from classical organic solvents or water by ILs in industry has still not yet been realized. In order for ILs to become more popular, the key is how to reduce the amount of the ILs used, while obtaining a certain benefit. We believe that the importance of ILs would be more highlighed not as solvents, but as the controlling agents of chemical reactions. In such cases, we need only small quantities of the ILs and no modification of the present reaction process. We can recover the ILs and repeatedly use them following a simple purification after the reaction. In fact, we always recycle our ionic liquids after the reaction and have not wasted any in the past. We are still using several ionic liquids that have more than an 18-year history. We hope this chapter may provide a hint for the reader's research studies.

This paper is dedicated to the late Professor Kenneth R. Seddon due to his sudden passing on January 21, 2018.

6.4   GLOSSARY

Cations
$[C_4mim]^+$: 1-butyl-3-methylimidazolium
$[C_4dmim]^+$: 1-butyl-1,3-dimethylimidazolium
$[[HMPmim]^+$: 1-hydroxy
$[N_{221ME}]^+$: N,N-diethyl-N-methyl-N-(2-methoxyethyl)ammonium
$[P_{4441}]^+$: tributyl(methyl)phosphonium
$[N_{221MEM}]^+$: N,N-diethyl-N-methyl-N-(2-methoxyethoxymethyl)ammonium
$[P_{444MEM}]^+$: tributyl((2-methoxyethoxy)methyl)phosphonium
$[Bpyr]^+$: Butylpyridinium
$[C_{16}tam]^+$: N-cetyl-N,N,N-trimethylammonium
$[BF_4]^-$: tetrafluoroborate
$[PF_6]^-$: hexafluorophosphate
$[OTf]^-$: trifluoromethanesulfonate
$[Tf_2N]^-$: bis(trifluoromethylsulfonyl)amide. $[Tf_2N]$ is widely known as "bis(trifluoromethylsulfonyl)imide". However, "imide" means "an amido compound which is connected to two carbonyl groups", therefore, $[Tf_2N]$ should be labelled as "bis(trifluoromethylsulfonyl)amide" according to the IUPAC rule.
IL1: 1-butyl-2,3-dimethylimidazolium cetyl-PEG10 sulfate
Tz1: 3-butyl-1-methyl-1H-1,2,3-triazol-3-ium cetyl-PEG10 sulfate
Cetyl-PEG$_{10}$ SO$_4$: 3,6,9,12,15,18,21,24,27,30-decaoxacetyltriacontyl sulfate

ee: enantiomer excess
PEG: polyethylene glycol
BINAP: (1,1'- Binaphthalene-2,2'-diyl)bis(diphenylphosphine)

REFERENCES

1. The first review to define the term "Ionic liquid": T. Welton, *Chem. Rev.*, 99, 2071–2084 (1999).
2. Number of publications in the field of ILs from 1975 to 2018 (searched by "Web of Science", on January 6, 2018). 82,992 references are listed for the topic of "ionic liquid*" in the database of Web of Science Core Collections of Thomson Reuters, Ltd. Since "Ionic Liquid" became popular after 1999, we also used "molten salt" and "room temperature" as key words and eliminated papers belonging to different topics.
3. Z. Lei, B. Chen, Y-M. Koo, and D. R. MacFarlane. *Chem. Rev.*, 117, 6633–6635 (2017).
4. J. P. Hallett and T. Welton. *Chem. Rev.*, 111, 3508–3576 (2011).
5. J. Ranke, S. Stolte, R. Störmann, J. Arning, and B. Jastorff. *Chem. Rev.*, 107, 2183–2206 (2007).
6. C. Dai, J. Zhang, C. Huang, and Z. Lei. *Chem. Rev.*, 117, 6929–6983 (2017).
7. T. Itoh. *Chem. Rev.*, 117, 10567–10607 (2017).
8. K. S. Egorova, E. G. Gordeev, and V. P. Ananikov. *Chem. Rev.*, 117, 7132–7189 (2017).
9. Z. Zhang, J. Song, and B. Han. *Chem. Rev.*, 117, 6834–6880 (2017).
10. J. A. Boon, J. A. Levisky, J. L. Pflug, and J. S. Wilkes. *J. Org. Chem.*, 51, 480–483 (1986).
11. P. A. Z. Suarez, J. E. L. Dullius, S. Einloft, R. F. de Souza, and J. Dupont. *Polyhedron*, 15, 1217 (1996).
12. A. L. Monteiro, F. K. Zinn, R. F. de Souza, and J. Dupont. *Tetrahedron: Asymmetry*, 8, 177–179 (1997).
13. (a) J. T. Link. *Org. React.*, 60, 157–534 (2002). (b) C. I. M. Santos, J. F. B. Barata, M. A. F. Faustino, C. Lodeiro, and M. G. P. M. S. Neves. *RSC Adv*, 3, 19219–19238 (2013).
14. J. Carmichael, M. J. Earle, J. D. Holbrey, P. B. McCormac, and K. R. Seddon. *Org. Lett.*, 1, 997–1000 (1999).
15. W. A. Herrmann and V. P. W. Bölm. *J. Organomet. Chem.*, 572, 141–145 (1999).
16. J. Dupont, G. S. Fonseca, A. P. Umpierre, P. F. P. Fischtner, and S. R. Tiexeira. *J. Am. Chem. Soc.*, 124, 4228–4229 (2002).
17. P. Migowski and J. Dupont. *Chem. Eur. J.*, 13, 32–39 (2007).
18. (a) T. Torimoto, K. Okazaki, T. Kiyama, K. Hirahara, N. Tanaka, and S. Kuwabata. *Appl. Phys. Lett.*, 89, 243117 (2006). (b) K. Okazaki, T. Kiyama, K. Hirahara, N. Tanaka, S. Kuwabata, and T. Torimoto. *Chem. Commun.*, 691–693 (2008). (c) T. Kameyama, Y. Ohno, T. Kurimoto, K. Okazaki, T. Uematsu, S. Kuwabata, and T. Torimoto. *Phys. Chem. Chem. Phys.*, 12, 1804–1811 (2010).
19. Y. Hamashima, H. Takano, D. Hotta, and M. Sodeoka. *Org. Lett.*, 5, 3225–3228 (2003).
20. T. Fukuyama, M. Shinmen, S. Nishitani, M. Sato, and I. Ryu. *Org. Lett.*, 4, 1691–1694 (2002).
21. T. Fukuyama, M. T. Rahman, H. Mashima, H. Takahashi, and I. Ryu. *Org. Chem. Front.*, 4, 1863–1866 (2017).
22. A. Punzi, D. I. Coppi, S. Matera, M. A. M. Capozzi, A. Operamolla, R. Ragni, F. Babudri, and G. M. Farinola. *Org. Lett.*, 19, 4754–4757 (2017).
23. (a) M. Fujiwara, M. Kawatsura, S. Hayase, M. Nanjo, and T. Itoh. *Adv. Synth. Catal.*, 351, 123–128 (2009). (b) M. Kawatsura, K. Kajita, S. Hayase, and T. Itoh. *Synlett.*, 1243–1246 (2010). (c) C. Ibara, M. Fujiwara, S. Hayase, M. Kawatsura, and T. Itoh. *Sci. China Chem.*, 55, 1627–1632 (2012).
24. N. Audic, H. Clavier, M. Mauduit, and J.-C. Guillemin. *J. Am. Chem. Soc.*, 125, 9248–9249 (2003).
25. (a) S. Doherty, P. Goodrich, C. Hardacre, J. G. Knight, M. T. Nguyen, V. I. Parvulescu, and C. Paun. *Adv. Synth. Catal.*, 349, 951–963 (2007). (b) S. Doherty, P. Goodrich, C. Hardacre, V. Parvulescu, and C. Paun. *Adv. Synth. Catal.*, 350, 295–302 (2008).

26. C. Baleizao, B. Gigante, H. Garcia, and A. Corma. *Tetrahedron Lett.*, 44, 6813–6816 (2003).

27. C. Li, J. Zhao, R. Tan, Z. Peng, R. Luo, M. Peng, and D. Yin. *Catal. Commun.*, 15, 27–31 (2011).

28. S. Koguchi, A. Mihoya, and M. Mimura. *Tetrahedron*, 72, 7633–7637 (2016).

29. (a) T. Nokami, N. Sasaki, Y. Isoda, and T. Itoh. *ChemElectroChem.*, 3, 2012–2016 (2016). (b) N. Sasaki, T. Nokami, and T. Itoh. *Chem. Lett.*, 46, 683–685 (2017).

30. Y. Wang, X. Zhao, Q. Kong, J. Yao, X. Meng, and Z. Li. *Tetrahedron Lett.*, 58, 1655–1658 (2017).

31. (a) T. L. Amyes, S. T. Diver, J. P. Richard, F. M. Rivas, and K. Toth. *J. Am. Chem. Soc.*, 126, 4366–4374 (2004). (b) A. M. Magill, K. J. Cavell, and B. F. Yates. *J. Am. Chem. Soc.*, 126, 8717–8724 (2004). (c) S. Tsuzuki, H. Tokuda, K. Hayamizu, and M. Watanabe. *J. Phys. Chem. B*, 109, 16474–16481 (2005).

32. M. Kawatsura, H. Nobuto, S. Hayashi, T. Hirakawa, D. Ikeda, and T. Itoh. *Chem. Lett.*, 40, 953–955 (2011).

33. (a) T. Ramnial, D. D. Ino, and J. A. C. Clyburne. *Chem. Commun.*, 325–326 (2005). (b) T. Ramnial, S. A. Taylor, J. A. C. Clybume, and C. J. Walsby. *Chem. Commun.*, 2066–2068 (2007).

34. (a) V. Jurcik and R. Wilhelm. *Green Chem.*, 7, 844–848 (2005). (b) S. T. Handy, *J. Org. Chem.*, 71, 4659–4662 (2006).

35. M. C. Law, K.-Y. Wong, and T. H. Chan. *Chem. Commun.*, 2457–2459 (2006).

36. T. Itoh, K. Kude, S. Hayase, and M. Kawatsura. *Tetrahedron Lett.*, 48, 7774–7777 (2007).

37. K. Kude, S. Hayase, M. Kawatsura, and T. Itoh. *Heteroatom Chem.*, 22, 397–404 (2011).

38. A. C. Cole, J. L. Jensen, I. Ntai, K. L. T. Tran, K. J. Weaver, D. C. Forbes, and J. H. Davis Jr. *J. Am. Chem. Soc.*, 124, 5962–5963 (2002).

39. (a) P. Wasserscheid, M. Sesing, W. Korth. *Green. Chem.*, 4, 134–138 (2002). (b) V. Singh, S. Kaur, V. Sapehiya, J. Singh, and G. L. Kad. *Catal. Commun.*, 6, 57–60 (2005). (c) J. J. Ma, C. Wang, Q. H. Wu, R. X. Tang, H. Liu, and Q. Li. *Heteratom. Chem.*, 19, 609–611 (2008). (d) H. Gao, C. Guo, J. Xing, J. Zhao, and H. Liu. *Green Chem.*, 12, 1220–1224 (2010).

40. D. Cao, Y. Zhang, C. Liu, B. Wang, Y. Sun, A. Abdukadera, H. Hu, and Q. Liu. *Org. Lett.*, 18, 2000–2003 (2016).

41. (a) T. L. Greaves and C. J. Drummond. *Chem. Rev.*, 115, 11379–11448 (2015). (b) A. S. Amarasekara. *Chem. Rev.*, 116, 6133–6183 (2016).

42. S. Luo, X. Mi, L. Zhang, S. Liu, H. Xu, and J.-P. Cheng. *Angew. Chem. Int. Ed.*, 45, 3093–3097 (2006).

43. C. M. R. Volla, I. Atodiresei, and M. Rueping. *Chem. Rev.*, 114, 2390–2431 (2014).

44. A. Obregón-Zúniga, M. Milán, and E. Juaristi. *Org. Lett.*, 19, 1108–1111 (2017).

45. M. Varyani, P. K. Khatri, and S. L. Jain. *Tetrahedron Lett.*, 57, 723–727 (2016).

46. (a) G. A. Grasa, R. M. Kissling, and S. P. Nolan. *Org. Lett.*, 4, 3583–3586 (2002). (b) G. W. Nyce, J. A. Lamboy, E. F. Connor, R. M. Waymouth, and J. H. Hedrick. *Org. Lett.*, 4, 3587–3590 (2002). (c) A. Sarkar, S. R. Roy, N. Parikh, and A. K. Chakraborti. *J. Org. Chem.*, 76, 7132–7140 (2011).

47. M. Zanatta, F. P. dos Santos, C. Biehl, G. Martin, G. Ebeling, P. A. Netz, and J. Dupont. *J. Org. Chem.*, 82, 2622–2629 (2017).

48. (a) H. Hagiwara, Y. Shimizu, T. Hoshi, T. Suzuki, M. Ando, K. Ohkubo, and C. Yokoyama. *Tetrahedron Lett.*, 42, 4349–4351 (2001). (b) H. Hagiwara, K. H. Ko, T. Hoshi, and T. Suzuki. *Chem. Comm.*, 2838–2840 (2007). (c) H. Hagiwar, H. Sasaki, T. Hoshi, and T. Suzuki. *Synlett*, 643–647 (2009). (d) H. Hagiwara, T. Kuroda, T. Hoshi, and T. Suzuki. *Adv. Synth. Catal.*, 352, 909–916 (2010).

49. C. P. Mehnert, R. A. Cook, N. C. Dispenziere, and M. Afeworki. *J. Am. Chem. Soc.*, 124, 12932–12933 (2002).

50. (a) A. Riisager, P. Wasserscheid, R. van Hal, and R. Fehrmann. *J. Catalysis*, 219, 452–455 (2003). (b) A. Riisager, R. Fehrmann, S. Flicker, R. van Hal, M. Haumann, and P. Wasserscheid. *Angew. Chem. Int. Ed.*, 44, 815–819 (2005). (c) A. Riisager, R. Fehrmann, M. Haumann, and P. Wasserscheid. *Eur. J. Inorg. Chem.*, 695–706 (2006). (d) M. Jakuttis, A. Schönweis, S. Werner, R. Franke, K.-D. Wiese, M. Haumann, and P. Wasserscheid. *Angew. Chem. Int. Ed.*, 50, 4492–4495 (2011). (e) C. Meyer, V. Hager, W. Schwieger, and P. Wasserscheid. *J. Mol. Catalysis*, 292, 157–165 (2012). (f) M. J. Schneider, M. Haumann, and P. Wasserscheid, *J. Mol. Catal. A: Chemical*, 376, 103–110 (2013).

51. M. Kusche, F. Enzenberger, S. Bajus, H. Niedermeyer, A. Bösmann, A. Kaftan, M. Laurin, J. Libuda, and P. Wasserscheid. *Angew. Chem. Int. Ed.*, 52, 5028–5032 (2013).

52. M. Abai, M. P. Atkins, A. Hassan, J. D. Holbrey, Y. Kuah, P. Nockemann, A. A. Oliferenko et al. *Dalton Trans.*, 44, 8617–8624 (2015).

53. J. Pavlinac, M. Zupan, K. K. Laali, and S. Stavber. *Tetrahedron*, 65, 5625–5862 (2009).

54. M. Hasegawa, H. Ishii, and T. Fuchigami. *Green Chem.*, 5, 512–515 (2003).

55. (a) D. W. Kim, C. E. Song, and D. Y. Chi. *J. Am. Chem. Soc.*, 124,10278–10279 (2002). (b) D. W. Kim and D. Y. Chi. *Angew. Chem. Int. Ed.*, 43, 483–485 (2004). (c) D. W. Kim, S. J. Oh, and D. Y. Chi. *J. Am. Chem. Soc.*, 128, 16394–16397 (2006).

56. A. Taber, K. C. Lee, H. J. Han, and D. W. Kim. *Org. Lett.*, 19, 3342–3345 (2017).

57. M. A. N. Mostafa, R. M. Bowley, D. T. Racys, M. C. Henry, and A. Sutherland. *J. Org. Chem.*, 82, 7529–7537 (2017).

58. (a) H. Ohara, H. Kiyokane, and T. Itoh. *Tetrahedron Lett.*, 43, 3041–3044 (2002). (b) T. Itoh, K. Kawai, S. Hayase, and H. Ohara. *Tetrahedron Lett.*, 44, 4081–4084 (2003).

59. D. S. Choi, D. H. Kim, U. S. Shin, R. R. Deshmukh, S.-G. Lee, and C. E. Song. *Chem. Commun.*, 3467–3469 (2007).

60. J. W. Lee, J. Y. Shin, Y. S. Chun, H. B. Jang, C. E. Song, and S.-G. Lee. *Acc. Chem. Res.*, 43, 985–994 (2010).

61. J. H. Kim, J. W. Lee, U. S. Shin, J. Y. Lee, S.-G. Lee, and C. E. Song. *Chem. Commun.*, 4683–4685 (2007).

62. R. R. Deshmukh, J. W. Lee, U. S. Shin, J. Y. Lee, and C. E. Song. *Angew. Chem. Int. Ed.*, 47, 8615–8617 (2008).

63. U. T. Bornscheuer, G. W. Huisman, R. J. Kazlauskas, S. Luts, J. C. Moore, and K. Robins. *Nature*, 485, 185–194 (2014).

64. K. Faber. *Biotransformations in Organic Chemistry*, A Textbook, 6th ed. Springer, Heidelberg, Germany (2011).

65. T. Itoh, E. Akasaki, K. Kudo, and S. Shirakami. *Chem. Lett.*, 30, 262–263 (2001).

66. T. Itoh, Y. Nishimura, N. Ouchi, and S. Hayase. *J. Mol. Catalysis B: Enzym.*, 26, 41–45 (2003).

67. (a) T. Itoh, S-H. Han, Y. Matsushita, and S. Hayase. *Green Chem.*, 6, 437–439 (2004). (b) T. Itoh, Y. Matsushita, Y. Abe, S.-H. Han, S. Wada, S. Hayase, M. Kawatsura, S. Takai, M. Morimoto, and Y. Hirose. *Chem. Eur. J.*, 12, 9228–9237 (2006).

68. K. Yoshiyama, Y. Abe, S. Hayse, T. Nokami, and T. Itoh. *Chem. Lett.*, 42, 663–665 (2013).

69. (a) T. Itoh, E. Akasaki, K. Kudo, and S. Shirakami. *Chem. Lett.*, 30, 262–263 (2001). (b) T. Itoh, Y. Nishimura, N. Ouchi, and S. Hayase. *J. Mol. Catal. B: Enzym.*, 26, 41–45 (2003).

70. (a) Y. Abe, K. Kude, S. Hayase, M. Kawatsura, K. Tsunashima, and T. Itoh. *J. Mol. Catal. B: Enzym.*, 51, 81–85 (2008). (b) Y. Abe, K. Yoshiyama, Y. Yagi, S. Hayase, M. Kawatsura, and T. Itoh. *Green Chem.*, 12, 1976–1980 (2010). (c) Y. Abe, Y. Yagi, S. Hayase, M. Kawatsura, and T. Itoh. *Ind. Eng. Chem. Res.*, 51, 9952–9958 (2012).

71. S. Kadotani, R. Inagaki, T. Nishihara, T. Nokami, and T. Itoh. *ACS Sustainable Chem. Eng.*, 5, 8541–8545 (2017).

72. T. Nishihara, A. Shiomi, S. Kadotani, T. Nokami, and T. Itoh. *Green Chem.*, 19, 5250–5256 (2017).

73. (a) R. A. Sheldon. *Chem. Eur. J.*, 22, 12984–12999 (2016). (b) M. N. Nawshad, Y. A. Elsheikh, M. I. A. Mutalib, A. A. Bazmi, R. A. Khan, H. Khan, S. Rafiq, Z. Man, and I. Khan. *J. Ind. Eng. Chem.*, 21, 1–10 (2015). (c) C.-Z. Liu, F. Wang, A. R. Stiles, and C. Guo. *Appl. Energy*, 92, 406–414 (2012). (d) A. H. M. Fauzi and N. A. S. Amin *Renew. Sust. Energ. Rev.*, 16, 5770–5786 (2012).

74. P. Lozano, J. M. Bernal, G. Sánchez-Gómez, G. López-López, and M. Vaultier. *Energy Environ. Sci.*, 6, 1328–1338 (2013).

Section II

Ionic Liquids in Biocatalysis and Biomass Processing

7 Biotransformations in Deep Eutectic Solvents

Vicente Gotor-Fernández and Caroline Emilie Paul

CONTENTS

7.1 INTRODUCTION

From the discovery of ionic liquids (ILs) to the development of deep eutectic solvents (DESs) (Abbott et al. 2004), and more recently with the progress toward natural DESs (NADESs) (Paiva et al. 2014), the use of neoteric solvents in chemical synthesis has attracted considerable attention due to their simple preparation, tuneable properties, applicability, and sustainability. As the latter has become of high

importance in green chemistry and in the development of biocatalysis, trying to
ensure the greenness of chemical processes (Domínguez de María and Hollmann
2015; Sheldon 2017), the interest in neoteric solvents has shifted little by little from
ILs to DESs.

DESs are the result of a mixture between a quaternary ammonium salt, so-called
hydrogen bond acceptor, such as choline chloride (ChCl) with a hydrogen bond
donor (HBD) such as alcohols, amides, amines, carboxylic acids, etc. (Figure 7.1).
Their main difference with ILs resides in their hydrogen-bonding nature, which low-
ers their overall melting point, and avoids the need for further purification (Zhang
et al. 2012). By the proper selection of their subunits (hydrogen bond acceptor and
HBD), such as choline chloride with urea [ChCl:U (1:2), freezing point of 12°C] or
glycerol [ChCl:Gly (1:2), freezing point of −40°C], DESs can be fine-tuned to obtain
variable properties in addition to their low volatility. Advantages include lower cost
of production due to inexpensive starting materials, atom efficiency, and, hence, no
purification required.

DESs are generally considered more sustainable than ILs (Radošević et al. 2015),
however, choline chloride-based DESs with glucose (Glu), glycerol (Gly), and oxalic acid
as HBD were evaluated for toxicity, ChCl:Glu and ChCl:Gly exhibited low cytotoxicity,
whereas ChCl:Ox had moderate cytotoxicity. Toxicity studies have therefore shown not
all DESs are equally sustainable, with choline chloride-based DESs being less toxic than
phosphonium-based DESs (Hayyan 2013a, 2013b; Hayyan et al. 2016). Applications
for DESs are widely spread in a wide variety of areas (Smith et al. 2014), examples
include their use in extraction processes (Maugeri et al. 2012; Krystof et al. 2013; Pena-
Pereira and Namieśnik 2014), in the production of materials (Wagle et al. 2014), and
their use as reaction media in synthetic transformations (Zhang et al. 2012), including
also biotransformations, which we will further describe (Domínguez de María and
Maugeri 2011; Guajardo et al. 2016; Xu et al. 2017). (NA)DESs were hypothesized to
mimic the metabolites present in cells forming another type of liquid, apart from water

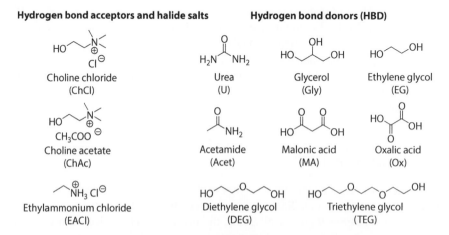

FIGURE 7.1   Selected structures of DESs and their corresponding abbreviations used in
this chapter.

and lipids, in which the protein structure may be preserved in the absence of water (Choi et al. 2011), leading to interesting media alternatives. In the next sections, we will discuss the use of DESs in biocatalytic processes with hydrolases, the most widely reported enzyme class used with these (co)solvents, followed by the current research on oxidoreductases, lyases, and whole-cell biocatalysis.

7.2 USE OF HYDROLASES IN CONVENTIONAL PROCESSES

Among the six classes of enzymes, hydrolases (EC.3) have traditionally received major attention due to the possibility to catalyze a wide number of hydrolytic (hydrolysis of esters, amides, epoxides, etc.) or the reversible bond formation reactions (esterification, amidation, transesterification, etc.) depending on the reaction medium and conditions. The use of hydrolases by organic chemists has been boosted due to their simplicity of handling, commercial availability, and lack of cofactor requirements. Satisfyingly, the stability of free and immobilized hydrolases has been demonstrated in aqueous medium, but also in organic and neoteric solvents including ILs, supercritical conditions, and DESs.

Next, we will focus on the main organic synthetic transformations reported using hydrolase-catalyzed reactions in DESs, rather than disclosing extensive discussions about the stability and kinetic studies of enzymes in model reactions. This will give to the reader a wider view about the synergy between enzymes and DESs than previous bibliographic revision (Durand et al. 2013b).

Kazlauskas and co-workers reported, for the first time, the use of hydrolytic enzymes, such as lipases, esterases, and epoxide hydrolases in DES, (Gorke et al. 2008). In this study a few biotransformations and enzyme classes were studied in order to explore new reaction media for enzyme-catalyzed reactions. Firstly, the lipase-catalyzed transesterification of ethyl valerate with 1-butanol as nucleophile was explored as model reaction (Figure 7.2a). *Candida antarctica* lipase type B (CAL-B), *Candida antarctica* lipase type A (CAL-A), and *Pseudomonas cepacia* lipase (PCL) were tested in eight DESs, finding excellent levels of activity with CAL-B in both free and immobilized forms. These activity values were comparable with those attained with a hydrophobic solvent, such as toluene, or even better when employing ChCl:Gly, ChCl:U, and EACl:Gly mixtures. The activity of immobilized CAL-B was also successfully tested in the aminolysis reaction of ethyl valerate

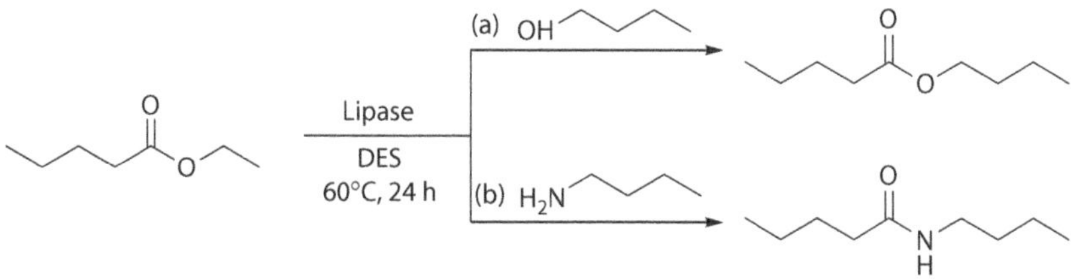

FIGURE 7.2 Lipase-catalyzed reactions of ethyl valerate in DESs: (a) transesterification with *n*-butanol and (b) aminolysis with butylamine. (Based on Gorke, J.T. et al., *Chem. Commun.*, 1235–1237, 2008.)

with a little excess of 1-butylamine as nucleophile, reaching over 90% conversion in ChCl:Gly, ChCl:U, and toluene (Figure 7.2b).

In addition, the role of DESs as co-solvents for biotransformations in aqueous media, such as ester or epoxide hydrolysis was explored. In this context, the hydrolysis of p-nitrophenyl acetate using pig liver esterase in the presence of ChCl:Gly (10% v/v) increased up to three-fold compared to the reaction in aqueous system, while the rate of epoxide hydrolase-catalyzed hydrolysis of styrene oxide dramatically increased until 20 times using ChCl:Gly (25% v/v), maintaining the stereoselectivity.

In the next sections, we will try to summarize the contributions reported in the literature that employed DESs as suitable media or co-solvents for conventional hydrolase-catalyzed reactions. This includes from synthetic reactions (transesterification, aminolysis, esterification, etc.) to the hydrolysis of esters and epoxides.

7.2.1 TRANSESTERIFICATION REACTIONS

The reaction between esters and alcohols is probably the most common hydrolase-catalyzed process excluding hydrolytic procedures. These transesterification processes are based on the exchange of an alkoxy group between both molecules, receiving different names depending on the literature sources, such as acylation or alcoholysis. In this context, the use of large excess of alcohol, organic solvents, or ionic liquids have been largely explored in the literature. Lipase-catalyzed transesterification reactions are normally developed by mixing an activated ester, such as vinyl esters, with aliphatic alcohols or phenols.

The main advantage of using enzymes for synthetic transformations is the possibility to carry out highly asymmetric processes based on the intrinsic chirality of the biocatalyst. Surprisingly, for the hydrolase-catalyzed transesterification reaction, only one example reports a kinetic resolution process. In this case, the use of organic solvents, such as cyclopentyl methyl ether and toluene represent a better option than DESs (EACl:Gly, ChCl:Gly, ChCl:Iso, ChCl:LA, ChCl:Ox, and ChCl:U) in terms of higher *Thermomyces lanuginosus* lipase activity (Figure 7.3). However, the poor solubility of the product, namely, (*S*)-benzoin butyrate, in ChCl:Iso compared to organic solvents, provide an elegant alternative for downstream processing (Petrenz et al. 2015).

In recent years, the use of DESs in transesterification reactions has opened new possibilities by using these neoteric solvents in three different varieties: (a) as unique

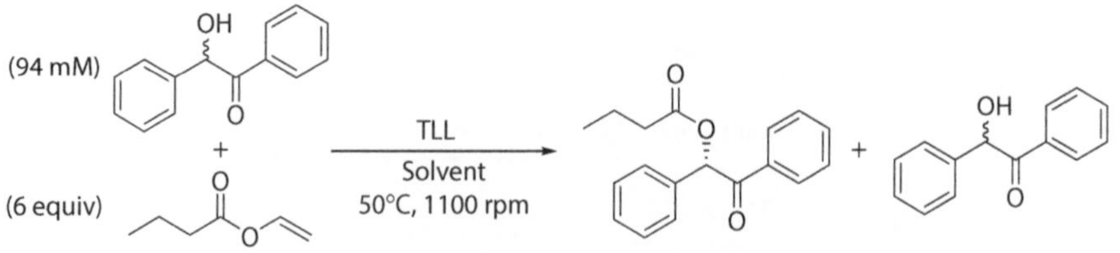

FIGURE 7.3 Lipase-catalyzed kinetic resolution of racemic benzoin in organic solvents and DESs. (Adapted from Petrenz, A. et al., *J. Mol. Catal. B: Enzym.*, 114, 42–49, 2015.)

solvents allowing a good solubility of the reactants; (b) using water as co-solvent that interferes in the hydrogen bond network between the DESs and the reactants, increasing the reactivity of the system; and (c) in combination with traditional organic solvents, then providing a higher stability to the enzyme.

7.2.1.1 Transesterification Reactions in Pure DES

The application of hydrolases, mainly lipases, has been attracting recent attention due to good solubility of polar compounds in these neoteric solvents and the great stability shown by hydrolytic enzymes in DESs. Pöhnlein and co-workers have reported the highly selective acylation of glucose with 2 equivalent (equiv.) of vinyl hexanoate using CAL-B as biocatalyst (Pöhnlein et al. 2015). Different DESs were tested, finding the best results in ChCl:U (1:2) and ChCl:Glu (1:1), which led to glucose-6-*O*-hexanoate as the major product after three days at 70°C and 1500 rpm (Figure 7.4). In this context, the high stability of new immobilized CAL-B preparations using hierarchical assembly techniques have been recently disclosed, enzymes that were later applied in the transesterification and esterification of glucose with vinyl laurate and lauric acid, respectively (Andler et al. 2017).

Biocatalyzed biodiesel production has received great attention in the last decades since lipases have displayed excellent activities in these transesterification protocols consisting in the reaction between natural triglycerides and aliphatic alcohols, mainly methanol and ethanol (Figure 7.5).

These enzymatic transformations offer, therefore, several advantages over traditional methods in terms of mild reaction conditions, use of renewable stocks, and very low waste production. The emergence of neoteric solvents, such as ionic liquids, liquid polymers, perfluorcarbons, supercritical conditions, and DESs has opened a myriad of possibilities to substitute the traditional organic volatile solvents, which have been recently summarized in the literature (Gutiérrez-Arnillas et al. 2016). These

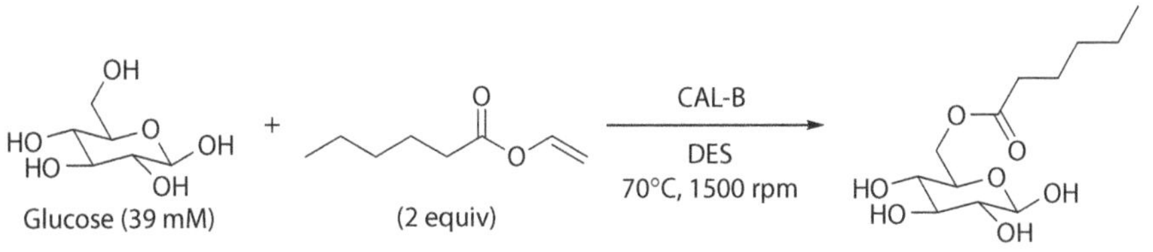

FIGURE 7.4 Lipase-catalyzed transesterification of glucose with vinyl hexanoate in pure DESs. (Adapted from Pöhnlein, M. et al., *Eur. J. Lipid. Sci. Technol.*, 117, 161–166, 2015.)

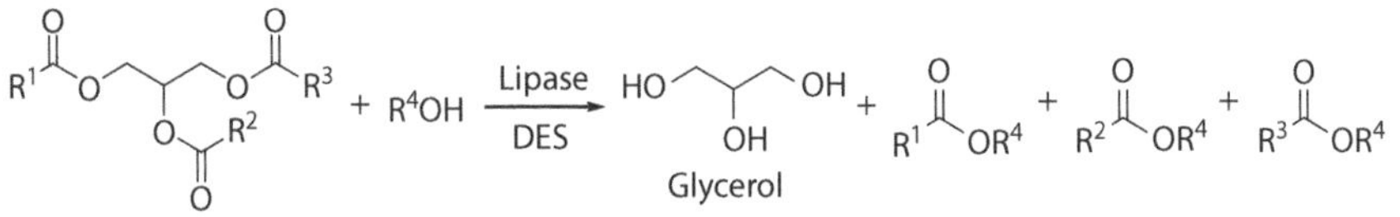

FIGURE 7.5 Enzyme-catalyzed production of biodiesel through transesterification processes.

biotransformations are based in the use of pure DESs that will be covered in this section, but also binary mixtures of DESs and water that will be discussed later in this contribution. For instance, the crude oil extracted from *Millettia pinnata* seeds was treated with methanol (MeOH) in ChAc:Gly (1:2) at 50°C and 220 rpm (Huang et al. 2014). From the six hydrolases tested that were CAL-B, immobilized lipase from *Thermomyces lanuginosus* (TLL IM), *Aspergillus niger* lipase (ANL), *Aspergillus oryzae* lipase, *Rhizopus chinesis* lipase, and *Penicillium expansum* esterase, only CAL-B (55%) and TLL IM (45%) gave significant conversion values after 48 h.

Yellow horn seed oil has also served as an ideal candidate for the production of biodiesel using CAL-B, pure DESs, and microwave conditions (Zhang et al. 2016). An exhaustive study including the use of 11 DESs, different enzyme loadings, methanol/oil ratios, and microwave parameters, led to the identification of ChCl:Gly (1:2), 6 equiv. of MeOH, and 50°C for the production of biodiesel in 95% conversion after 120 h, the possibility of reusing the enzyme being possible although with gradual deactivation of the enzyme until 70% conversion after the fifth recycling experiment.

7.2.1.2 Transesterification Reactions in Mixtures of DESs and Water

Water plays a key role in synthetic hydrolase-catalyzed processes usually providing additional stability to the biocatalyst, despite that the reverse hydrolytic reaction can occur in a concomitant mode. In the case of DESs, the enzyme stability and selectivity are highly dependent on the water content, the possibility to carry out transesterification being possible with usually less than 10% v/v of water to avoid the undesired hydrolytic reaction.

For instance, Villeneuve and co-workers have studied the transesterification of vinyl laurate with different alcohols (1-butanol, 1-octanol, and 1-octadecanol) using water contents below 1% and immobilized CAL-B (Durand et al. 2012). In this case, the influence of the hydrogen bond donor components of the DESs was analyzed, finding that ChCl-based DESs containing malonic acid (MA), oxalic acid, or ethylene glycol (EG) react with the ester competing with the alcohol, which caused a significant decrease in the reaction selectivity (Figure 7.6). In addition, not only the formation of by-product was observed, but also the destruction of the DESs over the time.

The same research group has reported the use of pure DESs or DES-water binary mixtures for the alcoholysis of phenolic esters using CAL-B, finding a great benefit when using water as co-solvent (Durand et al. 2013a). In fact, in some cases the variation ranged from very low conversions (<2% after four days) in pure DESs (ChCl:U and ChCl:Gly mixtures) to quantitative conversions when using only 8% v/v–10% v/v of water without significant concomitant ester hydrolysis. It is worth mentioning

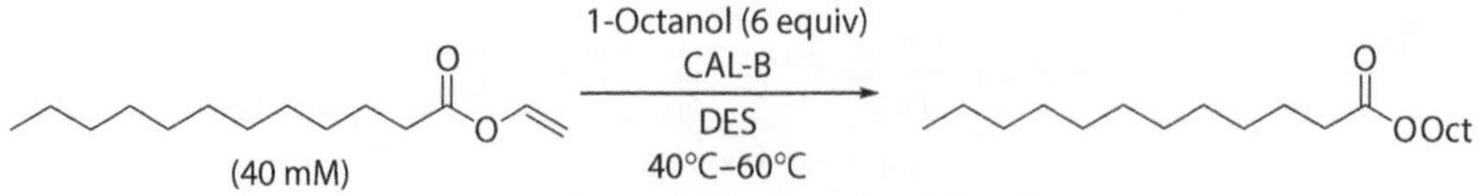

FIGURE 7.6 CAL-B catalyzed transesterification of vinyl laurate with 1-octanol. (Adapted from Durand, E. et al., *Process Biochem.*, 47, 2081–2089, 2012.)

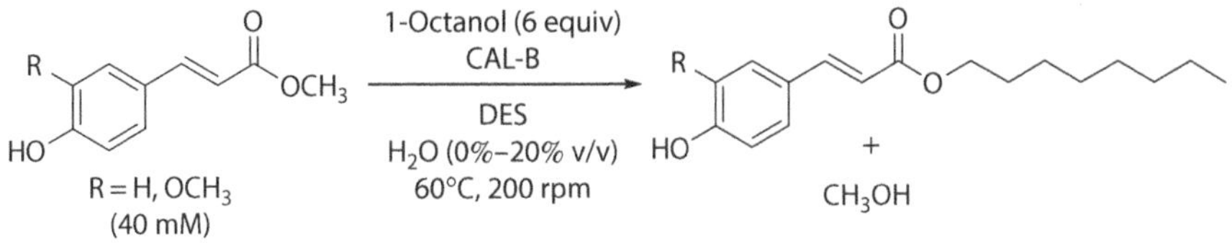

FIGURE 7.7 Lipase-catalyzed transesterification of phenolic esters with octanol using DESs or DES-water binary mixtures. (Adapted from Durand, E. et al., *Green Chem.*, 15, 2275–2282, 2013a.)

that higher water contents (10%–20%) led also to complete conversions, but the side-hydrolytic reaction starts to be observed in an appreciable extent. As a practical application, the CAL-B catalyzed process between methyl *p*-coumarate and 6 equiv. of 1-octanol was successfully scaled-up to 3 g of substrate, yielding 97% of the octyl ester after 72 h and a 92% isolated yield (Figure 7.7).

Similarly, CAL-B has displayed good activities in the transesterification of propyl gallate with methanol to produce methyl gallate (Figure 7.8). Gallic acid alkyl esters present a wide range of biological properties ranging from antitumoral to antioxidant activities, however, their synthesis through lipase-catalyzed esterification of gallic acid in alcohols, organic solvents, or DESs led to no conversion, probably caused by phenolic acid inhibition (Ülger and Takaç 2017). Then, the transesterification process was extensively analyzed using CAL-B as biocatalyst in terms of water content (0%–20%), enzyme concentration (10 g/L–80 g/L), methanol concentration (1–8 equiv.), temperature (35°C–60°C), and agitation speed (75 rpm–250 rpm), finding conversions around 60% after long reaction times under optimal conditions, although with almost negligible formation of gallic acid as hydrolytic by-product.

Lipases and mainly CAL-B are the most common hydrolases used for transesterification purposes, however, also proteases, such as immobilized subtilisin and α-chymotrypsin have displayed excellent activities in the reaction between *N*-acetyl-L-phenylalanine ethyl ester and 1-propanol (Zhao et al. 2011b). In this case, low water contents (2%–4%) were employed, finding the best activity with the cross-linked subtilisin in ChCl:Gly (1:2) and, in the presence of 3% water, biotransformations that occur also with complete selectivity (Figure 7.9).

Previously, the production of biodiesel has been reported in pure DESs, although in some cases the use of low water contents has provided great benefits. For instance,

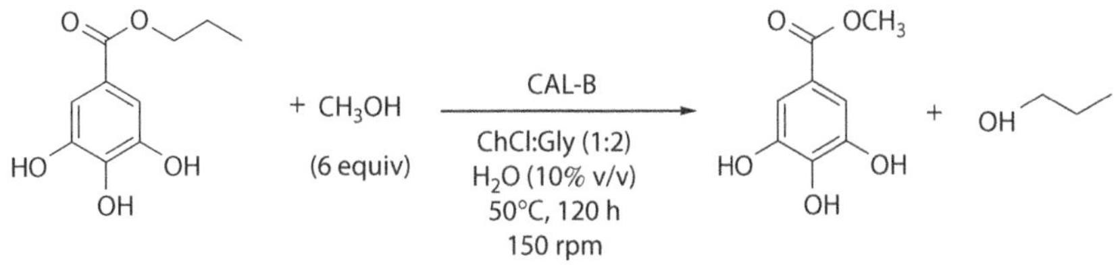

FIGURE 7.8 Lipase-catalyzed transesterification of propyl gallate with methanol using ChCl:Gly (1:2) and 10% water. (Adapted from Ülger, C. and Takaç, S., *Biocatal. Biotransformation*, 35, 407–416, 2017.)

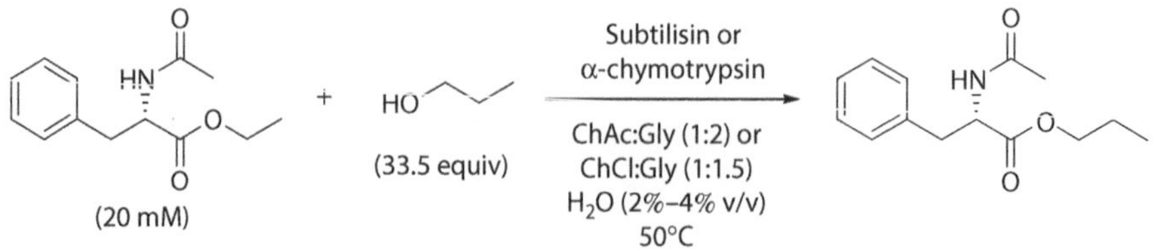

FIGURE 7.9 Protease-catalyzed transesterification of *N*-acetyl-L-phenylalanine ethyl ester with 1-propanol in pure choline-based DESs. (Adapted from Zhao, H. et al., *J. Mol. Catal. B: Enzym.*, 72, 163–167, 2011b.)

ChAc or ChCl coupled with glycerol have been used as adequate solvents for the transesterification of the triglyceride Miglyol® oil 812 with methanol using CAL-B and very low contents of water (Zhao et al. 2011a). High conversions were obtained (82%–97%) in short reaction times (1 h–3 h) under optimal conditions, this is 50°C, ChAc:Gly (1:1.5), and just 1% water v/v. The same research group has also reported the production of biodiesel from soybean oil through its CAL-B catalyzed transesterification with methanol in similar conditions. In this case, the use of ChCl:Gly (1:2) and 50°C led to the biodiesel after 24 h in 81% or 88% conversion when using 1% or just only 0.2% of water, respectively (Zhao et al. 2013). Satisfyingly, at the lowest water concentration the reusability of the enzyme was demonstrated for four cycles (79%–88% conversion). More recently, Schörken and co-workers have reported a two-step one-pot approach for the production of biodiesel from cooking and acidic oils (Kleiner et al. 2016). This involves the use of two free lipases, first the TLL catalyzed a regioselective transesterification in the presence of ethanol and aqueous solutions for 24 h at 30°C, requiring later the transesterification of residual monoglycerides by the action of CAL-B in DESs, such as ChCl:Gly mixtures at 45°C.

7.2.1.3 Transesterification Reactions in Mixtures of DESs and Organic Solvent

In other cases, the development of hydrolase-catalyzed reaction in organic solvents can be improved by the addition of DESs as co-solvents. This is the case of the monoacetylation of dihydromyricetin (DMY), which is a natural flavanonol with a broad spectrum of biological properties including anti-inflammatory, antibacterial, and antitumoral activities (Figure 7.10). The reaction of DMY with 10 equiv. of vinyl acetate

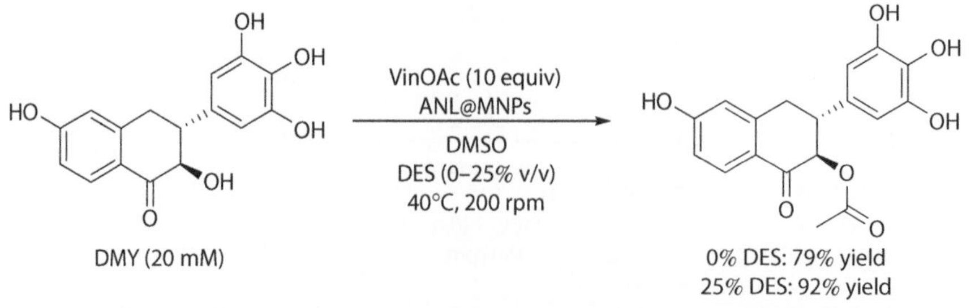

FIGURE 7.10 Lipase-catalyzed acetylation of DMY using DESs as co-solvent. (Based on Cao, S.L. et al., *J. Agric. Food. Chem.*, 65, 2084–2088, 2017.)

(VinOAc) catalyzed by *Aspergillus niger* lipase immobilized onto magnetic nanoparticles (ANL@MNP) allowed the production of the monoacetate in the C-16 position, while the use of ChCl:Gly (25% v/v) enhanced the reactivity until 92% (Cao et al. 2017). On the one hand, dimethyl sulfoxide allows the complete solubility of the reactants, and on the other hand the use of DESs prevent the inactivation of the enzyme that usually occurs in a high concentration of strong polar solvents.

7.2.2 ESTERIFICATION REACTIONS

In this section, we will cover the examples reported in the literature regarding the reaction between carboxylic acids and alcohols to form esters as main products and water as by-product. Since the reaction of alcohol with other more reactive carboxylic acid derivatives, such as anhydrides and acyl chlorides usually occur under mild chemical conditions, the esterification processes of these organic compounds in combination with hydrolases and DESs systems has been scarcely reported. However, Bubalo and co-workers described the reaction between acetic anhydride and 1-butanol to give butyl acetate using ILs or DESs as solvents (Figure 7.11). In this context, the use of water was crucial, as the employed ChCl-based solvents with different hydrogen donors (Gly, EG, and U, ratio 1:2) were identified as poor media with yields below 5%, probably caused by the entrapment of substrates within the DESs (Bubalo et al. 2015b). The addition of water dramatically enhanced the enzyme activity and stability, providing up to 80% conversion.

Regarding the reaction between carboxylic acids and alcohols, the successful esterification of oleic acid and benzoic acid in the presence of CAL-B will be later discussed. Unfortunately, negligible reactivity was found in the reaction between levulinic acid and 5-hydromethylfurfural (HMF) for the production of the fuel additive HMF levulinate in three different eutectic mixtures (Qin et al. 2016), although the use of the biomass-derived 2-methyltetrahydrofuran as solvent provides an eco-friendly synthetic solution.

Zeng and co-workers reported the lipase-catalyzed esterification of oleic acid with glycerol, acting the latest as substrate, but also as part of the solvent (Zeng et al. 2015). With that purpose, two quaternary ammonium salts (ChCl and betaine) were considered as hydrogen bond acceptors, studying the influence of different reaction parameters, such as the screening of five lipases, water content (0%–4%), reaction time, and acceptor:donor ratio (1 to 1–2.5). The best results were found with the ChCl:Gly (1:2) mixture and CAL-B at 50°C, reaching a 43% conversion

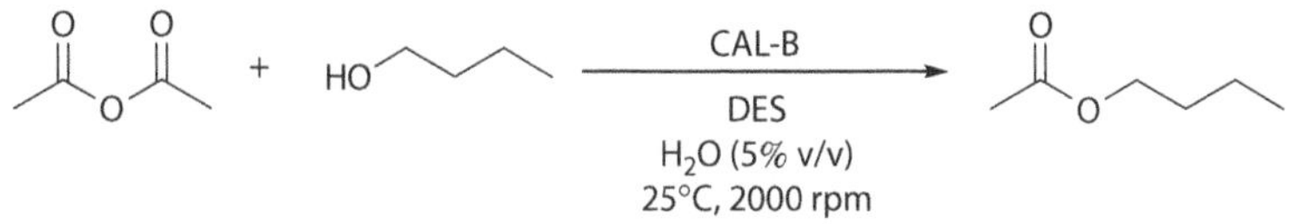

FIGURE 7.11 CAL-B catalyzed reaction of acetic anhydride with 1-butanol in the presence of water. (Adapted from Bubalo, M.C. et al., *J. Mol. Catal. B: Enzym.*, 122, 188–198, 2015b.)

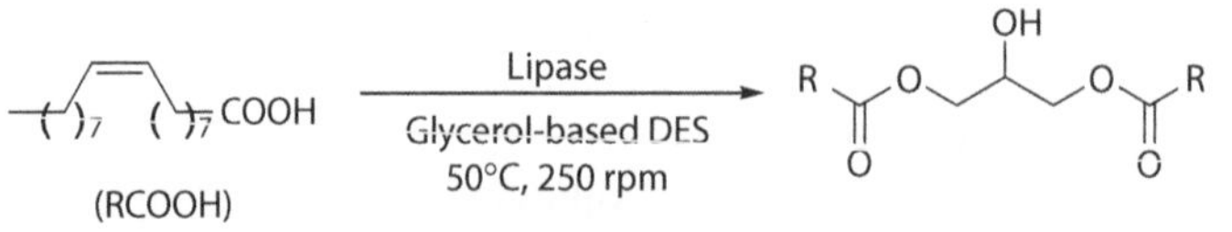

FIGURE 7.12 CAL-B catalyzed reaction of acetic anhydride with 1-butanol in the presence of water. (Based on Zeng, C.-X. et al., *Bioprocess Biosyst. Eng.*, 38, 2053–2061, 2015.)

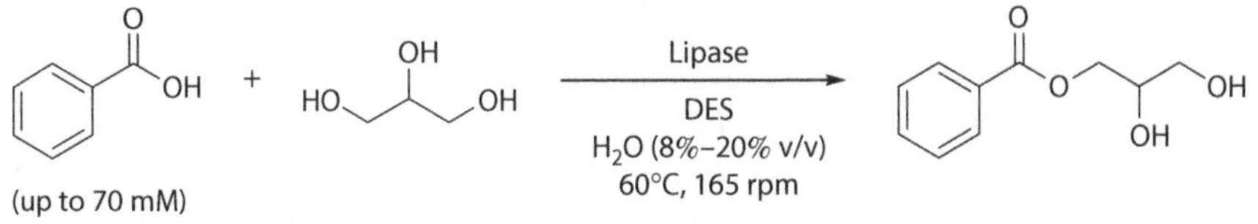

FIGURE 7.13 Esterification of benzoic acid with glycerol in DES-water binary mixtures using CAL-B as enzyme. (Adapted from Guajardo, N. et al., *ChemCatChem*, 9, 1393–1396, 2017.)

into the corresponding 1,3-diacylglycerol after just 1 h (Figure 7.12). Oleic acid has also been subjected to study in its reaction with 1-decanol catalyzed by CAL-B in DESs (Kleiner and Schörken 2015). Interestingly, a two-phase reaction system is proposed where the lipase catalyzed the esterification process in the interface, shifting the equilibrium toward decyl oleate production by entrapment of the water molecules, by-product of the reaction, in the DESs.

Very recently, the great benefits of using water as co-solvent have been demonstrated in the esterification of benzoic acid with glycerol catalyzed by CAL-B (Guajardo et al. 2017). After testing four different DESs, full conversions into the α-monobenzoate glycerol were attained after 24 h at 60°C when using ChCl:Gly (1:2) and water as co-solvent in a range between 8% v/v and 20% v/v (Figure 7.13). Percentages of water below 5% led to conversion values below 80%, while the use of 30% highly favored the appearance of by-products due to hydrolytic side reactions.

7.2.3 Hydrolase-Catalyzed Reactions of Esters with Nitrogenated Nucleophiles

Carboxylic acid and esters are susceptible to react with nitrogenated nucleophiles via non-enzymatic and enzymatic aminolysis processes. Particularly attractive is the use of lipases, mainly CAL-B (Gotor-Fernández et al. 2006), and proteases, as they allow the production of chiral amines, amides, carbamates, and amino acid derivatives under mild reaction conditions. In this context, α-chymotrypsin displayed high activity in the reaction between racemic *N*-acetyl-L-phenylalanine ethyl ester and glycinamide hydrochloride (Maugeri et al. 2013). From four tested ChCl-based DESs, ChCl:Gly was selected as ideal media to later optimize the water content, enzyme loading, and reactant concentrations (Figure 7.14). The desired peptide was finally obtained with excellent conversions at low water concentrations (10%–25%), while higher contents in water (i.e., 50%) led to significant amounts of the corresponding carboxylic acid resulting from the ester or the final amide hydrolysis reaction.

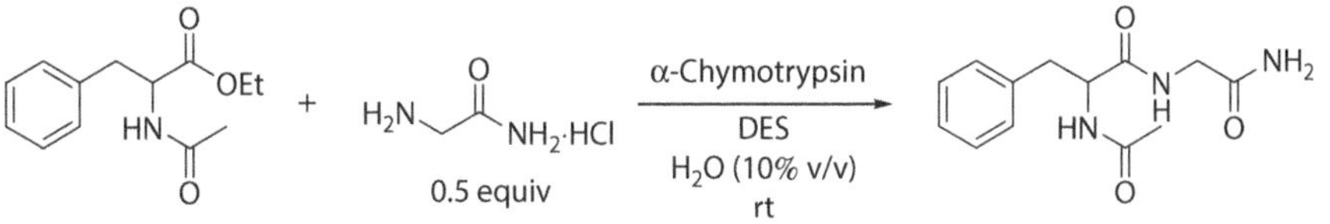

FIGURE 7.14 Chymotrypsin-catalyzed peptide synthesis in DES:water binary mixtures. (Adapted from Maugeri, Z. et al., *Eur. J. Org. Chem.*, 2013, 4223–4228, 2013.)

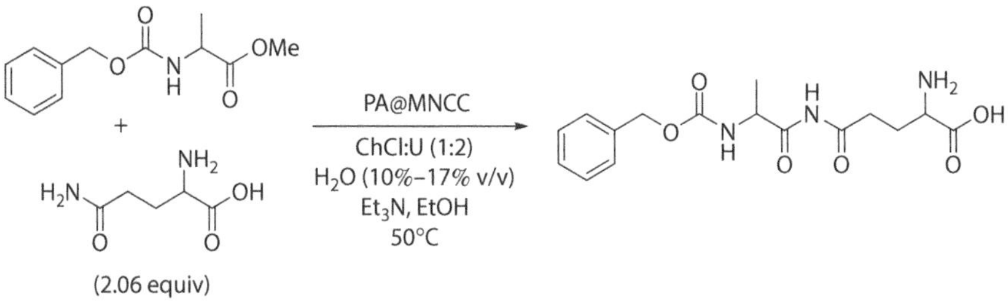

FIGURE 7.15 Synthesis of *N*-(benzyloxycarbonyl)-alanyl-glutamine catalyzed by immobilized papain in DESs. (Adapted from Cao, S.-L. et al., *ACS Sustain. Chem. Eng.*, 3, 1589–1599, 2015.)

Finally, the reusability of the enzyme was also analyzed, finding a significant loss of activity after the fourth cycle.

Similarly, Cao and co-workers reported the use of papain immobilized onto magnetic nanocrystalline cellulose for the synthesis of *N*-(benzyloxycarbonyl)-alanyl-glutamine dipeptide in ChCl:U (1.2) as solvent (Cao et al. 2015). The use of free or immobilized papain and the influence of water content, ratio ester:amide, and temperature were studied, finally providing a selective access toward *N*-(benzyloxycarbonyl)-alanyl-glutamine in around 65% yield (Figure 7.15). The use of triethylamine was necessary for the complete dissolution of the reactants, gratifyingly observing the high operational stability of the biocatalyst, as more than 88% of its activity was retained after reusing it for five times.

7.2.4 HYDROLYTIC REACTIONS

The use of DESs as solvents or co-solvents in enzymatic hydrolytic procedures has allowed the syntheses of high-added value products and biologically active molecules. This section covers the action of different hydrolases, such as lipases, glucosidases, cellulases, and epoxide hydrolases, which usually have acted with exquisite selectivity. Thus, DESs can be used as the main solvent of the reaction using waster as nucleophile, or the most common approach is the use of a buffer as solvent and DESs as co-solvent (5% v/v–40% v/v) to favor the solubility of the starting material and the final product.

Ester hydrolysis has received great attention in aqueous systems, but also in organic and neoteric solvents. For instance, Hayyan and co-workers reported the hydrolysis of

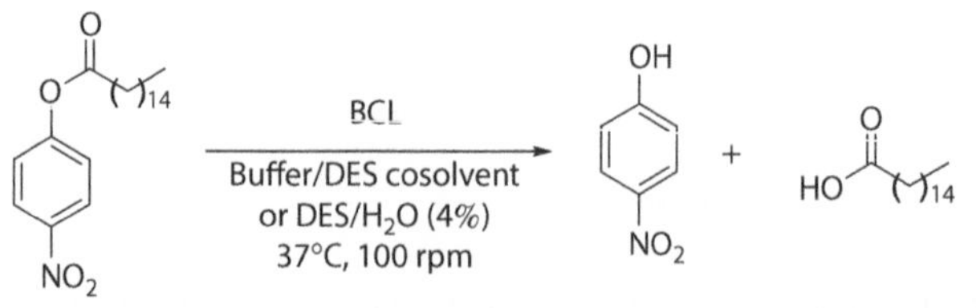

FIGURE 7.16 Lipase-catalyzed hydrolysis of *p*-nitrophenyl palmitate in DES-aqueous solution mixtures. (Based on Juneidi, I. et al., *Biochem. Eng. J.*, 117, 129–138, 2017.)

p-nitrophenyl palmitate using *Burkholderia cepacia* lipase (BCL, also known as PCL) in buffer using DESs (ChCl:Gly, ChCl:U, ChCl:EG, ChCl:DEG, EACl:Gly, EACl:EG, and EACl:TEG 1:2) as cosolvent or, alternatively, with DESs and main solvent and low contents of water (Juneidi et al. 2017). Best results were found in buffer medium and 40% ChCl:EG mixture, while the enzyme was mostly deactivated in pure DESs (Figure 7.16). However, the addition of 4% water allowed the hydrolysis of *p*-nitrophenyl palmitate, which occurred in better shape than using buffer as unique medium or when methanol or 1-ethyl-3-methylimidazolium tetrafluoroborate were considered as co-solvents.

The kinetic resolution of tetrahydropyrazolo[1,5-α]pyrimidine derivatives was successfully developed by hydrolysis of the corresponding ethyl esters in buffer (Fernández-Álvaro et al. 2014), using dimethyl sulfoxide, 1,4-dioxane or ChCl-based DESs (Gly, U, and Xyl) as co-solvents (10% v/v). For the three substrates tested, CAL-B provided high to excellent enantioselectivities ($E > 80$), while pig and rabbit liver esterases were found as very active, but poorly selective enzymes (Figure 7.17). Overall, this enzymatic strategy offers an eco-friendly alternative to existing chiral high performance liquid chromatography (HPLC) preparative separations of this family of compounds.

The use of DESs for flavonoid chemistry is particularly appealing due to the low solubility of this class of organic compounds in traditional solvents. Between the different synthetic possibilities, enzymatic hydrolysis of carbohydrates and glycoconjugates has attracted recent attention, based on the compatibility of glycosidases and DESs. This is the case of the hydrolysis of daidzin to prepare daidzein, both from the isoflavone class and present in plants (Figure 7.18). After exhaustive optimization of reaction parameters (temperature, reaction time, and water content), the best conversion (97.5%) was found for the ChCl:EG (1:2) from 16 tested DESs mixtures (Cheng and Zhang 2017). Interestingly, the almond β-D-glucosidase was reused for six times, decreasing the conversion down to 50%. In a similar manner,

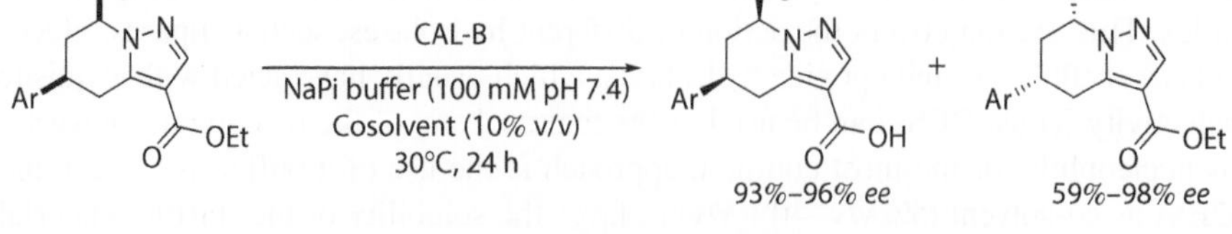

FIGURE 7.17 Kinetic resolution of tetrahydropyrazolo[1,5-α]pyrimidines in buffer using organic solvents or DESs as co-solvents. (Adapted from Fernández-Álvaro, E. et al., *J. Mol. Catal. B: Enzym.*, 100, 1–6, 2014.)

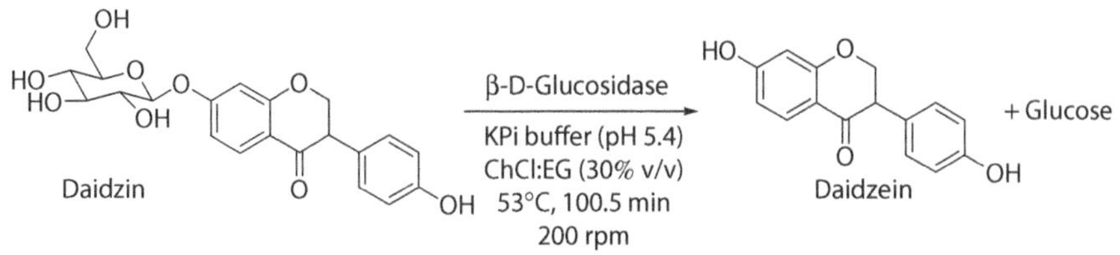

FIGURE 7.18 Kinetic resolution of tetrahydropyrazolo [1,5-α]pyrimidines in buffer using organic solvents or DESs as co-solvents. (Based on Cheng, Q.-B. and Zhang, L.-W., *Molecules*, 22, 186, 2017.)

6-*O*-α-L-rhamnosyl-β-D-glucosidase was employed for the hydrolysis of hesperidin to produce hesperitin, obtaining rutinose as by-product (Weiz et al. 2016). Three ChCl-based solvents were used containing U, Gly, and EG as hydrogen bond donor, which interestingly dissolved up to 90 mM of hesperidin. However, high contents of DESs led to the complete deactivation of the enzyme.

The use of DESs in cellulose-catalyzed hydrolytic reactions has also been demonstrated in a couple of examples. On one hand, cellulases serve as environmentally friendly tools for the depolymerization of cellulose under mild reaction conditions, while on the other hand, DESs allow the solubilization of substrates which are hardly soluble in aqueous systems. Schwaneberg and co-workers reported the design of a novel fluorescence high-throughput screening for the identification of active cellulases in mixtures of water and salty effluents (ILs, DESs, and seawater), previously evolved by directed evolution methods (Lehmann et al. 2012). In this case, ChCl:Gly (1:2) was used as co-solvent in the hydrolysis of the fluorogenic substrate 4-methylumbelliferyl-β-D-cellobioside (Figure 7.19). In general, a decrease in the activity of the cellulase mutants was observed even at low DESs concentrations (5%), although a cellulase variant 4D1 displayed a significant 7.5-fold increase in activity compared to the wild-type enzyme. The same research group published two years later the design of a nearly inactive cellulase in buffer for the degradation

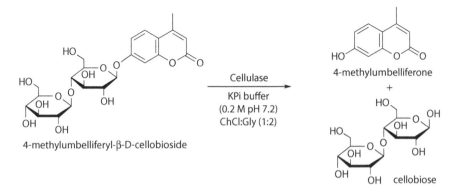

FIGURE 7.19 Use of evolved cellulases in the degradation of 4-methylumbelliferyl-β-D-cellobioside in DES-water binary mixtures. (Adapted from Lehmann, C. et al., *Green Chem.*, 14, 2719–2726, 2012.)

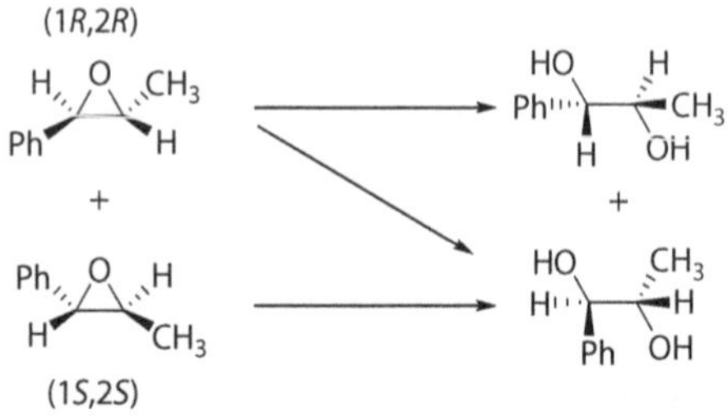

(1R,2R)

(1S,2S)

FIGURE 7.20 Epoxide hydrolase-catalyzed hydrolysis reaction of 1,2-*trans*-2-methylstyrene oxide in buffer and DESs as co-solvents. (Adapted from Lindberg, D. et al., *J. Biotechnol.*, 147, 169–171, 2010.)

of 4-methylumbelliferyl-β-D-cellobioside and 4-*p*-nitrophenyl-β-D-cellobioside, but highly active in high salt concentrations media including ILs, DESs (ChCl:Gly 1:2), and NaCl solutions as co-solvents (Lehmann et al. 2014).

The enzymatic hydrolysis of epoxides has received great attention in the literature, as epoxide hydrolases enable a high control of the regio- and stereoselectivity for the production of vicinal diols. Widersten and co-workers reported the use of potato epoxide hydrolase from *Solanum tuberosum* in the hydrolysis of 1,2-*trans*-2-methylstyrene oxide (Lindberg et al. 2010) using different buffer: DESs ratios, including ChCl:EG, ChCl:Gly and ChCl:U (1:2 mixtures). DESs provided beneficial effects accelerating the reaction, improving the regioselectivity toward the epoxide ring opening at the benzylic carbon, and facilitating the use of higher substrate concentrations (Figure 7.20). Similarly, the asymmetric hydrolysis of 1,2-epoxyoctane to (*R*)-1,2-octanediol (41% yield, 81% enantiomeric excess, *ee*) was described by immobilizing the soybean epoxide hydrolase and its later application in a DES-containing system (15% v/v ChCl:U) (Cao et al. 2016).

7.2.5 OTHER CONVENTIONAL CATALYZED REACTIONS

In this last section, a few conventional catalyzed reactions are covered which occurred in DES-containing systems. Recently, the phospholipase D-catalyzed transphosphatidylation of phosphatidylcholine has been reported with 6 equiv. of L-serine for the synthesis of phosphatidylserine, phospholipid with interesting properties to rejuvenate brain cell membranes and increase acetylcholine levels in the brain (Yang and Duan 2016). The biotransformation was attempted in various DESs with 0.5% v/v water, yielding over 90% yield employing ChCl:EG (1:2, 90.3% after 7 h) or ChCl:Gly (1:2, 92.1% after 12 h) at 40°C. In addition, 81% of the original enzymatic activity was maintained after nine reuses when the reaction was carried under optimal conditions in ChCl:EG.

Microbial haloalkane dehalogenases are valuable enzymes catalyzing the hydrolytic cleavage of carbon-halogen bonds, however, their practical applications are usually hampered due to the poor solubility of their substrates in aqueous medium. With this challenge, DES-water binary mixtures were employed in two model dehalogenation reactions, such as the one over 1-iodohexane and the kinetic resolution of racemic 2-bromopentane (Stepankova et al. 2014). Interestingly, the three haloalkane dehalogenases tested were active in ChCl:Gly (1:2)-glycine buffer pH 8.6 systems at 37°C.

7.3 USE OF HYDROLASES IN NON-CONVENTIONAL AND TANDEM PROCESSES

Enzymes are highly selective catalysts, which are able to catalyze a myriad of transformations depending on their enzyme class and subclass. However, biocatalytic promiscuity has emerged as a recent trend in biocatalysis, illustrating the potential of enzymes to mediate unexpected transformations (López-Iglesias and Gotor-Fernández 2015; Busto et al. 2010). Particularly attractive is the use of hydrolases that are involved in two main types of unconventional transformations: (a) bi- or multicomponent transformations that are far away from the scope of hydrolytic enzymes such as C-C and C-N bond formation reactions and (b) tandem processes that involve firstly the development of a conventional enzymatic transformation forming an active intermediate, which is the final responsible reagent for the success of the main reaction, this includes epoxidation and Baeyer-Villiger reactions among others.

7.3.1 HYDROLASE-CATALYZED PROMISCUOUS REACTIONS

Dandekar and co-workers reported the methylation of chitosan using the combination of a lipase with dimethyl carbonate for the synthesis of methylated chitosan species (Figure 7.21), biopolymers with interesting properties in the pharmaceutical and textile industrial sector (Bangde et al. 2016). ChCl:Gly and ChCl:U (1:2) were used in combination with aqueous systems or/and dimethylformamide, finding BCL as a highly active enzyme for the synthesis of *N*- and *O*-methylated chitosan derivatives, while no reaction was observed with CAL-B.

The aldol reaction is one of the most powerful strategies for the formation of C-C bonds, and the application of porcine pancreas lipase and DESs has been recently demonstrated in the synthesis of hydroxyl ketones starting from a series of aromatic

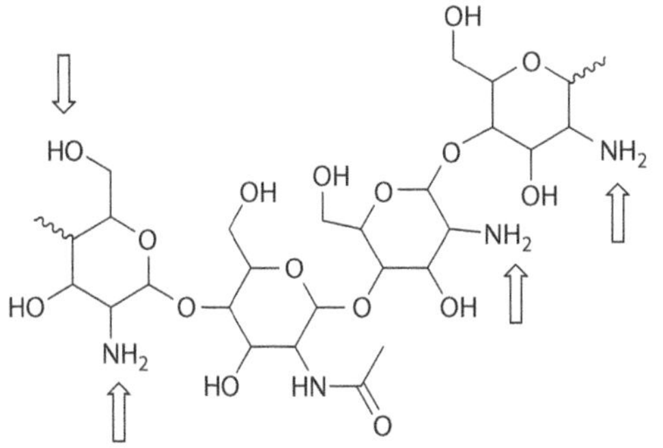

FIGURE 7.21 Chitosan positions susceptible of reaction in its lipase-catalyzed methylation. (Adapted from Bangde, P.S. et al., *ACS Sustain. Chem. Eng.*, 4, 3552–3557, 2016.)

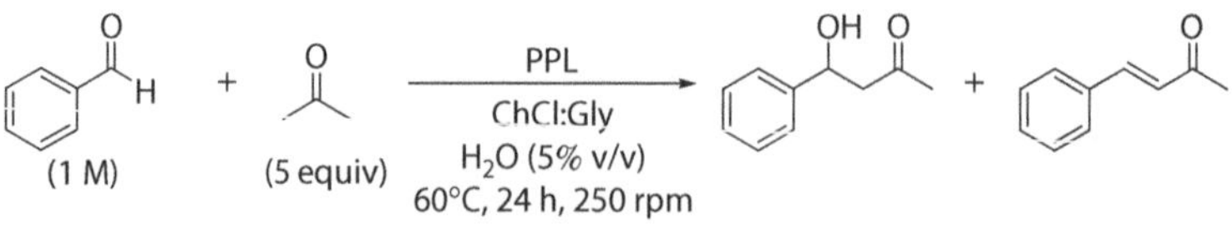

FIGURE 7.22 Porcine pancreas lipase-catalyzed the formation of aldol and aldol-dehydration products by reaction benzaldehyde and acetone in DESs. (Adapted from González-Martínez, D. et al., *Eur. J. Org. Chem.*, 1513–1519, 2016.)

and aliphatic aldehydes and acetone (González-Martínez et al. 2016). After optimization of reaction parameters that included benzaldehyde:acetone molar ratio, enzyme loading, and ChCl:Gly composition, full conversions toward the formation of 4-hydroxy-4-phenylbutan-2-one were achieved after 24 h at 60°C (Figure 7.22). The appearance of the corresponding aldol-dehydration compound as minor by-product was also observed in all the cases, although in different extents depending on the eutectic mixture composition. This reaction was finally successfully extended to the use of substituted benzaldehydes and cyclic ketones, such as cyclopentanone and cyclohexanone.

The *Aspergillus niger* lipase was found to be a highly active catalyst for the Henry reaction between nitromethane and different benzaldehydes in pure DESs (Tian et al. 2016). The biotransformation between 4-nitrobenzaldehyde and 10 equiv. of nitromethane was studied as model reaction using ChCl-based DESs including Gly, EG, and U as hydrogen bond donors in different ratios. Later, the influence of water content, reactant molar concentrations, enzyme loading, and temperature were analyzed, finding higher conversions for the reactions carried out in ChCl:Gly with 30% of water rather than using only pure water (Table 7.1). Overall, reaction yields were very high with benzaldehydes containing electron-withdrawing groups like the nitro in different positions at the aromatic ring, lower with halogen substitutions, and poor with electro-donating rests.

Satisfyingly, in the same contribution, the aza-Henry reaction catalyzed by a hydrolase was reported for the first time in DESs. The biotransformation between (*E*)-*N*-[(4-methylbenzene-1-sulfonyl)oxy]-1-phenylmethanimine and 15 equiv. of nitromethane in ChCl:Gly (1:2) at 25°C, led to the aza-Henry product in 41%–44% conversion using 30%–50% water content (Figure 7.23). In both cases, for the Henry and aza-Henry reactions, no enantioselectivity was observed.

The synthesis of tricyanovinyl substituted aniline and indole derivatives was carried out by reacting equimolecular amounts of *N*-alkyl anilines or indole with tetra-cyanoethylene in DESs and the absence of enzymes (Sanap and Shankarling 2014). Among different DESs and DES-organic solvent mixtures, ChCl:U (1:2) provided the best conversion (Figure 7.24), while hydrolases, such as lipase from *Pseudomonas* species and protease from *Bacillus subtilis* were also able to catalyze this process in dichloromethane. Unfortunately, the reactions in the presence of both DESs and enzyme were not developed.

Finally, in a more complex approach involving three reactants, a series of dihydropy-rimidin-2(1*H*)-one derivatives were synthesized starting from an aromatic aldehyde,

TABLE 7.1

ANL-Catalyzed Henry Reaction between Benzaldehydes and Nitromethane

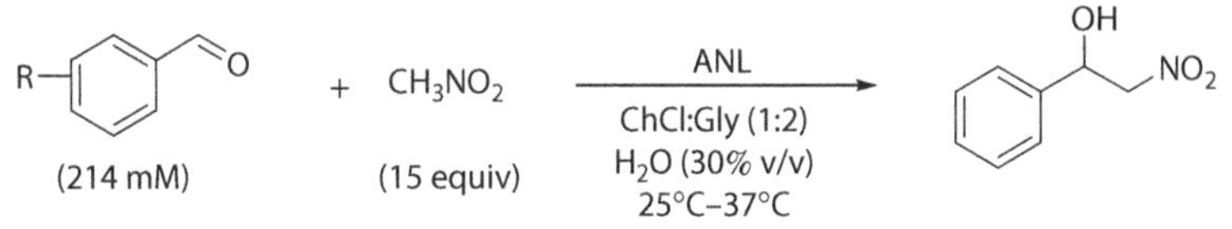

		100% H_2O		ChCl:Gly (1:2) and 30% H_2O	
Entry	R	Time (h)	Yield (%)	Time (h)	Yield (%)
1	H	120	—	48	—
2	$2-NO_2$	10	81	4	87.7
3	$3-NO_2$	10	85	4	88.5
4	$4-NO_2$	10	87	4	92.2
5	$2,4-(NO_2)_2$	10	70	4	91.7
6	4-F	120	46[a]	48	12.3
7	4-Cl	120	80[a]	24	62
8	4-Br	120	91	24	89
9	4-OMe	120	37[a]	48	9.6
10	4-Me	120	14[a]	48	—

Source: Tian, X. et al., 2016. J. Microbiol. Biotechnol., 26, 80–88, 2016.

[a] These biotransformations were carried out at 37°C. For all the others, 25°C were employed.

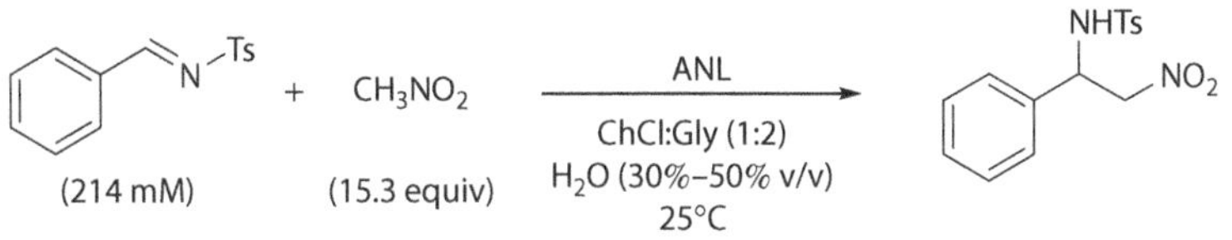

FIGURE 7.23 ANL-catalyzed aza-Henry reaction between a tosyl imine and nitromethane in ChCl:Gly (1:2) and water. (Based on Tian, X. et al., J. Microbiol. Biotechnol., 26, 80–88, 2016.)

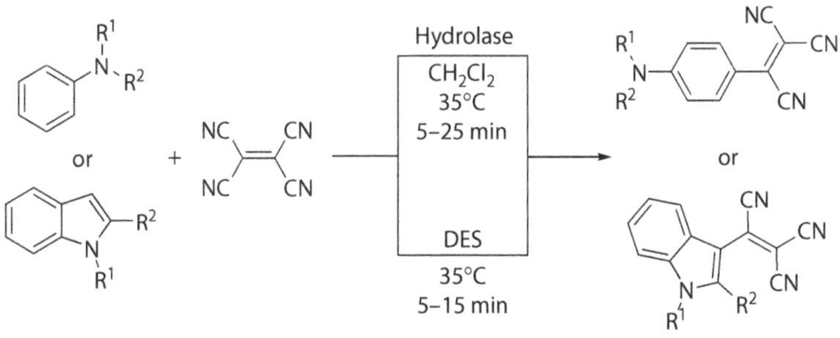

FIGURE 7.24 Synthesis of tricyanovinyl substituted anilines and indoles carried out in DESs or by the action of hydrolases in dichloromethane. (Adapted from Sanap, A.K. and Shankarling, G.S., Catal. Commun., 49, 58–62, 2014.)

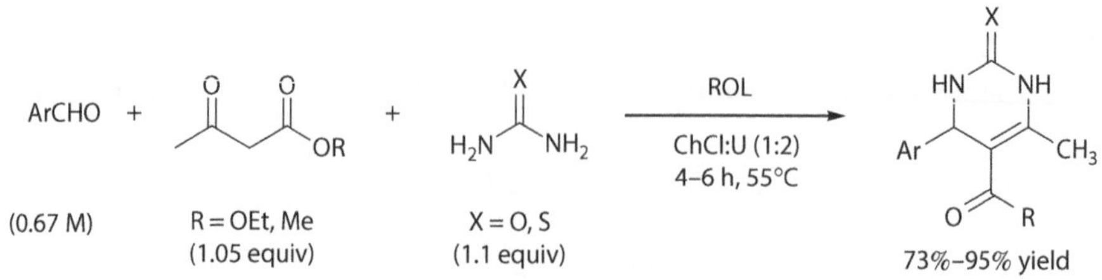

FIGURE 7.25 Lipase-catalyzed Biginelli reaction using ROL and a ChCl:U mixture. (Adapted from Borse, B.N. et al., *Curr. Chem. Lett.*, 1, 59–68, 2012.)

a dicarbonyl compounds (ethyl acetoacetate or pentane-2,4-dione), and urea or thiourea in ChCl:U (1:2) (Borse et al. 2012). In particular, the Biginelli reaction between 6-methoxynaphthalene-2-carbaldehyde, ethyl acetoacetate, and urea was highly accelerated by the *Rhizopus oryzae* lipase (ROL), as in the absence of the enzyme only 20% yield was reached after 7.5 h, while 95% was attained after 4 h in the presence of ROL (Figure 7.25). The recyclability of the DESs and lipase system was also demonstrated in this model reaction, obtaining a range of 75%–95% yield after four recycling uses. Interestingly, the use of ChCl:U (1:2) as solvent gave better results in comparison with the use of water, methanol, 1,4-dioxane, or dimethylformamide.

7.3.2 TANDEM UNCONVENTIONAL REACTIONS MEDIATED BY HYDROLASES

In the previous section, examples of the promiscuous activity of hydrolases have been disclosed, mainly focused in the C-C and C-N bond formation reactions. Alternatively, hydrolases can be involved in tandem processes catalyzing an individual conventional reaction (for instance, transesterification or perhydrolysis), although the global transformation could be initially unexpected (aldol or epoxidation reactions).

Taking advantage of the high activity displayed by CAL-B in DESs, Domínguez de María and co-workers developed the *in situ* formation of acetaldehyde through the transesterification between vinyl acetate with 2-propanol in ChCl:Gly (1:2) at room temperature (Figure 7.26). The reaction between the resulting acetaldehyde with 4-nitrobenzaldehyde in their own reaction medium catalyzed by a highly substituted proline derivative led to 1-(4-nitrophenyl)propane-1,3-diol in 92% conversion, 70% isolated yield, and 95% ee after reduction of the aldehyde intermediate with sodium borohydride (Müller et al. 2014). The recycling of the DESs and CAL-B system was satisfyingly achieved after extraction in ethyl acetate (EtOAc), without requiring the addition of the organocatalyst for two cycles. This approach was successfully extended to seven aldehydes, the conversion being highly dependent on the aromatic pattern substitution, but occurring in all cases with high selectivity. The same research group proposed the chemical synthesis and application of a structurally similar organocatalyst, but containing in this case an extra hydroxyl group (Müller et al. 2015). This additional hydrogen bond donor group allows a stronger interaction with the DESs, improving the conversion and recycling possibilities of the evolved catalytic system.

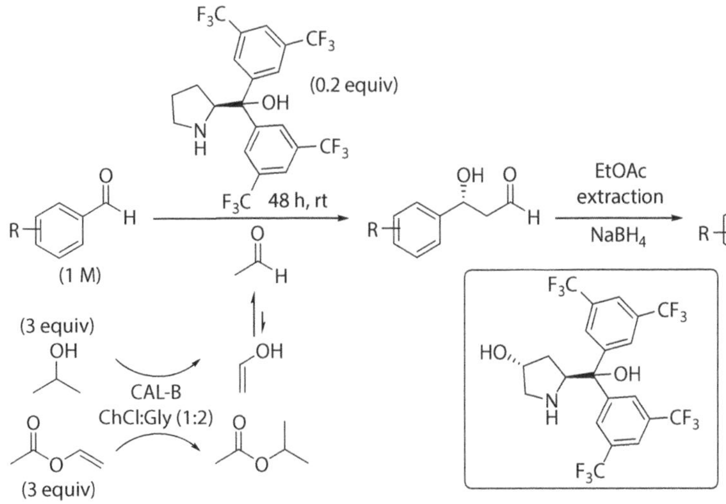

FIGURE 7.26 Tandem lipase-diaryl prolinol catalyst approach for cross-aldol reactions between aromatic aldehydes and acetaldehyde generated *in situ*. (Adapted from Müller, C.R. et al., *RSC Adv.*, 4, 46097–46101, 2014.)

The perhydrolysis activity of CAL-B in DESs has been demonstrated in the transformation of octanoic acid into peroctanoic acid using an aqueous solution of hydrogen peroxide (30% H_2O_2) (Wang et al. 2017). Besides, this peracid intermediate has been used in the same reaction medium for the Baeyer-Villiger oxidation of cyclohexanone allowing the formation of ε-caprolactone and 6-hydroxyhexanoic acid, the latest corresponding to the lipase-catalyzed hydrolysis of the lactone (Figure 7.27).

Different ratios of products were found depending of the type of CAL-B, and, interestingly, in a mixture of hexane-water (2:1), the lactone was observed as major product for the CAL-B mutant, while the hydrolysis activity of CAL-B was very significant, favoring the formation of the carboxylic acid as main product. Pleasingly, when the reaction was carried out in DESs, such as ChCl:sorbitol (1:1), better selectivities were found, favoring the formation of the Baeyer-Villiger products starting from cyclohexanone and other aliphatic ketones (Table 7.2).

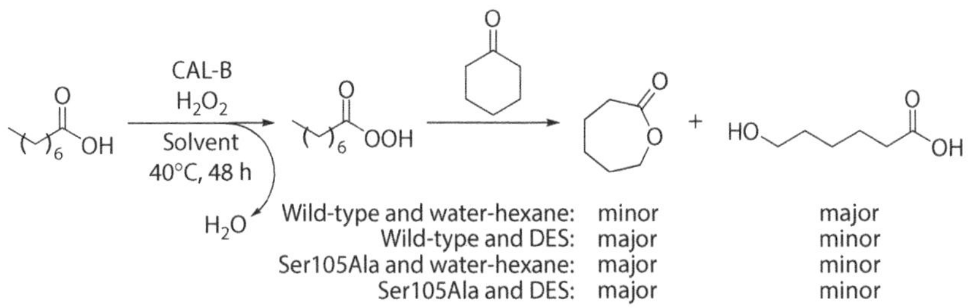

FIGURE 7.27 Baeyer-Villiger oxidation of cyclohexanone through the lipase-catalyzed formation of peroctanoic acid as oxidizing agent. (Adapted from Wang, X.-P. et al., *Sci. Rep.*, 7, 44599, 2017.)

TABLE 7.2

Baeyer-Villiger Oxidation of Ketones with 2 Equivalents of H₂O₂ Mediated by CAL-B Wild-Type and Ser105Ala Mutant in ChCl:sorbitol (1:1) after 48 h at 40°C

		Wild-Type CAL-B		Ser105Ala	
Entry	R	Conversion (%)	Selectivity (%)	Conversion (%)	Selectivity (%)
1	Cyclobutanone	99	93	99	100
2	Cyclopentanone	95	48	51	97
3	Cyclohexanone	92	46	47	99
4	4-Heptanone	79	35	38	96

Source: Wang, X.-P. et al., *Sci. Rep.*, 7, 44599, 2017.

Taking advantage of the perhydrolytic activity of some hydrolases, a series of global epoxidation reactions mediated by lipases have been developed in 2017 using DESs as solvents. For instance, the epoxidation of seven alkenes was reported using equimolar amounts of H₂O₂ (30%) and octatonic acid as peracid precursor (Zhou et al. 2017a). After screening nine eutectic mixtures, ChCl:sorbitol (1:1) was found to be the most effective medium, yielding the corresponding epoxides in 72%–97% conversion after 11 h–24 h at 30°C or 40°C (Figure 7.28). A similar strategy was reported by Sieber and co-workers for the CAL-B-mediated epoxidation of monoterpenes, including 3-carene, limonene, α-pinene, and camphene in ChCl-based DESs (Ranganathan et al. 2017). Octanoic acid was also used as peracid intermediate, employing in this case the urea-hydrogen peroxide complex as oxidant, recovering the resulting epoxides in quantitative conversions, and good to high yields after short reaction times and a simple extraction protocol.

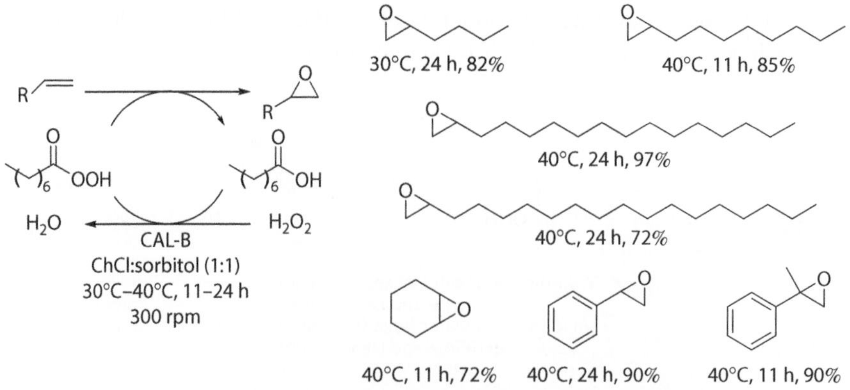

FIGURE 7.28 Epoxidation of alkenes through CAL-B-catalyzed formation of peroctanocid acid as oxidizing agent. (Adapted from Zhou, P. et al., *RSC Adv.*, 7, 12518–12523, 2017a.)

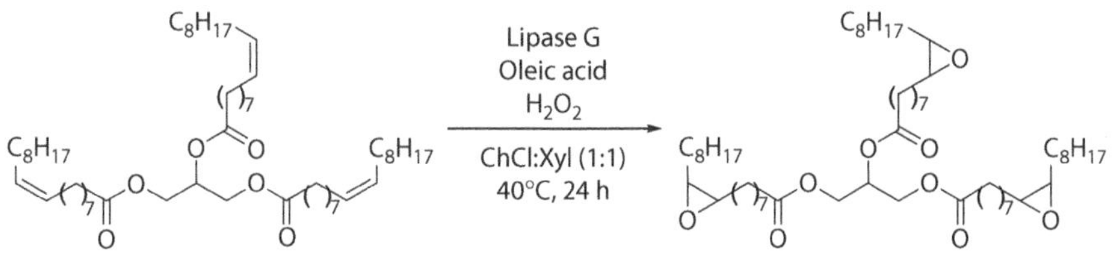

FIGURE 7.29 Epoxidation of glyceryl triolate using the lipase G from *Penicillium camembertii*. (Adapted from Zhou, P. et al., *ChemCatChem*, 9, 934–936, 2017b.)

Wang and co-workers reported the lipase-mediated epoxidation of glyceryl trioleate, the lipase G from *Penicillium camembertii* displaying a higher activity than CAL-B (Zhou et al. 2017b). After optimization, the best conditions were found with the ChCl:Xyl (1:1), 3 equiv. of hydrogen peroxide and 40°C (Figure 7.29). Oleic acid was used as peracid precursor without requiring its external addition, as it was present in the vegetable oil samples used as starting materials. Similarly, the same research group has recently described the epoxidation of this soybean oil using PCL, finding the best results in a biphasic system composed of the oil and water where the ChCl:sorbitol (1:1) served to reduce the surface tension of water (Lan et al. 2017).

7.4 USE OF OXIDOREDUCTASES AND LYASES

As described in the previous sections, hydrolases have attracted the main focus for biotransformations carried out in DESs. Nevertheless, an increasing number of biocatalytic reactions have been reported using other classes of enzymes, namely oxidoreductases (EC 1), especially ketoreductases, and lyases (EC 4). In the following sections, the use of whole-cells and isolated enzymes with DESs is described for these enzyme families.

7.4.1 USE OF LYASES

The group of Domínguez de María reported the use of ThDP-dependent enzyme benzaldehyde lyase (BAL) in DES-water mixtures for the carboligation of various aldehydes (Maugeri and Domínguez de María 2014a). For the enzyme-catalyzed self-condensation of two aldehydes molecules, butyraldehyde, valeraldehyde, benzaldehyde, and 2-furaldehyde were considered using ChCl:Gly (1:2) as model DESs medium (Figure 7.30). The behavior of BAL was determined to be very sensitive to the amount of DESs used, noting a decrease in the enzyme activity at higher DESs concentrations, which was almost negligible at 90% DESs content. Interestingly, high conversions (75%–98%) for the formation of the enantioenriched (*R*)-products (27% *ee*–99% *ee*) were found after 24 h at room temperature when using a 60:40 mixture of ChCl:Gly (1:2)-phosphate buffer at pH 8. These results were similar to those obtained with 2-methyl-tetrahydrofuran as co-solvent. Additionally, the BAL-catalyzed carboligation of valeraldehyde was tested in the presence of ChCl:Gly (1:2), but also ChCl:U (1:2) and ChCl:Xyl (1:1) at 70% v/v in phosphate buffer, ChCl:U was

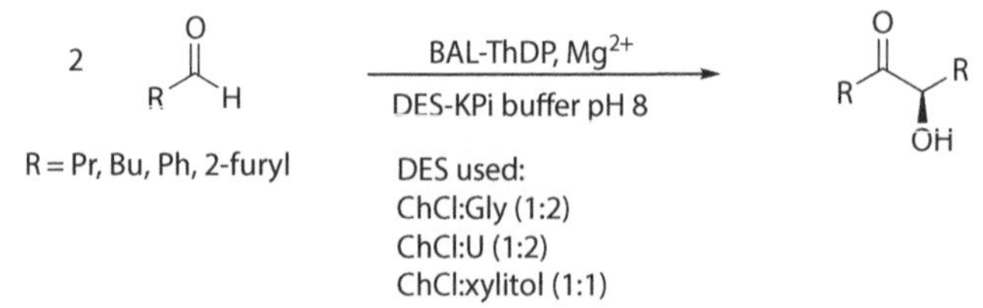

FIGURE 7.30 BAL-catalyzed enantioselective carboligation of aldehydes in DES-buffer mixtures. (Adapted from Maugeri, Z. and Domínguez de María, P., *J. Mol. Catal. B: Enzym.*, 107, 120–123, 2014a.)

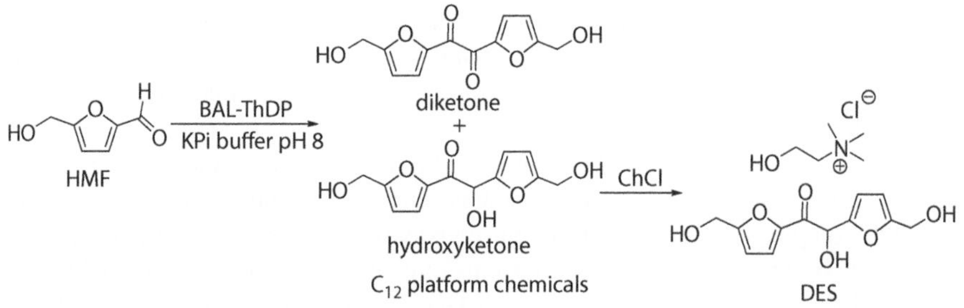

FIGURE 7.31 BAL-catalyzed condensation of HMF to C_{12} platform chemicals and formation of DESs with choline chloride. (Adapted from Donnelly, J. et al., *Green Chem.*, 17, 2714–2718, 2015.)

shown to afford full conversion, while the others DESs reached only around 70% conversion. The use of DESs with ThDP-dependent enzymes, lyases, could therefore be used with careful consideration on the type of HBD included (such as urea) in the choline chloride-based DESs.

The same research group has reported the umpolung carboligation of furfural, HMF, and mixtures of them for the formation of C_{10}–C_{12} platform chemicals using BAL, investigating later the use of the formed hydroxyketone as a HBD for the formation of DESs (Figure 7.31). This opens up a wider perspective for the production of DESs (Donnelly et al. 2015).

The chondroitin ABC lyase (cABCl, EC 4.2.2.4) is an important clinical enzyme used in the treatment of spinal lesions, however, its stability is an issue. For that reason, the group of Khajeh studied the stability of cABCl from *Proteus vulgaris* in different DESs (trimethylglycine and choline chloride-based DESs) (Daneshjou et al. 2017). The DESs were reported to facilitate the effective contact between the substrate and cABCl, thus affecting the activity, structure, and stability of the enzyme. Twenty percent of the DESs ChCl:Gly (1:2) in phosphate buffer at pH 6.8 was found to improve the thermal stability of cABCl at −20°C, 4°C, and 37°C. In buffer, the enzyme only retained 20% of its original activity after 2 h at 37°C, however, it retained 82% with the ChCl (1:2) added. The same effect was observed at the other temperatures, for instance, at −20°C, the enzyme lost activity after five days, but retained 95% of it with ChCl:Gly after 15 days. The authors used fluorescence studies to observe conformational changes induced by the DESs, displaying higher fluorescence intensity and, hence, a more compact structure with ChCl:Gly. Therefore, both the stability and activity of cABCI in aqueous solution were improved by the addition of DESs. Further studies are needed to fully understand the interactions involved between the DESs and lyases to explain these outcomes.

7.4.2 USE OF OXIDOREDUCTASES

7.4.2.1 Ketoreductases in Whole-Cells

The compatibility of isolated alcohol dehydrogenases (ADHs) with DESs as non-conventional media has not yet been reported, as it occurs with ILs (Paul et al. 2014). Whole-cells are generally used to take advantage of the co-factor recycling system, as well as preserving the enzyme environment against a solvent, while saving costs in terms of purification and co-factor. In general, the use of DESs allows working with higher substrate loadings and avoids inhibition. *E. coli* whole-cells were reported to retain their integrity in DES-buffer mixtures. However, a disadvantage of using whole-cells is the presence of other enzymes with overlapping substrate scope and differing enantioselectivities, thus lacking in orthogonal activity, such as the many ketoreductases present in baker's yeast, leading to low *ee*. The following section will describe the use of DESs:water mixtures with whole-cells for oxidoreductase-catalyzed reactions.

7.4.2.1.1 Reduction Reactions

The stereoselective bioreduction of ethyl acetoacetate catalyzed by baker's yeast (BY), *Saccharomyces cerevisiae*, has been reported in different mixtures of water and DESs (Maugeri and Domínguez de María 2014b). Remarkably, BY remained active at rather long reaction times (>200 h), finding an interesting stereoinversion of the selectivity when increasing the amount of water in the DES-water mixture. Thus, the (*S*)-ethyl 3-hydroxybutyrate was preferred in pure water, while the (*R*)-alcohol was mostly recovered in pure ChCl:Gly (1:2) (Figure 7.32). Interestingly, the racemic alcohol was obtained using a 30:70 DES-water mixture. For these unexpected results, it was hypothesized the potential inhibition in DESs of (*S*)-oxidoreductases present in baker's yeast, while the (*R*)-selective enzymes remained active.

One year later, Redovniković and co-workers studied in more detailed the same BY-catalyzed bioreduction of ethyl acetoacetate (Figure 7.33), employing in this case, a wide number of ChCl-based DESs (Bubalo et al. 2015a). Various HBDs were employed such as ChCl:Gly (1:2), EG (1:2), Glu (2:1), fructose (3:2), xylose (2:1), MA (1:1), OA (1:1), and U (1:2), and variable amounts of buffer. The yield of

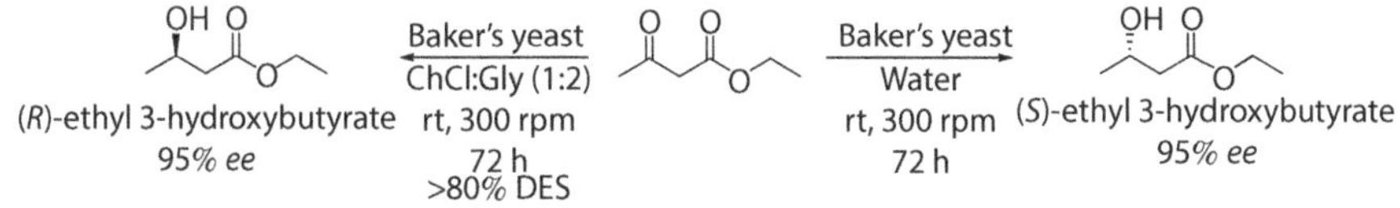

FIGURE 7.32 Stereoselective baker's yeast-catalyzed bioreduction of ethyl acetoacetate in DESs or water. (Adapted from Maugeri, Z. and Domínguez de María, P., *ChemCatChem*, 6, 1535–1537, 2014b.)

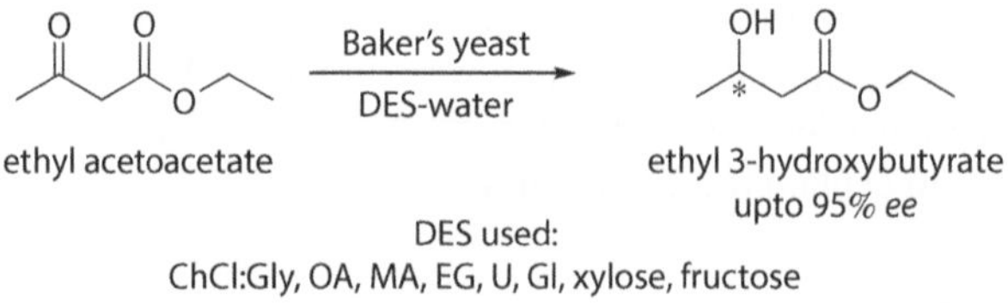

FIGURE 7.33 Baker's yeast-catalyzed reduction of ethyl acetoacetate in DESs.

the reaction was affected by the type of HBD and the water content; for example, 50% w/w of water in sugar and alcohol DESs gave similar yields as with phosphate buffer (>93%). With acid or amide-based DESs lower yields were obtained (<49%), implying the pH of the reaction medium is influenced by the structure of the DESs. As previously observed by Maugueri and Domínguez de María, the amount of DESs could change the enantioselectivity of the bioreduction, but also the selectivity was heavily dependent of the DESs source, and, for instance, ChCl:fructose was shown to afford the opposite stereoselectivity. Sugar containing DESs (glucose, xylose) were found to be more biocompatible with yeast cells as a non-conventional reaction medium for biotransformations.

Following the idea that a possible inhibition of (*S*)-selective ketoreductases occurs in the presence of DESs (Vitale et al. 2017), the bioreduction of arylpropanones was studied in various ChCl-based DESs (Figure 7.34). The structural components of the DESs, as well as the water content of the medium, were crucial to observe an inversion of the stereoselectivity, the access to the corresponding alcohols being possible in high optical purity. For instance, using the ChCl:Gly (2:1) as DESs component, the (*R*)-alcohol was obtained up to 96% *ee* in DES-aqueous mixture, while the use of pure water favored the formation of the (*S*)-enantiomer with up to 96% *ee*.

Domínguez de María and co-workers investigated the reduction of aromatic ketones catalyzed by recombinant whole-cells expressing well-known ketoreductases in various aqueous DESs media (Müller et al. 2015). Recombinantly expressed ADHs in *E. coli*, such as from *Ralstonia* sp. and from horse liver remained active when varying amounts of DESs in buffer were added (DES-buffer, up to 80% v/v, Figure 7.35). The enantiomeric excess of the resulting chiral aromatic alcohols were also found to significantly increase in proportion to DESs addition, from <20% *ee* with 10% DES-buffer to >90% *ee* with 80% DES-buffer, using ChCl:Gly, ChCl:U, ChCl:EG mixtures as DESs.

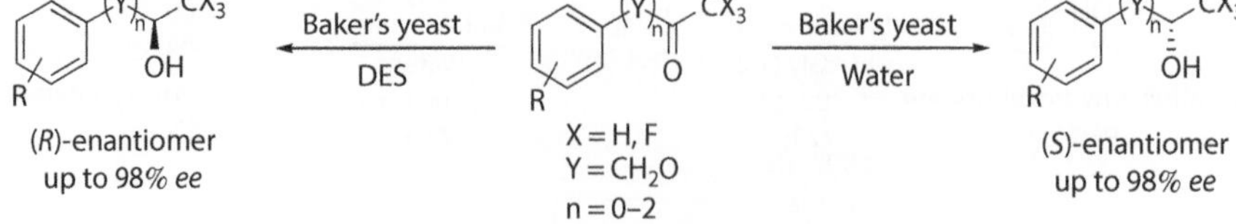

FIGURE 7.34 Baker's yeast-catalyzed reduction of arylpropanones in DES-water mixtures. (Adapted from Vitale, P. et al., *Adv. Synth. Catal.*, 359, 1049–1057, 2017.)

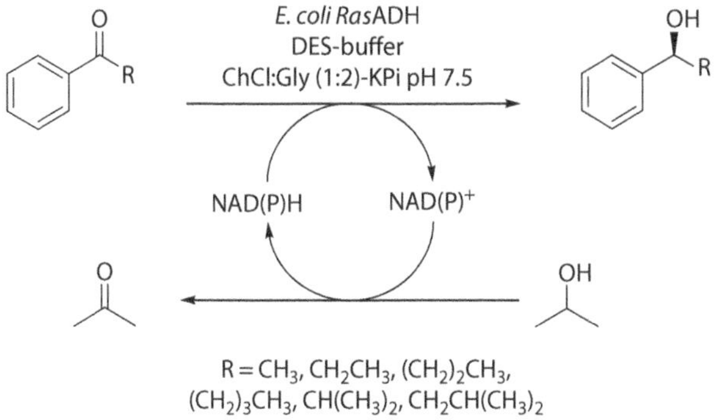

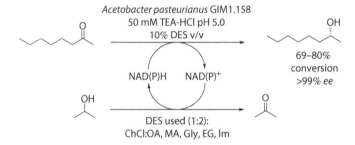

R = CH$_3$, CH$_2$CH$_3$, (CH$_2$)$_2$CH$_3$,
(CH$_2$)$_3$CH$_3$, CH(CH$_3$)$_2$, CH$_2$CH(CH$_3$)$_2$

FIGURE 7.35 Whole-cell-catalyzed reduction of ketones to chiral alcohols. (Adapted from Müller, C.R. et al., *ChemCatChem*, 7, 2654–2659, 2015.)

FIGURE 7.36 Asymmetric reduction of 2-octanone catalyzed by *Acetobacter pasteurianus* GIM1.158 cells using various DESs as co-solvents. (Adapted Xu, P. et al., *Sci. Rep.*, 6, 26158, 2016.)

The asymmetric reduction of 2-octanone catalyzed by *Acetobacter pasteurianus* GIM1.158 cells was reported in a biphasic system with DESs and water-immiscible ILs in 10% v/v (Figure 7.36). A series of choline chloride-based DESs were used (Xu et al. 2016), with either urea, glycerol, ethylene glycol, oxalic acid, malonic acid, or imidazole as hydrogen bond donor. ChCl:EG exhibited good biocompatibility and could moderately increase the cell membrane permeability, thus leading to the best results. The addition of ChCl:EG increased the optimal substrate concentration from 40 mM to 60 mM and the *ee* of the alcohol product to >99%. Additionally, the cells manifested good operational stability in the reaction system. Thus, ChCl:EG (1:2) and alternatively an imidazole-based IL (1-butyl-3-methylimidazolium hexafluorophosphate) were found as promising co-solvents for the efficient biocatalytic synthesis of (*R*)-2-octanol.

In a recent contribution, whole-cells of *E. coli* CCZU-T15 have been used with ChCl:Gly (1:2) for the biocatalytic reduction of ethyl 4-chloro-3-oxobutanoate (COBE) into ethyl (*S*)-4-chloro-3-hydroxybutyrate (CHBE, Figure 7.37). After optimization of the reaction conditions and adding L-glutamine or D-ribose for the synthesis of co-factors instead of directly using NAD⁺, the (*S*)-CHBE was obtained in >99% *ee* from a 2 M COBE solution with L-glutamine (150 mM) and ChCl:Gly-water (12.5% v/v at

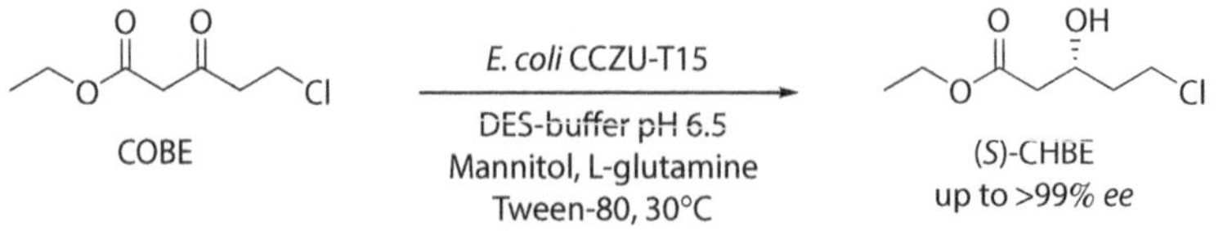

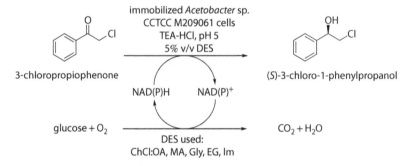

FIGURE 7.37　*E. coli* CCZU-T15 whole-cell-catalyzed reduction of COBE to (*S*)-CHBE. (Adapted from Dai, Y. et al., *Appl. Biochem. Biotechnol.*, 181, 1347–1359, 2017.)

immobilized *Acetobacter* sp.
CCTCC M209061 cells
TEA-HCl, pH 5
5% v/v DES

3-chloropropiophenone

(*S*)-3-chloro-1-phenylpropanol

NAD(P)H　　NAD(P)$^+$

glucose + O$_2$

CO$_2$ + H$_2$O

DES used:
ChCl:OA, MA, Gly, EG, Im

FIGURE 7.38　Whole-cell-catalyzed reduction of 3-chloropropiophenone, with immobilized *Acetobacter* sp. CCTCC M209061. (Adapted from Xu, P. et al., *ACS Sustain. Chem. Eng.*, 3, 718–724, 2015b.)

pH 6.5) as solvent (Dai et al. 2017). Once again, studies showed that the cell membrane was more permeable in the presence of ChCl:Gly and a surfactant as Tween-80.

The whole-cells of *Acetobacter* sp. CCTCC M209061 (CCTCC for China Center for Type Culture Collection) were used to reduce 3-chloropropiophenone to (*S*)-3-chloro-1-phenoylpropanol (Figure 7.38). The cells were immobilized on polyvinyl alcohol-sodium sulfate for higher stability and used with DESs as co-solvents (Xu et al. 2015b). The DES ChCl:U (1:2) displayed the best results in terms of biocompatibility, flow cytometry studies showing that the cell membrane permeability was slightly increased, affording the enantiopure (*S*)-alcohol with a productivity of 1.37 mM/h at 30°C after 6 h. This efficient bioreduction process was proved to be feasible in a 500 mL preparative scale.

7.4.2.1.2　Oxidation Reactions

The whole-cells of *Acetobacter* sp. CCTCC M209061 were also used for the resolution of racemic 1-(4-methoxyphenyl)ethanol (MOPE) through asymmetric oxidation (Figure 7.39). Several DESs were employed, achieving the best resolution with ChCl:Gly and its concentration exhibiting a significant influence on the reaction, with the optimal content being 10% v/v (Xu et al. 2015a). In the ChCl:Gly-buffer mixture, the substrate concentration was substantially increased (55 mM versus 30 mM) as compared with the ChCl:Gly-free aqueous (buffer only) reaction, while the residual substrate enantiomeric excess was kept higher than 99%. The good biocompatibility of ChCl:Gly with the cells and the improved cell membrane permeability in the ChCl:Gly-buffer mixture could partly account for the clearly enhanced reaction efficiency.

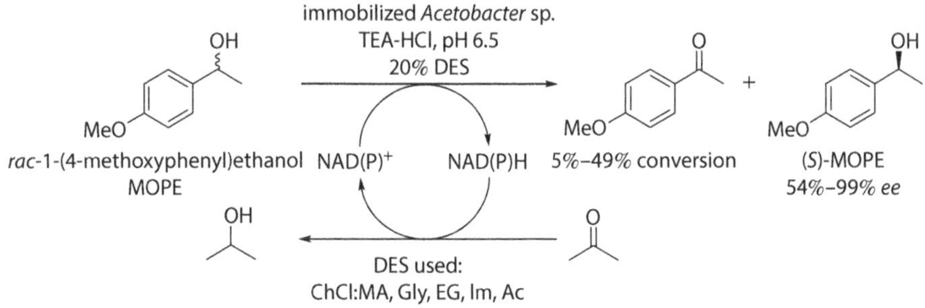

FIGURE 7.39 Resolution of *rac*-1-(4-methoxyphenyl)ethanol to the (*S*)-alcohol with immobilized *Acetobacter* sp. CCTCC M209061 cells through asymmetric oxidation. (Adapted from Xu, P. et al., *RSC Adv.*, 5, 6357–6364, 2015a.)

This resolution was further investigated by the same group (Wei et al. 2016), significantly improving the asymmetric oxidation by using DESs in a two-phase system. Thus, the asymmetric oxidation of MOPE to (*S*)-MOPE catalyzed by *Acetobacter* sp. CCTCC M209061 cells was studied in the presence of an IL and DES in a biphasic system looking for an improvement in the reaction efficiency (Figure 7.40). The IL 1-butyl-3-methylimidazolium hexafluorophosphate gave good results in combination with the buffer (97.8 μmol/min initial rate, 50.5% conversion and >99% *ee* after 10 h), however, when the DES ChCl:Gly was added as 10% v/v in buffer, the reaction was significantly improved (124.0 μmol/min initial rate, 51.3% conversion and >99% *ee* after 7 h), enhancing also the substrate concentration from 50 mM until 80 mM. When the cells were immobilized, 72% of their activity was retained after nine cycles in the IL/DES-buffer mixture, and the reaction could be upscaled to 500 mL.

Finally, DESs were evaluated as co-solvents for the dehydrogenation of steroids using *Arthrobacter simplex* (Mao et al. 2016). The 1,2-dehydrogenation of cortisone acetate to prednisone acetate showed that the DES ChCl:U as co-solvent could

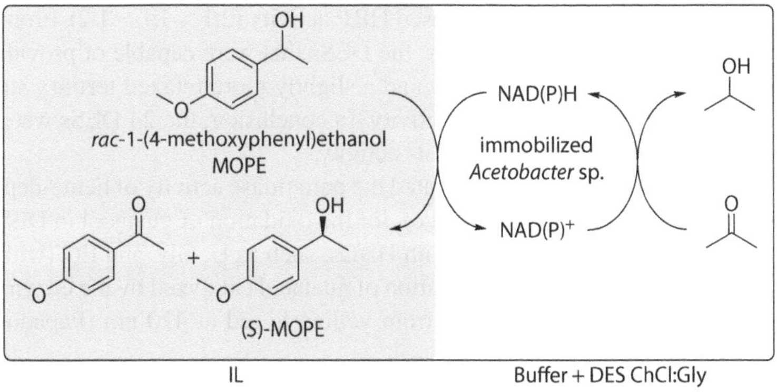

FIGURE 7.40 Oxidative resolution of racemic 1-(4-methoxyphenyl)ethanol with immobilized *Acetobacter* sp. CCTCC M209061 cells in a biphasic system. (Adapted from Wei, P. et al., *Microb. Cell Fact.*, 15, 5, 2016.)

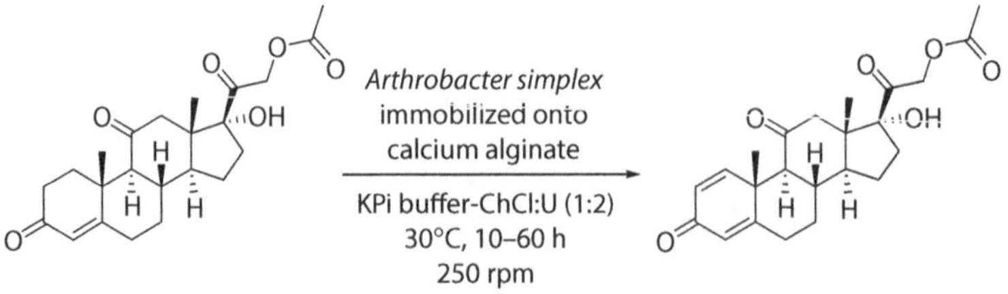

FIGURE 7.41   Immobilized whole-cell *Arthrobacter simplex* 1,2-dehydrogenation of corti-sone acetate to prednisone acetate. (Based on Mao, S.H. et al., *J. Chem. Technol. Biotechnol.*, 91, 1099–1104, 2016.)

improve the conversion of the reaction (Figure 7.41). A remarkable 93% conver-sion was achieved using substrate concentrations of 5 g/L, immobilized cells of *Arthrobacter simplex* on calcium alginate (4 g/L), and 6% of ChCl:U (1:2). The addi-tion of DESs to the reaction mixture allowed a better substrate solubility. Additionally, the reusability of the enzyme was demonstrated after recovery of the immobilized cells and reuse through five batches, which led to 82%–93% conversion values.

To conclude, the use of DESs with ketoreductases, so far only whole-cells have been used as biocatalysts, and remarkable increase in enantioselectivity, even inver-sion of selectivity, were observed in the reduction of ketone substrates. Further inves-tigation with isolated purified enzyme would be interesting to investigate the type of effects DESs have on ketoreductases, such as inhibition and conformational changes.

7.4.2.2   Peroxidases in DESs

The use of DESs was reported to improve the activity and stability of horseradish peroxidase (HRP) according to experimental and structural studies (Wu et al. 2014). The results were dependent on the two choline salts employed (ChCl and ChAc) and four HBDs (U, Gly, Acet, and EG) at three different molar ratios (1:2, 1:1, and 2:1). In total, 24 DESs were studied and various trends were observed. First, ChAc-based DESs were found to reduce peroxidase activity with respect to ChCl-based DESs. Second, for each DES composed of the same HBD, a higher concentration of the choline chloride salt resulted in increased HRP activity (2:1 > 1:1 > 1:2). From circu-lar dichroism and fluorescence studies, the DESs that were capable of providing the enzyme with a higher α-helix content and a slightly more relaxed tertiary structure were reported to facilitate the HRP activity. In conclusion, the 24 DESs were found to stabilize the HRP, but not increase its activity.

The group of Stamatis also investigated the peroxidase activity of heme-dependent enzymes, HRP, and cytochrome c (cyt c), for the oxidation of guaiacol in DESs con-sisting of either ChCl or EACl salts with HBDs, such as U, Gly, and EG (with ratios of 1:1.5 or 1:2, Figure 7.42). The oxidation of guaiacol catalyzed by the enzymes was measured according to color change from yellow to red at 470 nm (Papadopoulou et al. 2016).

The activity of the peroxidase depended on the nature of the choline salt, either ChCl or EACl, on the HBD, and on the amount of DESs used in the water mix-ture. The peroxidase activity of cyt c was increased in the DES-based media compared to the activity in pure buffer. This activity was reported to correlate

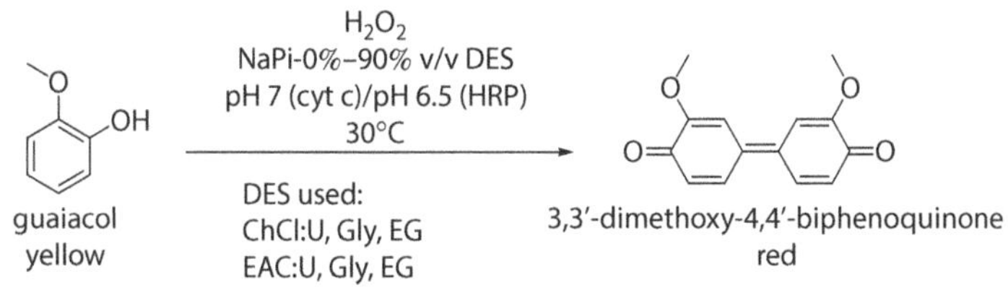

FIGURE 7.42 Peroxidase activity assay for cytochrome c and HRP, determined by the color change of guaiacol (yellow) during oxidation by hydrogen peroxide observed at 470 nm. (Adapted from Papadopoulou, A.A. et al., *Ind. Eng. Chem. Res.*, 55, 5145–5151, 2016.)

with the microenvironment perturbations of the heme, displaying higher stability with EACl-based DESs versus ChCl-based DESs. Especially when a mixture of 30% v/v ChCl:U-buffer was used, the activity increased eight-fold compared to that in pure buffer, and with EACl:U, the increase was 100-fold. This significant increase in activity was not observed for HRP. Regarding the stability of cyt c, a 35% decrease was noted after 24 h in buffer only, cyt c retained most of its activity in the presence of DESs. ChCl- and EACl-based DESs enhanced the stability of cyt c depending on the HBD, in the order U > Gly > EG. The stabilizing effect of urea as a HBD was also seen with HRP. Thus, lower stability was observed with ChCl:Gly and ChCl:EG compared to ChC:U, which correlated with conformational changes of the heme prosthetic group. The authors also investigated the cyt c-catalyzed decolorization of the pinacyanol chloride dye, which has industrial applications, and report enhanced cyt c activity with DESs, especially with 50% v/v of EACl:U and EACl:EG in buffer, giving a 3.3-fold increase. These DESs were also studied for their recyclability, showing the same enzyme activity after four cycles. Therefore, the two reported studies on heme-dependent enzymes up to now have shown that DESs were beneficial for the stability of HRP and cyt c and for the activity of cyt c.

Peroxidases are also able to promote free-radical polymerization reactions as occurs in the case of acrylamide, which have been efficiently developed in nearly non-aqueous Ch:U and ChCl:Gly (Sánchez-Leija et al. 2016). Although the HRP activity and thermal stability were found to be lower at high DESs concentration in comparison with phosphate buffer pH 7, taking advantage of the low freezing point of ChCl:Gly, the production of polyacrylamide was possible at 4°C in 80% DESs v/v. Remarkably, no polymerization was observed at this temperature in pure buffer, so these results open up a series of synthetic possibilities toward the production of important materials by using novel enzymatic cryo-polymerization reactions.

7.4.2.3 Catalases in DESs

Very recently, the effect of DESs has been investigated on the catalytic function and structure of the bovine liver catalase (Harifi-Mood et al. 2017). Catalases (EC 1.11.1.6) can convert hydrogen peroxide to water, and mixtures of DESs [ChCl:U (1:2) called reline, ChCl:Gly (1:2) called glyceline] in buffer were used. The binding affinity of the hydrogen peroxide substrate to the catalase enzyme increased with a glyceline (ChCl:Gly) concentration of 100 mM. The structure of the enzyme was reported to be influenced, hence, the K_M and k_{cat} values of the catalase enzyme changed upon addition of the DESs. Activity was retained in both glyceline and reline at 70% and 80%, respectively.

7.4.2.4 Other Oxidoreductases in Whole-Cells

The group of Yang studied the oxidative cleavage of isoeugenol to vanillin catalyzed by *Lysinibacillus fusiformis* cells from the China General Microbiological Culture Collection Center (CGMCC) called CGMCC1347 (Figure 7.43), using 24 DESs and 21 NADESs as co-solvents in varying amounts (Yang et al. 2017). The exact nature of the enzymes catalyzing the reaction is not known, but suspected to be monooxygenases (Hua et al. 2007). The DESs used were either ChCl- or ChAc-based with various HBDs (Gly, U, EG, and Acet) with molar ratios of 1:2, 1:2, or 2:1, while the NADESs were composed of choline chloride and the HBDs LA, MA, OA, maleic acid, malic acid, citric acid, tartaric acid, EG, Gly, propylene glycol, xylitol, sorbitol, xylose, glucose, fructose, mannose, galactose, sucrose, maltose, lactose, or raffinose, in varying proportion (ranging from 1:1 to 11:2) (Yang et al. 2017).

Both the DESs and NADESs improved the conversion for the oxidative cleavage of isoeugenol to vanillin by 142% and 132%, respectively, with respect to the reactions in buffer only. In particular, the ChAc-based DESs gave higher yields than ChCl-based DESs, which was explained by their greater influence on cellular membranes. The NADESs added with 20% v/v could improve the yield, with ChCl:lactose (4:1) and ChCl:raffinose (11:2) giving the highest conversions of 132% and 131%, respectively, with respect to the activity in buffer only. The choice of the HBD affected the yield with sugars > alcohols > organic acids. The use of organic acids also lowered the pH of the mixture, thus exhibiting lower conversions with ChCl:OA (1:1), for example, also confirmed when using other whole-cells (Xu et al. 2015b, 2016; Bubalo et al. 2015a). When the cells were immobilized in poly(vinyl alcohol)-alginate beads, increased conversion with DESs and NADESs was observed, with best results using ChAc-based DESs, with activity retained for at least 13 cycles showing good operational stability. Therefore, the whole-cell-catalyzed oxidative alkene cleavage benefitted from the addition of (NA)DESs for increased yields in the production of vanillin, a world-wide valuable compound.

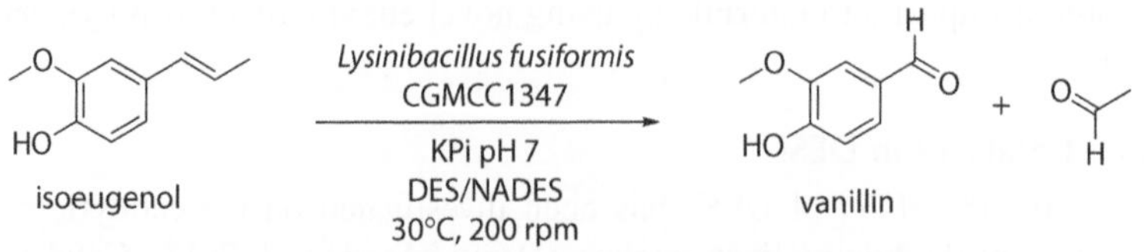

FIGURE 7.43 Whole-cell catalyzed oxidative cleavage of isoeugenol to vanillin. (Adapted from Yang, T.X. et al. *ACS Sustain. Chem. Eng.*, 5, 5713–5722, 2017.)

7.5　CONCLUSIONS

ILs have been widely used in biocatalysis, from proof-of-concept to direct applications, their use now being state-of-the-art. The aim of this chapter was to provide an overview of how DESs are currently emerging as an alternative medium or co-solvents for various enzyme classes to develop more sustainable processes. These recently discovered neoteric solvents are simple and cost-effective to produce, retain low volatility and flammability, and are less toxic than classical ILs.

The addition of DESs to enzymatic reactions leads to interesting properties and outcomes. Epoxide hydrolases and oxidoreductases showed in general higher enantioselectivities, haloalkane dehalogenase displayed increased performance, and the activity of lyases was improved in DES-water mixtures. We expect further developments in the near future with respect to the use of DESs as promising (co)solvents with oxidoreductases and more direct industrial applications with hydrolases.

REFERENCES

Abbott, A. P., Boothby, D., Capper, G., Davies, D. L., and Rasheed, R. K. 2004. Deep eutectic solvents formed between choline chloride and carboxylic acids: Versatile alternatives to ionic liquids. *J. Am. Chem. Soc.* 126:9142–9147.

Andler, S. M., Wang, L.-S., Rotello, V. M., and Goddard, J. M. 2017. Influence of hierarchical interfacial assembly on lipase stability and performance in deep eutectic solvent. *J. Agric. Food Chem.* 65:1907–1914.

Bangde, P. S., Jain, R., and Dandekar, P. 2016. Alternative approach to synthesize methylated chitosan using deep eutectic solvents, biocatalyst and "green" methylating agents. *ACS Sustain. Chem. Eng.* 4:3552–3557.

Borse, B. N., Borude, V. S., and Shukla, S. R. 2012. Synthesis of novel dihydropyrimidin-2(1H)-ones derivatives using lipase and their antimicrobial activity. *Curr. Chem. Lett.* 1:59–68.

Bubalo, M. C., Mazur, M., Radošević, K., and Redovnikovic, I. R. 2015a. Baker's yeast-mediated asymmetric reduction of ethyl 3-oxobutanoate in deep eutectic solvents. *Process Biochem.* 50:1788–1792.

Bubalo, M. C., Tušek, A. J., Vinkovic, M., Radošević, K., Srcek, V. G., and Redovniković, I. R. 2015b. Cholinium-based deep eutectic solvents and ionic liquids for lipase-catalyzed synthesis of butyl acetate. *J. Mol. Catal. B: Enzym.* 122:188–198.

Busto, E., Gotor-Fernández, V., and Gotor, V. 2010. Hydrolases: Catalytically promiscuous enzymes for non-conventional reactions in organic synthesis. *Chem. Soc. Rev.* 39:4504–4523.

Cao, S.-L., Xu, H., Li, X.-H., Lou, W.-Y., and Zong, M.-H. 2015. Papain@magnetic nanocrystalline cellulose nanobiocatalyst: A highly efficient biocatalyst for dipeptide biosynthesis in deep eutectic solvents. *ACS Sustain. Chem. Eng.* 3:1589–1599.

Cao, S.-L., Yue, D.-M., Li, X.-H. et al. 2016. Novel ano-/micro-biocatalyst: Soybean epoxide hydrolase immobilized on UiO-66-NH$_2$ MOF for efficient biosynthesis of enantiopure (R)-1,2-octanediol in deep eutectic solvents. *ACS Sustain. Chem. Eng.* 4:3586–3595.

Cao, S. L., Deng, X., Xu, P. et al. 2017. Highly efficient enzymatic acylation of dihydromyricetin by the immobilized lipase with deep eutectic solvents as cosolvent. *J. Agric. Food. Chem.* 65:2084–2088.

Cheng, Q.-B. and Zhang, L.-W. 2017. Highly efficient enzymatic preparation of daidzein in deep eutectic solvents. *Molecules* 22:186.

Choi, Y. H., van Spronsen, J., Dai, Y. T. et al. 2011. Are natural deep eutectic solvents the missing link in understanding cellular metabolism and physiology? *Plant Physiol.* 156:1701–1705.

Dai, Y., Huan, B., Zhang, H. S., and He, Y. C. 2017. Effective biotransformation of ethyl 4-chloro-3-oxobutanoate into ethyl (*S*)-4-chloro-3-hydroxybutanoate by recombinant *E. coli* CCZU-T15 whole cells in [ChCl][Gly]-water media. *Appl. Biochem. Biotechnol.* 181:1347–1359.

Daneshjou, S., Khodaverdian, S., Dabirmanesh, B. et al. 2017. Improvement of chondroitinases ABCI stability in natural deep eutectic solvents. *J. Mol. Liq.* 227:21–25.

Domínguez de María, P. and Hollmann, F. 2015. On the (un)greenness of Biocatalysis: Some challenging figures and some promising options. *Front. Microbiol.* 6:1257.

Domínguez de María, P. and Maugeri, Z. 2011. Ionic liquids in biotransformations: From proof-of-concept to emerging deep-eutectic-solvents. *Curr. Opin. Chem. Biol.* 15:220–225.

Donnelly, J., Müller, C. R., Wiermans, L., Chuck, C. J., and Domínguez de María, P. 2015. Upgrading biogenic furans: Blended C_{10}–C_{12} platform chemicals via lyase-catalyzed carboligations and formation of novel C_{12} – choline chloride-based deep-eutectic-solvents. *Green Chem.* 17:2714–2718.

Durand, E., Lecomte, J., Baréa, B., Dubreucq, E., Lortie, R., and Villeneuve, P. 2013a. Evaluation of deep eutectic solvent-water binary mixtures for lipase-catalyzed lipophilization of phenolic acids. *Green Chem.* 15:2275–2282.

Durand, E., Lecomte, J., Baréa, B., Piombo, G., Dubreucq, E., and Villeneuve, P. 2012. Evaluation of deep eutectic solvents as new media for *Candida antarctica* B lipase catalyzed reactions. *Process Biochem.* 47:2081–2089.

Durand, E., Lecomte, J., and Villeneuve, P. 2013b. Deep eutectic solvents: Synthesis, application, and focus on lipase-catalyzed reactions. *Eur. J. Lipid Sci. Technol.* 115:379–385.

Fernández-Álvaro, E., Esquivias, J., Pérez-Sánchez, M., Domínguez de María, P., and Remuiñán-Blanco, M. J. 2014. Assessing biocatalysis for the synthesis of optically active tetrahydropyrazolo[1,5-α]pyrimidines (THPPs) as novel therapeutic agents. *J. Mol. Catal. B: Enzym.* 100:1–6.

González-Martínez, D., Gotor, V., and Gotor-Fernández, V. 2016. Application of deep eutectic solvents in promiscuous lipase catalysed aldol reactions. *Eur. J. Org. Chem.*:1513–1519.

Gorke, J. T., Srienc, F., and Kazlauskas, R. J. 2008. Hydrolase-catalyzed biotransformations in deep eutectic solvents. *Chem. Commun.* 1235–1237.

Gotor-Fernández, V., Busto, E., and Gotor, V. 2006. *Candida antarctica* lipase B: An ideal biocatalyst for the preparation of nitrogenated organic compounds. *Adv. Synth. Catal.* 348:797–812.

Guajardo, N., Domínguez de María, P., Ahumada, K. et al. 2017. Water as cosolvent: Nonviscous deep eutectic solvents for efficient lipase-catalyzed esterifications. *ChemCatChem* 9:1393–1396.

Guajardo, N., Müller, C. R., Schrebler, R., Carlesi, C., and Domínguez de María, P. 2016. Deep eutectic solvents for organocatalysis, biotransformations, and multistep organo-catalyst/enzyme combinations. *ChemCatChem* 8:1020–1027.

Gutiérrez-Arnillas, E., Álvarez, M. S., Deive, F. J., Rodríguez, A., and Sanromán, M. Á. 2016. New horizons in the enzymatic production of biodiesel using neoteric solvents. *Renew. Energ.* 98:92–100.

Harifi-Mood, A. R., Ghobadi, R., and Divsalar, A. 2017. The effect of deep eutectic solvents on catalytic function and structure of bovine liver catalase. *Int. J. Biol. Macromol.* 95:115–120.

Hayyan, M., Hashim, M. A., Al-Saadi, M. A., Hayyan, A., AlNashef, I. M., and Mirghani, M. E. S. 2013a. Assessment of cytotoxicity and toxicity for phosphonium-based deep eutectic solvents. *Chemosphere* 93:455–459.

Hayyan, M., Hashim, M. A., Hayyan, A. et al. 2013b. Are deep eutectic solvents benign or toxic? *Chemosphere* 90:2193–2195.

Hayyan, M., Mbous, Y. P., Looi, C. Y. et al. 2016. Natural deep eutectic solvents: Cytotoxic profile. *Springerplus* 5:913.

Hua, D., Ma, C., Lin, S. et al. 2007. Biotransformation of isoeugenol to vanillin by a newly isolated Bacillus pumilus strain: Identification of major metabolites. *J. Biotechnol.* 130:463–470.

Huang, Z.-L., Wu, B.-P., Wen, Q., Yang, T.-X., and Yang, Z. 2014. Deep eutectic solvents can be viable enzyme activators and stabilizers. *J. Chem. Technol. Biotechnol.* 89:1975–1981.

Juneidi, I., Hayyan, M., Hashim, M. A., and Hayyan, A. 2017. Pure and aqueous deep eutectic solvents for a lipase-catalysed hydrolysis reaction. *Biochem. Eng. J.* 117:129–138.

Kleiner, B., Fleischer, P., and Schörken, U. 2016. Biocatalytic synthesis of biodiesel utilizing deep eutectic solvents: A two-step-one-pot approach with free lipases suitable for acidic and used oil processing. *Process Biochem.* 51:1808–1816.

Kleiner, B. and Schörken, U. 2015. Native lipase dissolved in hydrophilic green solvents: A versatile 2-phase reaction system for high yield ester synthesis. *Eur. J. Lipid. Sci. Technol.* 117:167–177.

Krystof, M., Pérez-Sánchez, M., and Domínguez de María, P. 2013. Lipase-catalyzed (trans) esterification of 5-hydroxy-methylfurfural and separation from HMF esters using deep-eutectic solvents. *ChemSusChem* 6:630–634.

Lan, D., Wang, X., Zhou, P., Hollmann, F., and Wang, Y. 2017. Deep eutectic solvents as performance additives in biphasic reactions. *RSC Adv.* 7:40367–40370.

Lehmann, C., Bocola, M., Streit, W. R., Martinez, R., and Schwaneberg, U. 2014. Ionic liquid and deep eutectic solvent-activated CelA2 variants generated by directed evolution. *Appl. Microbiol. Biotechnol.* 98:5775–5785.

Lehmann, C., Sibilla, F., Maugeri, Z. et al. 2012. Reengineering CelA2 cellulase for hydrolysis in aqueous solutions of deep eutectic solvents and concentrated seawater. *Green Chem.* 14:2719–2726.

Lindberg, D., Revenga, M. D., and Widersten, M. 2010. Deep eutectic solvents (DESs) are viable cosolvents for enzyme-catalyzed epoxide hydrolysis. *J. Biotechnol.* 147:169–171.

López-Iglesias, M. and Gotor-Fernández, V. 2015. Recent advances in biocatalytic promiscuity: Hydrolase-catalyzed reactions for nonconventional transformations. *Chem. Rec.* 15:743–759.

Mao, S. H., Yu, L., Ji, S. X., Liu, X. G., and Lu, F. P. 2016. Evaluation of deep eutectic solvents as co-solvent for steroids 1-en-dehydrogenation biotransformation by *Arthrobacter simplex. J. Chem. Technol. Biotechnol.* 91:1099–1104.

Maugeri, Z. and Domínguez de María, P. 2014a. Benzaldehyde lyase (BAL)-catalyzed enantioselective C-C bond formation in deep-eutectic-solvents-buffer mixtures. *J. Mol. Catal. B: Enzym.* 107:120–123.

Maugeri, Z. and Domínguez de María, P. 2014b. Whole-cell biocatalysis in deep-eutectic-solvents/aqueous mixtures. *ChemCatChem* 6:1535–1537.

Maugeri, Z., Leitner, W., and Domínguez de María, P. 2012. Practical separation of alcohol-ester mixtures using deep-eutectic-solvents. *Tetrahedron Lett.* 53:6968–6971.

Maugeri, Z., Leitner, W., and Domínguez de María, P. 2013. Chymotrypsin-catalyzed peptide synthesis in deep eutectic solvents. *Eur. J. Org. Chem.* 2013:4223–4228.

Müller, C. R., Lavandera, I., Gotor-Fernández, V., and Domínguez de María, P. 2015. Performance of recombinant-whole-cell-catalyzed reductions in deep-eutectic-solvent-aqueous-media mixtures. *ChemCatChem* 7:2654–2659.

Müller, C. R., Meiners, I., and Domínguez de María, P. 2014. Highly enantioselective tandem enzyme-organocatalyst crossed aldol reactions with acetaldehyde in deep-eutectic-solvents. *RSC Adv.* 4:46097–46101.

Müller, C. R., Rosen, A., and Domínguez de María, P. 2015. Multi-step enzyme-organocata-lyst C–C bond forming reactions in deep-eutectic-solvents: towards improved performances by organocatalyst design. *Sustain. Chem. Proc.* 3:12.

Paiva, A., Craveiro, R., Aroso, I., Martins, M., Reis, R. L., and Duarte, A. R. C. 2014. Natural deep eutectic solvents—Solvents for the 21st century. *ACS Sustain. Chem. Eng.* 2:1063–1071.

Papadopoulou, A. A., Efstathiadou, E., Patila, M., Polydera, A. C., and Stamatis, H. 2016. Deep eutectic solvents as media for peroxidation reactions catalyzed by heme-dependent biocatalysts. *Ind. Eng. Chem. Res.* 55:5145–5151.

Paul, C. E., Lavandera, I., Gotor-Fernández, V., and Gotor, V. 2014. Imidazolium-based ionic liquids as non-conventional media for alcohol dehydrogenase-catalysed reactions. *Top. Catal.* 57:332–338.

Pena-Pereira, F. and Namieśnik, J. 2014. Ionic liquids and deep eutectic mixtures: Sustainable solvents for extraction processes. *ChemSusChem* 7:1784–1800.

Petrenz, A., Domínguez de María, P. D., Ramanathan, A., Hanefeld, U., Ansorge-Schumacher, M. B., and Kara, S. 2015. Medium and reaction engineering for the establishment of a chemo-enzymatic dynamic kinetic resolution of *rac*-benzoin in batch and continuous mode. *J. Mol. Catal. B: Enzym.* 114:42–49.

Pöhnlein, M., Ulrich, J., Kirschhöfer, F. et al. 2015. Lipase-catalyzed synthesis of glucose-6-*O*-hexanoate in deep eutectic solvents. *Eur. J. Lipid. Sci. Technol.* 117:161–166.

Qin, Y.-Z., Zong, M.-H., Lou, W.-Y., and Li, N. 2016. Biocatalytic upgrading of 5-hydroxy-methylfurfural (HMF) with levulinic acid to HMF levulinate in biomass-derived solvents. *ACS Sustain. Chem. Eng.* 4:4050–4054.

Radošević, K., Bubalo, M. C., Srček, V. G., Grgas, D., Dragičević, T. L., and Redovniković, I. R. 2015. Evaluation of toxicity and biodegradability of choline chloride based deep eutectic solvents. *Ecotox. Environ. Safe.* 112:46–53.

Ranganathan, S., Zeitlhofer, S., and Sieber, V. 2017. Development of a lipase-mediated epoxidation process for monoterpenes in choline chloride-based deep eutectic solvents. *Green Chem.* 19:2576–2586.

Sanap, A. K. and Shankarling, G. S. 2014. Eco-friendly and recyclable media for rapid synthesis of tricyanovinylated aromatics using biocatalyst and deep eutectic solvent. *Catal. Commun.* 49:58–62.

Sánchez-Leija, R. J., Torres-Lubián, J. R., Reséndiz-Rubio, A., Luna-Bárcenas, G., and Mota-Morales, J. D. 2016. Enzyme-mediated free radical polymerization of acrylamide in deep eutectic solvents. *RSC Adv.* 6:13072–13079.

Sheldon, R. A. 2017. The E factor 25 years on: The rise of green chemistry and sustainability. *Green Chem.* 19:18–43.

Smith, E. L., Abbott, A. P., and Ryder, K. S. 2014. Deep eutectic solvents (DESs) and their applications. *Chem. Rev.* 114:11060–11082.

Stepankova, V., Vanacek, P., Damborsky, J., and Chaloupkova, R. 2014. Comparison of catalysis by haloalkane dehalogenases in aqueous solutions of deep eutectic and organic solvents. *Green Chem.* 16:2754–2761.

Tian, X., Zhang, S., and Zheng, L. 2016. Enzyme-catalyzed Henry reaction in choline chloride-based deep eutectic solvents. *J. Microbiol. Biotechnol.* 26:80–88.

Ülger, C. and Takaç, S. 2017. Kinetics of lipase-catalysed methyl gallate production in the presence of deep eutectic solvent. *Biocatal. Biotransformation* 35:407–416.

Vitale, P., Abbinante, V. M., Perna, F. M., Salomone, A., Cardellicchio, and C., Capriati, V. 2017. Unveiling the hidden performance of whole cells in the asymmetric bioreduction of aryl-containing ketones in aqueous deep eutectic solvents. *Adv. Synth. Catal.* 359:1049–1057.

Wagle, D. V., Zhao, H., Baker, and G. A. 2014. Deep eutectic solvents: Sustainable media for nanoscale and functional materials. *Acc. Chem. Res.* 47:2299–2308.

Wang, X.-P., Zhou, P.-F., Li, Z.-G., Yang, B., Hollmann, F., and Wang, Y.-H. 2017. Engineering a lipase B from *Candida antactica* with efficient perhydrolysis performance by eliminating its hydrolase activity. *Sci. Rep.* 7:44599.

Wei, P., Liang, J., Cheng, J., Zong, M. H., and Lou, W. Y. 2016. Markedly improving asymmetric oxidation of 1-(4-methoxyphenyl) ethanol with *Acetobacter* sp. CCTCC M209061 cells by adding deep eutectic solvent in a two-phase system. *Microb. Cell Fact.* 15:5.

Weiz, G., Braun, L., Lopez, R., Domínguez de María, P., and Breccia, J. D. 2016. Enzymatic deglycosylation of flavonoids in deep eutectic solvents-aqueous mixtures: Paving the way for sustainable flavonoid chemistry. *J. Mol. Catal. B: Enzym.* 130:70–73.

Wu, B. P., Wen, Q., Xu, H., and Yang, Z. 2014. Insights into the impact of deep eutectic solvents on horseradish peroxidase: Activity, stability and structure. *J. Mol. Catal. B: Enzym.* 101:101–107.

Xu, P., Cheng, J., Lou, W. Y., and Zong, M. H. 2015a. Using deep eutectic solvents to improve the resolution of racemic 1-(4-methoxyphenyl)ethanol through *Acetobacter* sp. CCTCC M209061 cell-mediated asymmetric oxidation. *RSC Adv.* 5:6357–6364.

Xu, P., Du, P.-X., Zong, M.-H., Li, N., and Lou, W.-Y. 2016. Combination of deep eutectic solvent and ionic liquid to improve biocatalytic reduction of 2-octanone with *Acetobacter pasteurianus* GIM1.158 cell. *Sci. Rep.* 6:26158.

Xu, P., Xu, Y., Li, X.-F., Zhao, B.-Y., Zong, M.-H., and Lou, W.-Y. 2015b. Enhancing asymmetric reduction of 3-chloropropiophenone with immobilized *Acetobacter* sp. CCTCC M209061 cells by using deep eutectic solvents as cosolvents. *ACS Sustain. Chem. Eng.* 3:718–724.

Xu, P., Zheng, G.-W., Zong, M.-H., Li, N., and Lou, W.-Y. 2017. Recent progress on deep eutectic solvents in biocatalysis. *Bioresour. Bioprocess.* 4:34.

Yang, S. L. and Duan, Z. Q. 2016. Insight into enzymatic synthesis of phosphatidylserine in deep eutectic solvents. *Catal. Commun.* 82:16–19.

Yang, T. X., Zhao, L. Q., Wang, J. et al. 2017. Improving whole-cell biocatalysis by addition of deep eutectic solvents and natural deep eutectic solvents. *ACS Sustain. Chem. Eng.* 5:5713–5722.

Zeng, C.-X., Qi, S.-J., Xin, R.-P., Yang, B., and Wang, Y.-H. 2015. Enzymatic selective synthesis of 1,3-DAG based on deep eutectic solvent acting as substrate and solvent. *Bioprocess Biosyst. Eng.* 38:2053–2061.

Zhang, Q. H., Vigier, K. D., Royer, S., and Jerome, F. 2012. Deep eutectic solvents: Syntheses, properties and applications. *Chem. Soc. Rev.* 41:7108–7146.

Zhang, Y., Xia, X., Duan, M. et al. 2016. Green deep eutectic solvent assisted enzymatic preparation of biodiesel from yellow horn seed oil with microwave irradiation. *J. Mol. Catal. B: Enzym.* 123:35–40.

Zhao, H., Baker, G. A., and Holmes, S. 2011a. New eutectic ionic liquids for lipase activation and enzymatic preparation of biodiesel. *Org. Biomol. Chem.* 9:1908–1916.

Zhao, H., Baker, G. A., and Holmes, S. 2011b. Protease activation in glycerol-based deep eutectic solvents. *J. Mol. Catal. B: Enzym.* 72:163–167.

Zhao, H., Zhang, C., and Crittle, T. D. 2013. Choline-based deep eutectic solvents for enzymatic preparation of biodiesel from soybean oil. *J. Mol. Catal. B: Enzym.* 85–86:243–247.

Zhou, P., Wang, X., Yang, B., Hollmann, F., and Wang, Y. 2017a. Chemoenzymatic epoxidation of alkenes with *Candida antarctica* lipase B and hydrogen peroxide in deep eutectic solvents. *RSC Adv.* 7:12518–12523.

Zhou, P., Wang, X., Zeng, C. et al. 2017b. Deep eutectic solvents enable more robust chemoenzymatic epoxidation reactions. *ChemCatChem* 9:934–936.

8 Ionic Liquids in Sustainable Carbohydrate Catalysis

Pilar Hoyos, Cecilia García-Oliva,
and María J. Hernáiz

CONTENTS

8.1 INTRODUCTION

There is considerable current interest in the utilization of carbohydrates as readily available, relatively inexpensive, and renewable feedstocks for the chemical, pharmaceutical, cosmetic, detergent, and food industries. While these compounds are mainly produced by chemical methods, the use of sustainable methods has been investigated over the past 20 years as a greener alternative to organic synthesis. Due to the low solubility of underivatized carbohydrates in traditional organic solvents, research has focused on the chemical and enzymatic synthesis of carbohydrates in polar green solvents, such as water and ionic liquids (ILs) (Farran et al., 2015).

It is necessary to define the concept of green chemistry, and the principles that govern it, in order to adapt the chemistry of carbohydrates to sustainable production and processing. A definition of green chemistry was proposed by Paul Anastas and John Warner in 1998 as the design of chemical products and processes that reduce or eliminate the use or generation of hazardous substances. Green chemistry has 12 principles that may be summarized as follows:

1. *Prevention*: Chemistry must avoid the production of toxic and hazardous waste, rather than to remove these wastes after they are formed.
2. *Atom economy*: In a synthesis, all components used should be incorporated to the maximum extent possible into a final product.
3. *Less hazardous chemical syntheses*: Wherever is possible, chemical synthesis should be designed to use and generate substances with low toxicity and with a low environmental impact.
4. *Designing safer chemicals*: Chemical products should be designed to exhibit their desired function with a minimum level of toxicity.
5. *Safer solvents and auxiliaries*: Auxiliary substances (e.g., solvents) should be avoided if possible and should be innocuous when used.
6. *Design for energy efficiency*: Energy requirements of chemical processes should be minimized to reduce their environmental impact and, if possible, should be performed at ambient temperature and pressure.
7. *Use of renewable feedstocks*: Raw materials or feedstock should be renewable.
8. *Reduce derivatives*: Whenever possible, unnecessary derivatization (e.g., protection/deprotection) should be reduced or avoided.
9. *Catalysis*: It is best to employ catalytic reagents (as selective as possible), rather than stoichiometric reagents.
10. *Design for degradation*: Chemical products should be designed so that at the end of their functional lifetime, they undergo an innocuous degradation process and do not persist in the environment.
11. *Real-time analysis for pollution prevention*: Analytical methodologies should be developed to allow for real-time analysis without the formation of hazardous substances.
12. *Inherently safer chemistry for accident prevention*: Substances involved in a chemical process should be chosen to minimize the potential for chemical accidents.

Traditional carbohydrate synthesis and modification frequently involves multiple protection and deprotection steps, the use of hazardous and harmful chemicals and solvents, and other harsh conditions that adversely impact the environment and human health (Chang et al., 2009; Crich and Li, 2007; Dasgupta and Nitz, 2011; Jing and Huang, 2004; Wang et al., 2007). There is a need to find new ways to produce carbohydrate-based products using more environmental friendly conditions. Green chemistry offers the tools to build a sustainable industrial and research effort. Renewable feedstocks, biocatalyzed reactions carried out under mild conditions, at room temperature, in water or other green solvents, and with high atom efficiency are among the strategies employed to achieve such goals.

ILs are currently the focus of increasing attention as reaction media for carbohydrate synthesis in general and enzymatic transformations in particular (Farran et al., 2015). Their potential as media for carbohydrate transformations was first pointed out by use and subsequently confirmed by Park and Kazkauskas (2001), who described the lipase-catalyzed acylation of glucose and maltose in ILs.

The term IL is now commonly accepted for salts with a melting point temperature below 100°C. They generally have negligible vapor pressure, high thermal and chemical stability, low volatility, are capable of dissolving both polar and non-polar compounds, and present tunable physical properties. Therefore, they are considered to be potential environmentally friendly alternatives to conventional volatile organic solvents. One of this impressive features is the wide variation in their physicochemical properties, including viscosity, polarity, density, melting point, and solubility. These properties are related to their ionic structures, which provide numerous possibilities to modify cation and anion structures for specific properties.

ILs are composed of a large organic cation (heterocycle, such as pyridine or imidazole) and organic and inorganic anion (halide, organic acid, isocyanate,...). The most commonly used ILs in carbohydrate chemistry, their abbreviated names, and their water solubilities are presented in Table 8.1.

These ionic solvents have the potential of replacing conventional volatile organic solvents in carbohydrate chemistry. At the same time, the number of studies on the toxicity, biodegradability, reusability, and environmental properties of ILs has increased (Capello et al., 2007; Coleman and Gathergood, 2010).

The first report of an IL, [Et$_3$N][NO$_3$], goes back 90 years, but its explosive nature precluded its widespread application. In the 1930s, an early patent application described cellulose dissolution by use of a molten pyridinium salt above 130°C (Graenacher, 1934). However, the use of ILs in chemical synthesis began in the 1990s (Seddon, 1997), affording first-generation ILs with unique physical-chemical properties, such as thermal and chemical stability, negligible volatility, flame retardancy, moderate viscosity, high polarity, low melting points, and high ionic conductivity. ILs that can dissolve many compounds without volatility, are thermally stable, and can be fine-tuned to show specific targeted behavior for an application are referred to as second-generation ILs.

Currently, there is a distinct trend toward the rational design of task-specific ILs, with small environmental footprints, that can be used for chemical or biocatalytic reactions (third-generation ILs).

Another green solvent characteristic of ILs is their reusability for the same chemical process repeatedly. Reaction products made in ILs can often be recovered by distillation from the non-volatile IL or extracted with water or hydrocarbon solvents that are IL-immiscible.

One of the major drawbacks in underivatized carbohydrate transformations is the low solubility of these high polar substrates in the conventional organic solvents. For this reason, polar solvents such as ILs, compromising the sustainability of the process, appear as greener alternative solvent systems, which can act not only as solvent media, but also as catalyst in different functionalization of carbohydrate processes (Farran et al., 2015).

On the other hand, ILs have also shown to be environmentally attractive reaction media for biocatalyzed carbohydrate transformation (Sheldon, 2016). Enzymes are valuable tools which allow the development of carbohydrate transformations under mild reaction conditions with high selectivity. In this respect, as it will be discussed

TABLE 8.1
Nomenclature of the Most Commonly Used ILs and Their Miscibility with Water

Cation	Anion	Common Notation	Water Miscibility
1-Butyl-3-methylimidazolium	Tetrafluoroborate	[BMIM][BF$_4$]	Yes
	Hexafluorophosphate	[BMIM][PF$_6$]	No
	Bis[(trifluoromethyl)sulfonyl] imide	[BMIM][Tf$_2$N]	—
	Trifluoromethanesulfonate	[BMIM][Tfms]	Yes
	Glycolate	[BMIM][OHCH$_2$COO]	—
	Octylsulfate	[BMIM][OctSO$_4$]	Yes
1-Butyl-2,3-dimethylimidazolium	Tetrafluoroborate	[BDMIM][BF$_4$]	Yes
	Hexafluorophosphate	[BDMIM][PF$_6$]	No
	Trifluoromethanesulfonate	[BDIM][Tfms]	
1-Butyl-4-methylpyridinium	Tetrafluoroborate	[BMP][BF$_4$]	Yes
1-Ethyl-3-methylimidazolium	Tetrafluoroborate	[EMIM][BF$_4$]	Yes
	Bis[(trifluoromethyl)sulfonyl] imide	[EMIM][Tf$_2$N]	No
	Trifluoromethanesulfonate	[EMIM][Tfms]	Yes
1-Ethylpyridinium	Tetrafluoroborate	[ETPY][BF$_4$]	—
	Trifluoroacetate	[ETPY][CF$_3$COO]	—
1-Butylpyridinium	Bis[(trifluoromethyl)sulfonyl] imide	[BUPY][Tf$_2$N]	—
1-Butyl-1-methylpyrrolidinium	Bis[(trifluoromethyl)sulfonyl] imide	[BMPY][Tf$_2$N]	Partly miscible
1-Hexyl-2,3-dimethylimidazolium	Tetrafluoroborate	[HDMIM][BF$_4$]	Partly miscible
1-Hexyl-3-methylimidazolium	Tetrafluoroborate	[HMIM][BF$_4$]	Partly miscible
	Hexafluorophosphate	[HMIM][PF$_6$]	No
1-(3-Hydroxypropyl)-3-methylimidazolium	Hexafluorophosphate	[HPMIM][PF$_6$]	
	Glycolate	[HPMIM][OHCH$_2$COO]	
	Chloride	[HPMIM][Cl]	
1,3-Dimethylimidazolium	Methylsulfate	[MMIM][MeSO$_4$]	Yes
3-Methyl-3-nonylimidazolium	Hexafluorophosphate	[MNIM][PF$_6$]	No
1-Methoxyethyl-3-methylimidazolium	Tetrafluoroborate	[MOEMIM][BF$_4$]	—
	Hexafluorophosphate	[MOEMIM][PF$_6$]	
1-Octyl-3-methylimidazolium	Tetrafluoroborate	[OMIM][BF$_4$]	Partly miscible

(*Continued*)

TABLE 8.1 (*Continued*)
Nomenclature of the Most Commonly Used ILs and Their Miscibility with Water

Cation	Anion	Common Notation	Water Miscibility
1-Octyl-3-nonylimidazolium	Hexafluorophosphate	[OMIM][PF$_6$]	No
1-Octyl-3-nonylimidazolium	Hexafluorophosphate	[ONIM][PF$_6$]	No
Methyltrioctylammonium	Bis[(trifluoromethyl)sulfonyl] imide	[MTOA][Tf$_2$N]	No
Butyltrimethylammonium	Bis[(trifluoromethyl)sulfonyl] imide	[BTMA][Tf$_2$N]	No
3-Hydroxypropyl-trimethylammonium	Bis[(trifluoromethyl)sulfonyl] imide	[HTMA][Tf$_2$N]	—
3-Cyanopropyl-trimethylammonium	Bis[(trifluoromethyl)sulfonyl] imide	[CPRTMATf$_2$N]	—
Butyltrimethylammonium	Bis[(trifluoromethyl)sulfonyl] imide	[BTMA][Tf$_2$N]	—
5-Cyanopentyl-trimethylammonium	Bis[(trifluoromethyl)sulfonyl] imide	[CPTMA][Tf$_2$N]	—
Hexyltrimethylammonium	Bis[(trifluoromethyl)sulfonylimide	[HTMA][Tf$_2$N]	—

later, ILs have been established as valuable reaction media for many chemical or enzyme-catalyzed processes (Itoh, 2017), and promising results have been reached in the field of carbohydrate transformations by the employ of biotransformation in ILs (García et al., 2017).

Carbohydrates are highly polar molecules that are generally soluble in water, strong acids, or organic solvents capable of forming hydrogen bonds, such as dimethyl sulfoxide, dimethylformamide, pyridine, or 2-methylpropanol. With the exception of water, these solvents have many undesirable properties, and there is a need to find new, environmentally benign green solvents to dissolve them and conduct their chemical or enzymatic synthesis.

In this chapter, the following two key points will be treated in depth:

- Chemical synthesis of carbohydrates in ionic liquids
- Enzymatic synthesis of carbohydrates in ionic liquids

8.2 IONIC LIQUIDS IN CARBOHYDRATES CHEMISTRY

8.2.1 CHEMICAL SYNTHESIS OF CARBOHYDRATES

Although great efforts have been devoted to design more efficient and greener alternatives for carbohydrate modifications, sustainability is still an open challenge. In this context, the introduction of ILs in sugar chemistry has become one of the most promising research areas, as it has allowed the development of new methodologies for carbohydrate modifications (Galan et al., 2013). Among their applications, it is worth of mention their use as polar solvents to dissolve high molecular weight carbohydrates (as cellulose) for further transformations and as solvent and recyclable catalysts in acetylation/deacetylation of saccharides for building block purposes and glycosylation reactions (Rosatella et al., 2011).

8.2.1.1 Polysaccharide-Based Materials as Substrates

Polysaccharides, such as cellulose, starch, hemicellulose, or chitin are very abundant biopolymers in nature, and they, and their derivatives, are widely used at industry as plastics, paper products, textiles, or pharmaceuticals among others (Liu et al., 2015; Llevot et al., 2016). However, one of the main drawbacks of polysaccharides processing is their low solubility in conventional media, especially cellulose and chitin, and the rigidity caused by intra- and intermolecular hydrogen bonding. In the last decade, the employ of ILs for polysaccharide processing has emerged as a very promising alternative. ILs are able to dissolve cellulose without derivatization, through the establishment of intermolecular interactions under relatively mild conditions. The group of Rogers described in 2002 the dissolution of cellulose in different ILs, such as 1-butyl-3-methylimidazolium chloride [BMIN] [Cl] (Figure 8.1), one of the most employed IL in the sugars field (Swatloski et al., 2002). Authors proposed that chloride containing ILs were the most effective solvent system due to the hydrogen bonding formed between the hydroxyl functions of cellulose and the anions of the solvent, disrupting the hydrogen bond network that keeps the cellulose together. Since then, many imidazolium-based ILs are recognized for their ability to dissolve cellulose, hemicellulose, or chitin. ILs applications have been also extended not only to cellulose dissolution, but also to serve as the reaction medium to functionalize cellulose and to prepare cellulose fibers, films, or beads (Wang et al., 2012b; Zhang et al., 2017). It should be noted the effectiveness in cellulose dissolution of some ILs, such as 1-allyl-3-methylimidazolium chloride [AMIM] [Cl] (Figure 8.1), which was developed as a new substituted imidazolium-based IL (Jiang et al., 2017; Zhang et al., 2005), as well as other non-halogen imidazolium ILs as 1-ethyl-3-methylimidazolium acetate ([EMIM] [AcO]) (Zhang et al., 2005) (Figure 8.1).

In last decade, other non-imidazolium ILs have been proven to dissolve cellulose as ILs based on tetramethylguanidine (King et al., 2011; Parviainen et al., 2013), 1,5-diazabicyclo-[4.3.0]non-5-ene, 1-hexylpyridinium chloride ([HPy] [Cl]) (Uju et al., 2013), and alkylalkyloxyammonium amino acids (Ohira et al., 2012).

Thus, ILs have been employed as the reaction media for the homogeneous functionalization of cellulose, allowing the preparation of a great variety of cellulose materials in an environmentally friendly way. Among them, cellulose esters are one of the most important cellulose derivatives.

Acetylation of cellulose mediated by acetic anhydride was successfully achieved by Wu and co-workers through the employ of [AMIM] [Cl] as reaction media. This

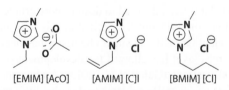

[EMIM] [AcO]   [AMIM] [C]l   [BMIM] [Cl]

FIGURE 8.1   ILs traditionally used in cellulose processing.

strategy allowed different advantages to the traditional acetylation processes, such as a DS-value (degree of substitution) controllable, the employ of recyclable solvent, or the absence of catalysts (Wu et al., 2004). The use of other reactants, such as propionic, butyric, or succinic anhydrides required the addition of 4-dimethylaminiopyridine, affecting the greenness of the process, but obtaining cellulose esters with high DS and soluble in different organic solvents (Luan et al., 2013). Acid chlorides have also been employed as acylating agents in ILs-mediated esterification of polysaccharides. For instance, the group of Meng prepared cellulose dinitrobenzoate, employing [BMIM][Cl] and 3,5-dinitrobenzoic acid chloride, obtaining a functionalized material as an oral adsorbent of creatinine in dialysate (Dai et al., 2015).

Esterification processes are promising ways to make up the properties of other class of polysaccharides for their specific applications. In this context, Wang et al. reported the preparation of lauroylated hemicelluloses in the presence of [BMIM] [Cl] and lauroyl chloride, without catalyst. The reaction mixture was incubated up to 90 min and then cooled and precipitate with ethanol, and the ILs could be further recycled (Wang et al., 2012a). On the other hand, [AMIM] [Cl] has been used as the reaction media for the homogeneous acetylation of chitosan using acetyl chloride, allowing the development of the process under mild conditions and simple work up (Figure 8.2) (Liu et al., 2013).

Cellulose is a linear polymer composed entirely of β-D-(1→4)-glucopyranose, and it is one important component of biomass (Figure 8.3). Pretreatment of biomass is a critical step for its effective utilization as a feedstock for the production of biofuels, specialty chemicals, and materials. Biomass typically consists of highly

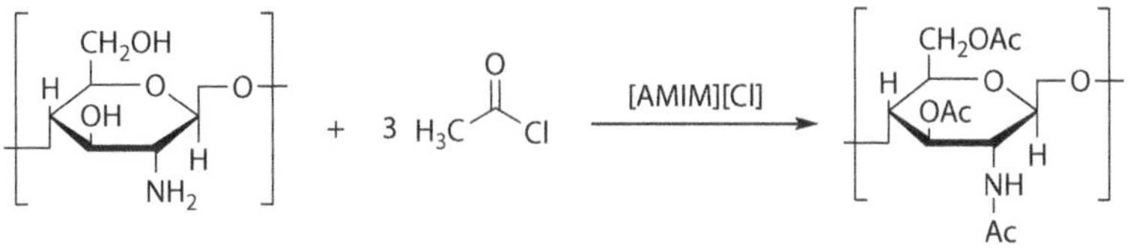

FIGURE 8.2 Acetylation of chitosan in [AMIM] [Cl]. (From Liu, L. et al., *J. Appl. Poly. Sci.*, 129, 28–35, 2013.)

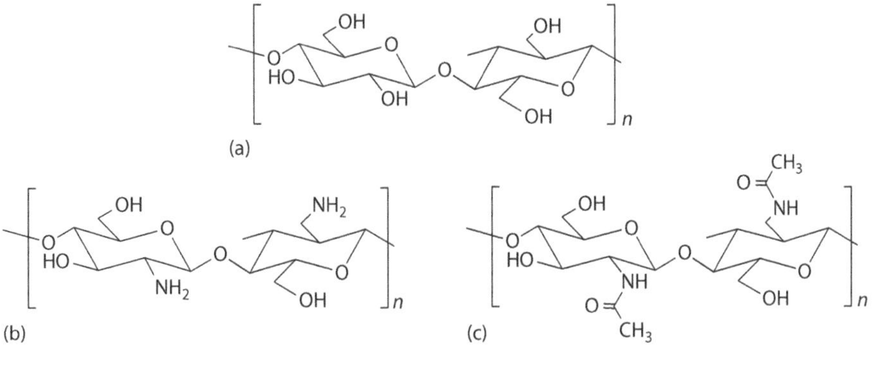

FIGURE 8.3 (a) Cellulose, (b) chitin, and (c) chitosan.

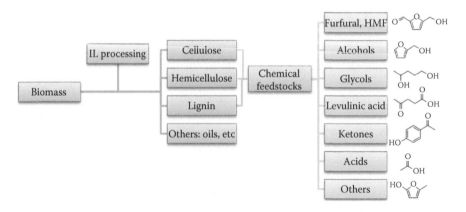

FIGURE 8.4 Biomass processing with ILs, obtaining simple compounds with different applications.

crystalline cellulose, hemicelluose, and xylan and lignin components (Figure 8.4). A major objective in biomass pretreatment is the separation of biomass into its individual components, their recovery, and the reduction of cellulose crystallinity. ILs have been proposed as green solvents for biomass treatment, as they are designer solvents, non-volatile, and should be completely recoverable and recyclable. The first stage of the catalytic cascade of cellulose processing in IL media is the hydrolysis of the natural polymer into oligomer glucans and ultimately into monomer sugar glucose and then can be transformed into a considerable range of chemical building blocks derived from biomass. Some of these are 5-hdroxymethylfurfural, levulinic acid, alcohols, ketones, and others (Bodachivskyi et al., 2018)

Chitin (Peniche et al., 2008; Rinaudo, 2006) is the world's second most important natural polymer and chitosan, partially *N*-acetylated chitin is the most important derivative of chitin. Chitin (Kim and Rajapakse, 2005) and chitosan show various functional properties that make them useful in many fields including food (No et al., 2007), cosmetics (Brode, 1991), biomedicine (Jayakumar et al., 2010), agriculture environmental protection (Badawy and Rabea, 2011), and waste water management (Yamada et al., 2005). However, the applications of chitin and chitosan have some limitations, mainly due to their lack of solubility in water and some organic solvents. Within this scenario, the dissolution of natural polymers, such as chitin and chitosan in ILs has been pointed out as an alternative strategy to enhance the dissolution and processability of these biomacromolecules. Similar to cellulose, these strategies combine to green chemistry principles, namely, the use of environmental solvents (ILs) and biorenewable feedstocks (chitin and chitosan, natural polymers). Many ILs have been reported in the literature with the ability to dissolve chitin and chitosan. Among them, [EMIM] [Ac] stands out as the most suitable IL to dissolve chitin. Chitin requires a more basic anion, such as acetate, due to the increased number of hydrogen bond donors and acceptors (Barber et al., 2013). Prasad et al. showed that the complete dissolution of chitin could also be achieved using [AMIM] [Br], to obtain a flowing solution or an ionic gel depending on chitin concentration (Prasad et al., 2009). The authors suggested that the reason behind chitin dissolution in [AMIM] [Br] can be related to the IL composition, which is a combination of an allyl substituent and a bromide counter anion on the imidazolium. Recent studies have shown that not only the structure of the anion or the cation is important, but also the features of chitin, namely, its polymorphic form, origin, molecular weight, and degree of acetylation (Wang et al., 2010).

In the case of chitosan, different ILs, such as [EMIM] [Cl], [BMIM] [Cl], [EMIM] [Ac], and [BMIM] [Ac] have been studied for their dissolution (Muzzarelli, 2011; Silva et al., 2012; Xie et al., 2006). In most of those studies, the solubility of chitosan in the ILs has been determined as a function of temperature (Chen et al., 2011; Feng et al., 2015).

8.2.1.2 Acetylation/Deacetylation of Carbohydrates

Per-O-acetylation of sugars is a very common and useful reaction to protect hydroxyl groups in multistep synthetic routes and also to facilitate the purification and the structural elucidation of natural products. Traditionally, this process has been performed using acetic anhydride and pyridine, but many other alternatives have been intended for this purpose, mainly based on Lewis acids such as $Sc(OTf)_3$ (Lee et al., 2002) or $(CH)_3SiOTf$ (Procopiou et al., 1998), among many others. On the basis of the properties displayed by ILs, concerning their carbohydrates dissolution capacity and their catalytic behavior, the possibility of the development of per-O-acetylation reactions in different ILs has been explored during the last decades.

Forsyth and co-workers described dicyanamide-based ionic liquids [BMIN] [DCA] and [EMIM] [DCA] (Figure 8.5) not only as effective solvents for saccharides, but also as active base catalysts for their O-acetylation (Forsyth et al., 2002). Different monosaccharides were acetylated under mild reactions conditions, employing acetic anhydride and those ILs as reaction media. Very high yields were achieved in all cases in very short reaction times. The protocol was successfully extended to the acylation of N-acetylneuraminic acid, sucrose, and the trisaccharide raffinose. The $^-N(CN)_2$ counterion seemed to be relevant, as the development of the process in other ILs with different anions, as bis(trifluoromethanesulfonyl)amide, ([BMIM] [Tfms]) failed to afford products.

The group of Linhardt reported the peracetylation and perbenzoylation of simple sugars and sulfated monosaccharides in dialkylimidazolium benzoates as ILs (Murugesan et al., 2003). Per-O-acetylation of α-D-glucose, β-D-glucose, α/β-D-mannose, and α/β-D-galactose was performed in 1-ethyl-3-methylimidazolium benzoate [EMIM] [BA] without the addition of any other catalyst and organic solvent, achieving high yields, ranging from 71% (in the case of α/β-D-galactose) to

[EMIM]   [BMIM]   DCA

FIGURE 8.5 Imidazolium cations and dicyanamide anion. (From Forsyth, S.A. et al., *Chem. Commun.*, 7, 714–715, 2002.)

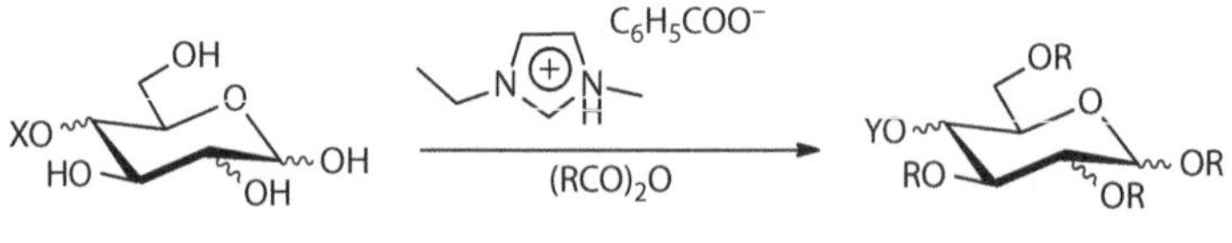

R = CH₃CO, PhCO
X = H or SO₃⁻ Na⁺; Y = H or SO₃⁻ Na⁺

FIGURE 8.6 Per-O-acetylation (or benzoylation) of α-D-glucose, β-D-glucose, α/β-D-mannose, α/β-D-galactose, and 4-O-sulfo-β-D-glucopyranoside. (From Murugesan, S. et al., *Synlett*, 9, 1283–1286, 2003.)

quantitative yields (β-D-glucose). Increasing the IL cation alkyl chain length to butyl ([BMIM] [BA]) or hexyl ([HMIM] [BA]) led to a decrease of the yields, attributed to the increase of the viscosity of the IL caused by the chain length, what would impede the mobility of the reactants. On the other hand, [EMIM] [BA] seemed to be unique for the complete dissolution of phenyl 4-O-sulfo-β-D-glucopyranoside and phenyl 6-O-sulfo-β-D-glucopyranoside, substrates only soluble in formamide and water. Thus, peracetylation of these compounds was carried out affording high yields (Figure 8.6).

Recently, Xiong and co-workers have reported the per-O-acetylation of a great variety of mono/disaccharides and sugar alcohols using acetic anhydride and a recyclable iodine/PEG (poly(ethyleneglycol)$_{400}$-based ionic liquid catalyst (I_2/IL_{400}) (Figure 8.7) under microwave irradiation. I_2 had been previously employed in the synthesis of peracetylated sugars, but its use at larger scales could not be possible without a support for this catalyst. Thus, the system developed by this group allowed the obtaining of excellent yields after just a few minutes and, in addition, the reuse of the I_2 in the following cycles (Xiong et al., 2015). After the reaction, products could be extracted from the reaction media by extractions with toluene, immiscible with IL_{400}, while the catalyst I_2 remained in IL_{400}. This system I_2/IL_{400}, could be recovered and reused at least six times without a significant loss of the catalytic activity. The excellent results conducted to the multigram-scale up synthesis of peracetylated glucopyranose, obtaining 90% yield after six cycles.

As another interesting green alternative, Lafuente and co-workers employed triethylammonium acetate as an efficient solvent and catalyst for the acetylation of alcohols, amines, oximes, and thiols in the presence of acetic anhydride at room temperature (Lafuente et al., 2016). The excellent yields prompted these authors to explore the per-O-acetylation of sugars, obtaining from moderate to high yields

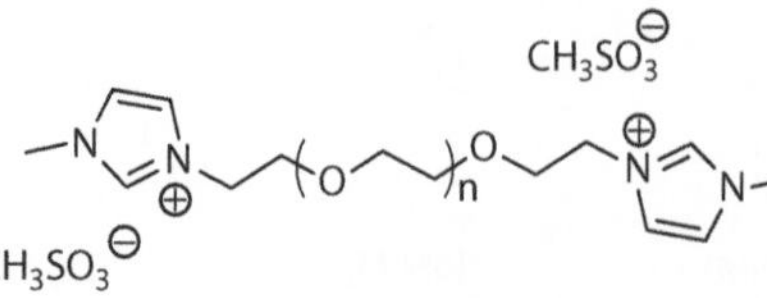

FIGURE 8.7 PEG-based ionic liquid. (From Xiong, X.Q. et al., *Chin. J. Catal.*, 36, 237–243, 2015.)

and high beta anomeric selectivity in the acetylation of glucose and galactose, without the addition of a catalyst. Furthermore, the ILs could be recycled and reused without apparent loss of activity until the fifth cycle, when a slight decrease of activity was detected.

8.2.1.3 Glycosylation Reactions

The formation of the glycosydic bond has been the target of multiple investigations, as there are many factors which influence the efficient synthesis of glycosides. The presence and selection of a leaving group on the anomeric position of the donor, the use of protecting groups on the donor and the acceptor, the choice of a promoter, and the solvent system are crucial for the success of the reaction (Nicolaou and Mitchell, 2001). Improving the regio- and stereoselectivity, as well as the yields, are still an open challenge in carbohydrate chemistry. Glycosilation of unprotected saccharides is a main goal since it implies the reduction of the reaction steps, but different drawbacks have to be saved, due to the poor solubility of carbohydrates in commonly employed solvents. The favorable properties of ILs make them promising reaction media for glycosylation reactions, and their possibilities on this area are being explored, not only as solvents, but also as catalysts of the reactions.

Park and co-workers described the glycosylation of simple, unprotected monosaccharides in the IL [EMIM] [BA]. D-glucose, D-mannose, D-galactose, and N-acetyl-D-galactosamine were used as inactivated donors, affording the correspondent benzyl glycosides with Amberlyst IR-120 (H+) resin in this IL (Park et al., 2007). Thus, a simple synthetic strategy was conducted to benzyl glycosides in fair to good yields of exclusively α-glycosides (Figure 8.8a). The same one step protocol was then explored in the preparation of disaccharides from unprotected donors and partially protected acceptors, employing p-TsOH as promoter. Excellent α-stereoselectivity was also achieved under moderate reaction conditions (Figure 8.8b).

The benzoate based-IL [EMIM] [BA] was also employed by Muñoz et al. in the green preparation of glycosides of D-glucose and N-acetylgalactosamine in a single step reaction, starting with unprotected and inactivated sugars donors (Munoz et al., 2009). The GalNAc derivative, containing a bifunctional linker, was immobilized

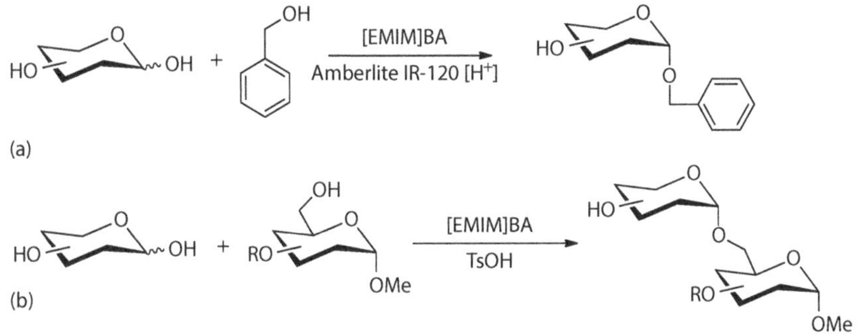

FIGURE 8.8 (a) Glycosylation of unprotected monosaccharides in [EMIM] [BA]. (b) Synthesis of disaccharides from unprotected donors in [EMIM] [BA]. (From Park, T.J. et al., *Carbohyd. Res.*, 342, 614–620, 2007.)

on a gold biosensor chip and SPR (Surface Plasmon Resonance)experiments were conducted to study the bounding of this glycoconjugate to specific lectins.

Galan and co-workers studied the applicability of the IL 1-butyl-3-methylimidazolium triflate [BMIM] [OTf] as co-solvent and promoter of glycosylation reactions of activated thioglycoside and trichloroacetimidate donors (Figure 8.9) at room temperature (Galan et al., 2009). Reactions were performed in anhydrous dichloromethane, adding [BMIM] [OTf] as co-solvent, which also influenced in the increase of substrates stabilities. Authors suggested that this IL worked as glycosylation promoter by the slow release of triflic acid, although less active peracetylated donors required the addition of catalytic amounts of triflic acid. Good yields were achieved in most of cases and the influence of the IL in the stereoselectivity was also demonstrated: when non-participating groups at C-2 of activated thioglycosides were present, an increase in α-glycoside products was observed compared with reactions catalyzed by TMSOTf (trimethylsilyl trifluoromethanesulfonate at low temperatures. In further studies, the same group explored the scope of imidazolium-based ILs as glycosylation promoters (Galan et al., 2010a). A wide range of imidazolium cations containing different alkyl chains, and a series of counter ions, were employed to study their influence on glycosilations of thioglycosides donors at room temperature in the presence of N-iodosuccinimide (NIS). Authors found that imidazolium ILs bearing triflate or triflimide counter ions could be employed as selective glycosylation promoters of that type of substrates. Substitutions

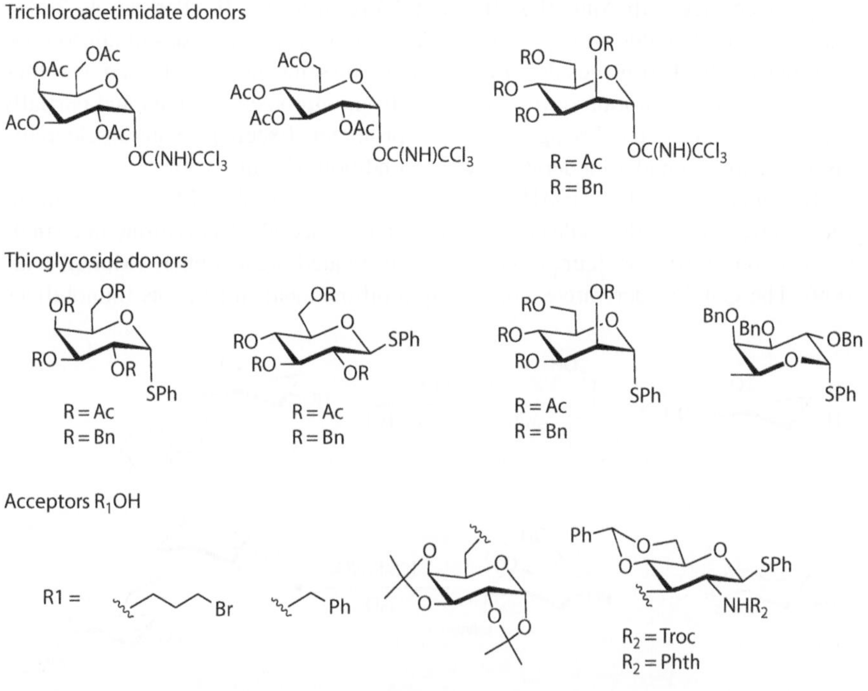

FIGURE 8.9 Trichloroacetimidates and thioglycosides donors employed in the study of the applicability of [BMIM] [OTf] as co-solvent/promoter of glycosylations. (From Galan, M.C. et al., *Tetrahedron Lett.*, 50, 442–445, 2009.)

at R_1 of the imidazolium ring did not have any effect on the process. However, susbstituents at R_2 considerably affected the rate of the reaction.

On the basis of previous results, Galan and co-workers developed an innovative one-pot protocol to synthesize oligosaccharides involving three components (Galan et al., 2010b), As they had formerly observed and extensively studied, [BMIM] [OTf] selectively activated benzylated thioglycosides at room temperature in the presence of acetylated donors, what could allow the development of *in situ* double glycosylations, to afford branched or linear disaccharides, depending on the components of the reaction. As an example, the linear trisaccharide shown on Figure 8.10 could be afforded as follows: a benzyl-protected thioglycoside donor, containing a free OH on C-2, reacted firstly with a protected acceptor, bearing a primary (more reactive) OH, in the presence of NIS and [BMIM] [OTf], to give the corresponding 1-6 linked disaccharide. Then a peracetylated thioglycoside was added to the reaction mixture and an excess of NIS and TMSOTf, affording the expected trisaccharide. Taking into account that peracetylated thioglycosides would not react until TMSOTf is added, a more attractive *one-pot* strategy was successfully developed, starting from a mixture of the three glycosides. After 3 h, allowing to form the disaccharide intermediate, an excess of NIS and TMSOTf was added and the trisaccharide was obtained.

On the other hand, Díaz and co-workers developed a stereoselective glycosylation of *endo*-glycals with different *O*-nucleophiles, using pTSA/[BMIM] [BF$_4$] catalytic system, as a green methodology to prepare α-2-deoxyglycosides (Díaz et al., 2012). Different alcohols were successfully employed in the glycosylation of perbenzyl-D-galactals and perbenzyl-D-glucals in the presence of this homogeneous catalytic system (Figure 8.11). The products could be extracted from the reaction media by extraction with organic solvents, allowing the recovering and reuse of the system [BMIM] [BF$_4$]/pTSA for four cycles without a significant loss of activity.

Talisman et al. employed imidazolium-based ILs to generate N-heterocyclic carbenes (NHCs) and reported *O*-glycosylation reactions promoted via silver NHC complexes formed *in situ* in ILs (Talisman et al., 2011). They showed that anion metathesis of the ILs with benzyltriethylamonium chloride resulted in silver-NHC formation and

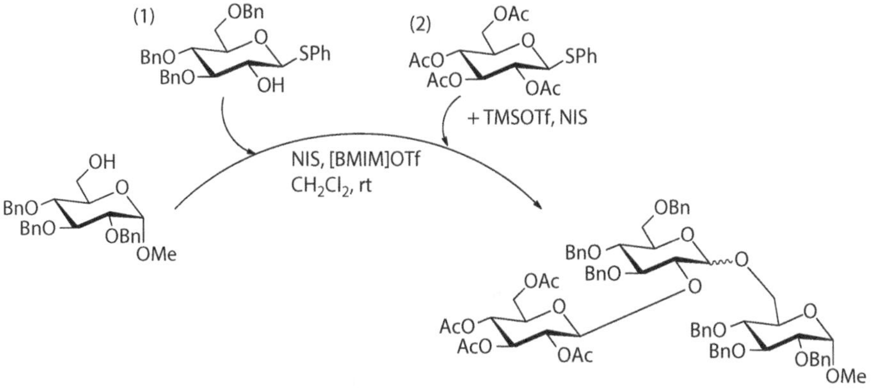

FIGURE 8.10 One-pot synthesis of trisaccharides in [BMIM] [OTf]. (From Galan, M.C. et al., *Chem. Commun.*, 46, 2106–2108, 2010b.)

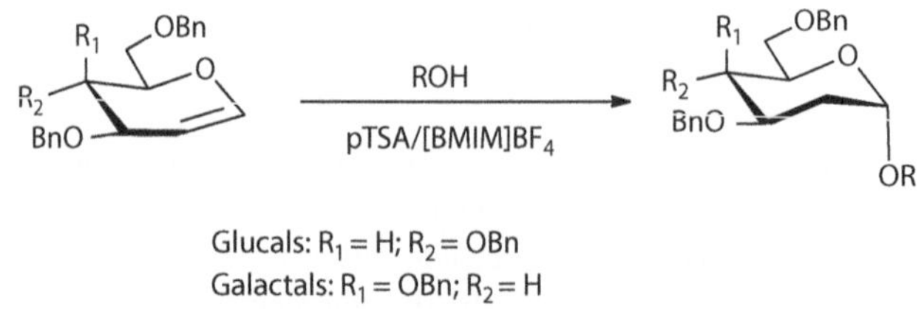

Glucals: R$_1$ = H; R$_2$ = OBn
Galactals: R$_1$ = OBn; R$_2$ = H

FIGURE 8.11 Synthesis of α-deoxyglycosides by pTSA/[BMIM]BF$_4$ mediated glycosylation of glycols. (From Díaz, G. et al., *Top. Catal.*, 55, 644–648, 2012.)

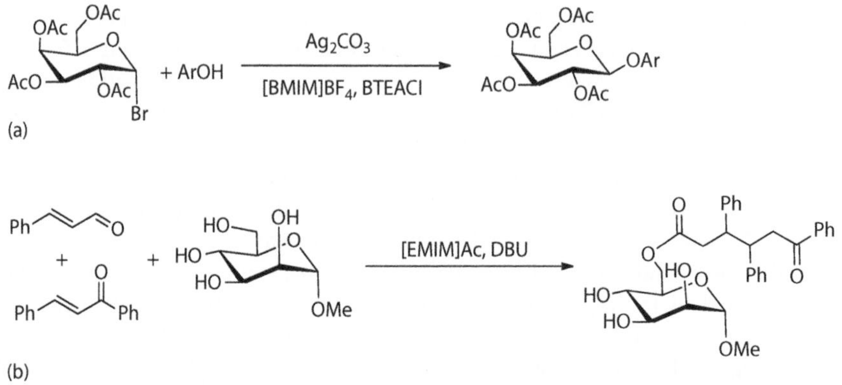

(a)

(b)

FIGURE 8.12 (a) O-Glycosidation reactions promoted by *in situ* generated silver NHC in IL. (From Talisman, I.J. et al., *Carbohyd. Res.*, 346, 883–890, 2011.) (b) Regioselective NHC-catalyzed functionalization of unprotected hexopyranosides in IL. (From Axelsson, A. et al., *Europ. J. Org. Chem.*, 20, 3339–3343, 2016.)

subsequent O-glycosylation in the presence of silver carbonate (Figure 8.12a). Thus, glycosylation of tetra-*O*-acetyl-α-D-galactopyranosyl bromide with phenol derivatives was investigated, achieving the corresponding products in good yields. Recently, Axelsson and co-workers have reported a regioselective NHC-catalyzed functionalization of unprotected hexopyranosides using ILs as NHC procatalysts (Figure 8.12b) (Axelsson et al., 2016). Unprotected carbohydrate reacted with chalcone and cinnamaldehyde in the presence of [EMIM] [Ac] and DBU (1,8-diazabicyclo[5.4.0]udec-7-ene), noticing in all cases total regioselectivity toward the primary alcohol.

8.3 ENZYMATIC SYNTHESIS OF CARBOHYDRATES

Enzyme-catalyzed processes offer attractive advantages in the synthesis and modification of carbohydrates because most of the relevant reactions require high degrees of chemo-, regio-, and stereoselectivity. In addition, these selective biocatalytic reactions are carried out under mild reactions conditions, allowing the development of sustainable processes. Different ILs have been shown to be compatible with enzyme activity, and, in addition, they have offered other benefits, such as an increase of enzyme stability and/or activity, resulting from conformational changes of the enzyme or an improvement of enzyme selectivity

(Sheldon, 2016). Thus, the combination of ILs and biocatalysts appears as a promising green alternative, and during the last decade, it has been increasingly implemented, especially in the conversion of polar substrates such as sugars (Galonde et al., 2012).

8.3.1 POLYSACCHARIDE-BASED MATERIALS AS SUBSTRATES

As it was previously mentioned, polysaccharides are promising renewable feedstocks for the production of a wide variety of products. Lignocellulose (composed of cellulose, hemicellulose, and lignin) is the most abundantly available biomass resource on earth, which can be converted into a fuel and chemical through chemical or biocatalytical routes. Conversion of lignocellulose usually is carried out in a three-step process: (a) pretreatment of the feedstock; (b) hydrolysis of the polysaccharides to monosaccharides; and (c) microbial fermentation to achieve the product (Wahlstrom and Suurnakki, 2015). Enzymatic hydrolysis is catalyzed by cellulases, usually applied as cocktails of different enzymatic activity: endoglucanases, which cleave the β-1,4-glycosidic bonds in cellulose chains; cellobiohydrolases, which catalyze the hydrolysis of cellobiose from the cellulose chain ends, and β-glucosidases, which catalyze the hydrolysis of the oligosachaccarides to glucose. Xylanases are hydrolytic enzymes that catalyze the hydrolysis of β-1,4-glycosidic bonds connecting anhydroxylose units in xylan. Their activity collaborates to make cellulose more accessible to cellulases. Thus, the combination of cellulases and xylanases has become a great promise for cellulose treatment. On the other hand, ILs have appeared as an attractive alternative for the lignocellulose pretreatment. On this basis, the *in situ* pretreatment and enzymatic hydrolysis in ILs is proposed as the best method for cellulose processing, but it is still an open challenge, as many cellulose dissolving ILs inactivate cellulases (Wahlstrom and Suurnakki, 2015). In this context, great efforts have been developed to find more ILs tolerant enzymes, and different ILs compatible cellulases have been studied from different extremotolerant sources (Datta et al., 2010; Gladden et al., 2014; Wahlstrom et al., 2013; Yu et al., 2016). As a recent sample, Rahikainen and co-workers have carried out a screening of glycoside hydrolases in cellulose dissolving ILs (Rahikainen et al., 2018). Authors reported that at 40% (w/v) IL concentration, several enzyme combinations retained more than 50% of activity. Endoglucanase from hyperthermophilic *Pyrococcus horikoshii* and GH10 *Thermopolyspora flexuosa* xylanase showed higher IL tolerance. ILs based on imidazolium cation and phosphate anion were found to be the least inactivating for the studied enzymes, especially those with short-chain alkyl substituents ([DMIM] [DMP]).

Metagenomic and directed evolution experiments have been conducted to more IL-tolerant cellulases (Ilmberger et al., 2012; Lehmann et al., 2014; Li et al., 2013), and the employ of immobilized enzymes is presented as a promising method to recycle both the IL and the biocatalyst (Ungurean et al., 2014). Lozano et al. established an effective protocol to improve the cellulose stability in [BMIM] [Cl] (Lozano et al., 2011): commercial cellulose from *T. ressei* immobilized on Amberlite XAD4 was further coated with the hydrophobic IL butyltrimethyl-ammonium bis(trifluoromethylsulfonyl)imide ([BTMA] [Tf$_2$N]), and this stabilized derivative was successfully employed for the saccharification of dissolved cellulose in [BMIM] [Cl] at 50°C and 1.5% (w/v) water content.

Cellulose derivatives, such as long-chain cellulose esters, can present improved properties in respect to non-modified cellulose, which makes them ideal substance to be employed as film materials, biodegradable plastics, packaging materials, or drug delivery systems, among others (Edgar et al., 2001). As explained later, enzymatic synthesis of mono/disaccharide esters has been extensively studied in ILs as reaction media, but enzyme catalyzed esterification of polysaccharides in ILs is still an emerging area, offering different advantages over traditional chemical methods, due to the high regioselectivity displayed by enzymes. The group of Li at al. has recently described the sustainable synthesis of long-chain cellulose esters in a one-step route mediated by the action of lipase from *Candida rugose* in a binary-IL system, composed by [BMIM] [Cl]:[BMIM] [PF$_6$] in different concentrations (Wang et al., 2017). Authors carried out the study of different reaction parameters (ILs proportion, acyl donor, temperature) in the synthesis of cellulose esters, and subsequently in the morphology, crystallinity and thermal properties of the products.

A similar approach has been developed to modify and improve the physic-chemical properties of chitosan, which presents a wide range of applications, such as immobilization support, surfactant, or in the medicine field (Muxika et al., 2017). Zhao and co-workers have developed an efficient green strategy for the synthesis of chitosan palmitate catalyzed by lipase Novozyme 435® in a binary-IL system (Zhao et al., 2017). Authors found best results employing a mixture of a hydrophilic IL with high chitosan-dissolving capability ([EMIM] [Ac] and a hydrophobic IL compatible with the enzyme activity ([BMIM] [PF$_6$], in a ratio 6:4, respectively. Once the chitosan is dissolved in the ILs, the lipase mediated the regioselective chitosan esterification, avoiding the protection-deprotection steps commonly employed in traditional chemical strategies.

8.3.2 ACYLATION OF CARBOHYDRATES

Enzymatic acylation of sugars is mediated by lipases (E.C. 3.1.1.3), which, in nature, catalyze the ester bond hydrolysis of triacylglycerols. In a non-aquous environment, lipases are able to promote the reverse reaction, dealing to the synthesis of esters. In this area, one of the main applications of lipases in carbohydrate is focused on the synthesis of sugar fatty acid esters (Khan and Rathod, 2015; van den Broek and Boeriu, 2013).

Sugar fatty acid esters are amphiphilic polysaccharides, very attractive non-ionic surfactants, which are becoming more important in food, beverage, cosmetic, and drug industries (Neta et al., 2015). They are derived from renewable resources and are considered biodegradable, non-toxic, and biocompatible and interesting medicinal properties have also been explored (Bezbradica et al., 2017). Commonly, these compounds are prepared by chemical methods, but enzymatic reactions offer a very attractive alternative due to the high enantio-, regio-, and chemoselectivity displayed by lipases. Furthermore, mild reaction conditions are required and more environmentally friendly processes. However, the main drawback of sugar esterification processes is the low solubility of carbohydrates in organic solvents. Besides the environmental benefits of ILs, these solvents have allowed a considerable increase of sugar substrate concentration. In addition, these ILs medium have also contributed to enhance lipase activity and regioselectivity (Galonde et al., 2012; Sheldon, 2016). Actually, different lipase-catalyzed synthesis of sugar esters in ILs have been described. Park and Kazlauskas reported the first lipase-catalyzed glucose acetylation in ILs (Park and Kazlauskas, 2001). They studied a diverse range of ILs to use as medium of lipase-catalyzed acylation of 1-phenylethanol and developed a regioselective 6-O-acetylation of glucose catalyzed by lipase B from *Candida antarctica* (CALB). 1-(2-methoxyethyl)-3-methylimidazolium derived IL, [MOEMIM] [BF$_4$], resulted to be the best choice for this reaction. In addition, it allowed a regioselective reaction, obtaining just the 6-O-acyl derivative, avoiding the formation of diacetylated products usually produced in conventional organic media (Figure 8.13). Authors suggested that the nature of the cation and the anion counterpart of the ILs may influence the rate of the reaction and the regioselectivity of the reaction, respectively. Similar studies have been further developed, as the one described by Galonde and co-workers, on the lipase catalyzed synthesis of mannosyl myristate (Galonde et al., 2013). This study showed that the lipase effectiveness was affected by the anion, obtaining best results with those ILs based on TFO⁻ anion.

Ganske and Bornscheuer reported the synthesis of glucose fatty acid esters catalyzed by immobilized CALB and using lauric and myristic acid vinyl ester as acyl donor in the presence of [BMIM] [PF$_6$] and [BMIM] [PF$_4$]. In order to improve the results, this group developed a binary system composed of any of those two ILs and 40% *t*-BuOH (Figure 8.14) (Ganske and Bornscheuer, 2005a; Ganske and

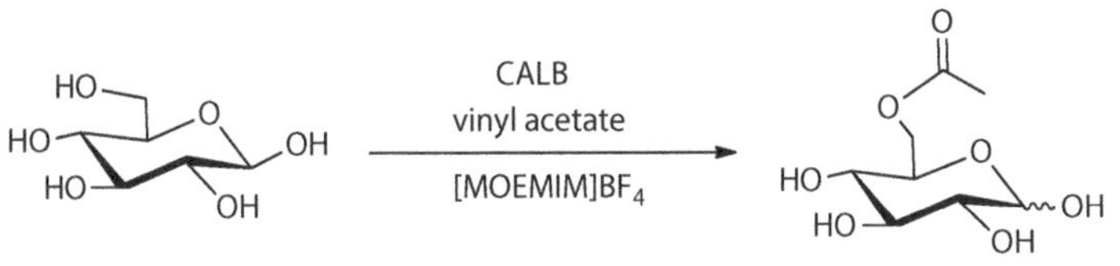

FIGURE 8.13 Glucose 6-O-acetylation catalyzed by CALB in IL. (From Park, S. and Kazlauskas, R.J., *J. Org. Chem.*, 66, 8395–8401, 2001.)

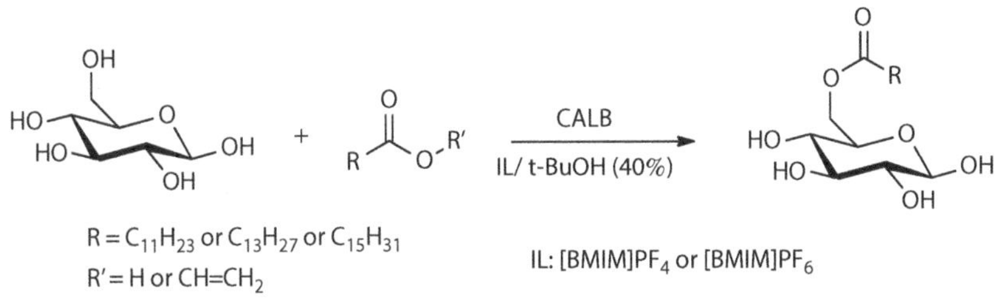

R = C$_{11}$H$_{23}$ or C$_{13}$H$_{27}$ or C$_{15}$H$_{31}$
R′ = H or CH=CH$_2$

IL: [BMIM]PF$_4$ or [BMIM]PF$_6$

FIGURE 8.14 Regioselective CALB catalyzed 6-O-esterification of glucose in a system composed of IL and 40% of t-BuOH. (From Ganske, F. and Bornscheuer, U.T., *Org. Lett.*, 7, 3097–3098, 2005a.)

Bornscheuer, 2005b). This strategy permitted diverse advantages: the improvement of glucose solubility; the use of commercial lipase CALB; the use not only of vinyl esters, but also the correspondent acids as acylating agents, the increase of the yields; and an excellent 6-O-regioselectivity.

The synthesis of 6-*O*-lauroyl-D-glucose catalyzed by immobilized CALB (Novozyme 435) in ILs was also investigated by the group of Koo (Lee et al., 2008b). The highest lipase activity was detected in [BMIM] [TfO], which could dissolve high concentrations of glucose, and the best lipase stability was found in hydrophobic [BMIM] [Tf$_2$N]. Then, authors successfully obtained best results in a 1:1 (v/v) mixture of both ILs. Subsequently this group developed a new strategy based on the preparation of supersaturated glucose solution in ILs (Lee et al., 2008a). Glucose was firstly solved in an aqueous medium and it was further transferred into an IL ([BMIM] [OTf]). Then, water was removed, leading to supersaturated IL, and the lipase-catalyzed direct esterification of glucose with lauric acid was efficiently carried out with the supersaturated glucose in the IL. In order to enhance lipase activity and stability, this strategy was optimized using the supersaturated sugar system in a mixture of ILs (Ha et al., 2010). The optimal stability and activity of Novozyme 435 were observed in a mixture [BMIM] [OTf]: [OMIM] [Tf$_2$N] 1:1 (v/v). Best conditions permitted the reuse of CALB and ILs, maintaining 78% of initial activity after five cycles. In similar optimization studies, this group also successfully employed ultrasounds to overcome the mass transfer limitations in viscous ILs, accelerating in this way, the lipase catalyzed sugar acylation in supersaturated ILs (Lee et al., 2008c).

It should be noted the lipase catalyzed synthesis of 6- and 6'-O-linoleyl-α-D-maltose described by Fischer et al. in a binary solvent system (Fischer et al., 2013). After a wide screening of lipases and reaction medium, lipases from *Pseudomonas cepacia* and *Candida antarctica* showed best catalytic behavior toward deesterification of thee disaccharide maltose with linoleic acid. Binary green solvents, [EMIM] [MeSO$_3$]/[BMPyr] [PF$_6$] and the acetone/[EMIM] [MeSO$_3$] mixture, tended to double maltose conversions in comparison to single solvent use (Figure 8.15).

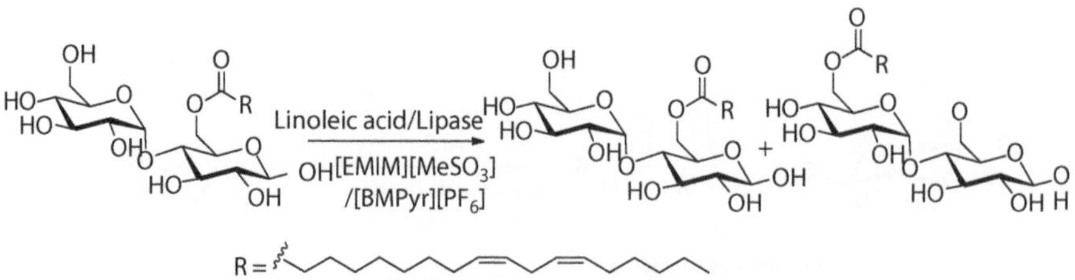

FIGURE 8.15 Lipase catalyzed synthesis of 6- and 6'-O-linoleyl-a-D-maltose in a binary IL system. (From Fischer, F. et al., *J. Mol. Catal. B-Enzymatic*, 90, 98–106, 2013.)

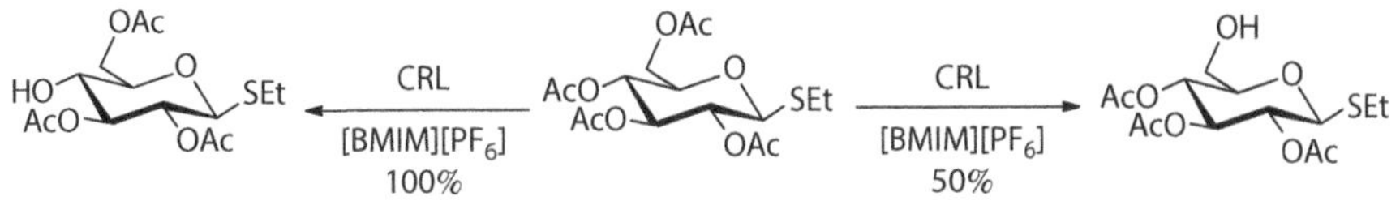

FIGURE 8.16 Lipase catalyzed regioselective hydrolysis of peracetylated ethyl thioglucopyranoside in [BMIM] [PF$_6$] systems.

It is worth mentioning that lipase catalysis in ILs has been also investigated in the regioselective hydrolysis of peracetylated sugars, leading to very useful synthetic intermediates. Gervaise and co-workers explored the regioselective hydrolysis of peracetylated ethyl thioglucopyranoside catalyzed by lipase from *Candida cylindracea* in a [BMIM] [PF$_6$]/buffer system (Gervaise et al., 2009). The amount of hydrophobic IL in the medium could modulate the lipase regioselectivity. Thus, different deacylated products could be isolated depending on the ratio IL:buffer employed in the reaction (Figure 8.16).

8.3.3 GLYCOSYLATION REACTIONS

Enzyme-catalyzed glycosylations reactions are becoming a very promising alternative to traditional chemical strategies, as the high selectivity displayed of enzymes allows to avoid many protection/deprotection steps. Glycosidases (E.C. 3.2.1) make up the group of enzymes commonly employed in oligosaccharides synthesis. In nature, this class of enzymes catalyzes the hydrolysis of glycosidic bonds, but, under appropriate conditions, many of them are able to mediate the synthesis of glycosidic linkages. In general, they are easily available enzymes, tolerant to organic solvents, present a broad range of substrates, and usually display high stereospecificity. Although the development of efficient enzymatic synthesis still remains challenging, different approaches have achieved desired glycosides in excellent yields and selectivities (Farran et al., 2015).

As glycosidases are a type of hydrolases, reactions in aqueous environments lead to low yields. Better results can be reached by increasing substrates concentrations or decreasing the amount of water in the media. In this context, those ILs with high capability to solve carbohydrates can offer the possibility of increase in the concentration of reactants and reduce the aqueous media (Hernaiz et al., 2010).

The synthesis of carbohydrates in the presence of ILs was firstly described by Kaftzik and co-workers (Kaftzik et al., 2002). This group employed the β-galactosidase from *Bacillus circulans* (*B. circulans*) as biocatalyst of the synthesis of *N*-acetyl-D-lactosamine from lactose and *N*-acetylglucosamine. The addition of 25% v/v of the water soluble IL 1,3-dimethylimidazolium methylsulphate [MMIM] [MeSO$_4$] could suppress the secondary hydrolysis of the

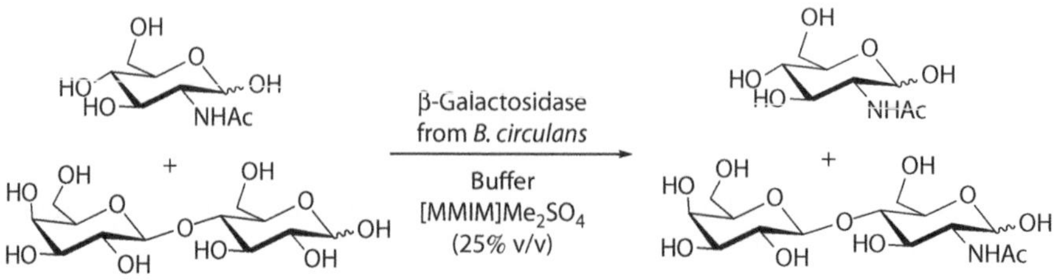

FIGURE 8.17 *B. circulans* β-galactosidase catalyzed synthesis of *N*-acetyllactosamine. (From Kaftzik, N. et al., *Org. Proc. Res. Devel.*, 6, 553–557.)

desired product and enhance the yield to 60% (Figure 8.17). Authors suggested that some interactions between the components of the IL and the enzyme could be also responsible of the enhancement of the yields. In this sense, Ferdjani et al. carried out a study about the relation of the thermostability of different glycosidases (β-glycosidase from *Thermus thermophilus* and α-galactosidases from *Thermotoga maritima* and *Bacillus stearothermophilus*) and their stability in hydrophilic ILs (Ferdjani et al., 2011). Authors concluded that there was a close correlation between the thermostability of the glycosidases and their stability in hydrophilic ILs. Thermostable enzymes present a more compact structure and rigidity, which could avoid the modification of the hydration environment caused by ILs. Then, it was preferable the use of thermostable glycosidases in water-miscible ILs.

Sandoval et al. (2012) found an important change from the classical regioselectivity of the transglycosylation reaction mediated by the β-galactosidase from *Thermus thermophilus* HB27 (TTP0042) in ILs. This thermophilic biocatalyst promotes the synthesis of *N*-acetyl-D-lactosamine, employing *p*-nitrophenyl-β-D-galactopyranoside (pNPGal) as donor. If the reaction was performed in a buffered medium, high amounts of self-condensation product derived from the donor were produced (Gal-β[1→3]Gal-β-*p*NP). The addition of different ILs (30%, v/v) considerably reduced the self-condensation product, and the reaction shifted toward the production of Gal-β[1→4]GlcNac or Gal-β[1→6]GlucNac depending on the IL employed. *N*-acetyl-D-lactosamine was produced in high yields and selectivity when [BMIM] [PF₆] was employed as co-solvent in the transglycosilation reaction. Authors carried out a deep molecular interaction study, and they concluded that the enzyme became more flexible in a water-IL mixture, what would allow the stabilization of the GlcNac molecule in the active center. In a similar manner, recombinant β-galactosidase from *B. circulans* ATCC 31382 (β-Gal-3-*N*Tag) was employed as the biocatalyst in an efficient and regioselective synthesis of β-(1→3)-galactosides in a reaction media based on ILs (Bayon et al., 2013). In the presence of 30% IL (v/v), the enzyme was able to synthesize the dissacharides Gal-β[1→3]GlcNAc or Gal-β[1→3] GalNAc, using pNPGal as donor, and GlcNAc or GalNAc as acceptor, respectively, in excellent yields (Figure 8.18). In addition, the employ of [BMIM]PF₆ permitted an easy isolation of the products, what allowed the developed of these procedures in a semi-preparative scale.

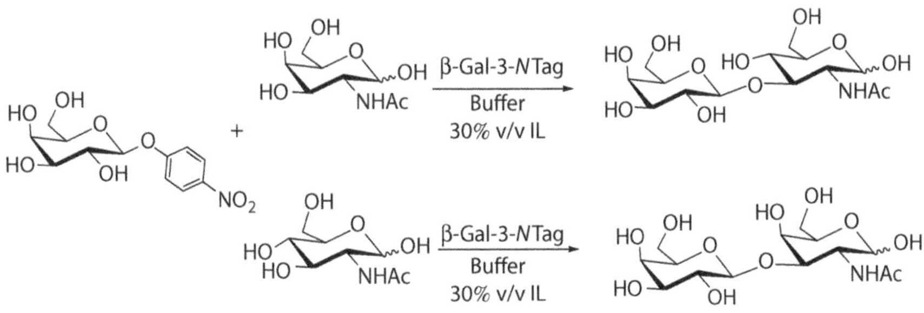

FIGURE 8.18 Synthesis of disaccharides catalyzed by β-galactosidase from *B. circulans.* (From Bayon, C. et al., *Tetrahedron*, 69, 4973–4978, 2013.)

The effect of [BMIM] [PF$_6$] over the transglycosylation reaction catalyzed by β-galactosidase from *Aspergillus oryzae* has also been explored (Brakowski et al., 2016). Authors tested the transglycosylation in buffer, in an emulsion buffer-IL, and in a fed-batch system, employing *o*-nitrophenyl-β-D-galactopyranoside as donor and glucose, mannose, or galactose as acceptor. Depending on the system and the acceptor concentration, the reaction could deal to the synthesis of di or trisaccharides.

Very recently, Yang and co-workers have reported the synthesis of different alkyl galactopyranosides in a sustainable process mediated by *Thermotoga naphthophila* RKU-10 β-galactosidase and milk processing waste lactose as galactosyl donor in ILs (Yang et al., 2017). Different alkyl alcohols were tested as nucleophiles (from *n*-butanol to *n*-tetradecanol) in systems containing different ILs. Most promising results were achieved with AMMOENG 102, a type of amphiphilic tetraammonium-ethylsulfate containing a C18 acyl chain and oligoethyleneglycol, which presented a better protective effect for the enzyme, allowing the reaction performance at 95°C. Interestingly, another IL from this amphiphilic group, AMMOENG 101, resulted to be compatible with the cellobiose phosphorylase from *Clostridium thermocellum*, permitting the glycosylation of aliphatic and aromatic alcohols (De Winter et al., 2015). Cellobiose phosphorylase catalyzes the reversible phosphorolysis of cellobiose into α-D-glucose 1-phosphate and D-glucose with inversion of the anomeric configuration. It also possesses synthetic activity, using α-D-glucose 1-phosphate as sugar donor and glucose as sugar acceptor. The immobilization of the enzyme and the use of this IL contributed to the stabilization of the biocatalyst, and its acceptor promiscuity was widely explored, obtaining the corresponding α-glucosides in different yields, depending on the alcohol employed.

One of the main drawbacks in the enzymatic synthesis of carbohydrate using glycosidases is that the newly formed product is a substrate for secondary hydrolysis with reduced final yields and makes their industrial application difficult. To overcome these limitations, glycosynthases, a new class of mutant glycosidases with no hydrolytic activity that synthesize glycans in enhanced yields, were obtained from β- (Mackenzie et al., 1998; Malet and Planas, 1998; Moracci et al., 1998) and α-aglycosidases (Cobucci-Ponzano et al., 2009; Cobucci-Ponzano and Moracci, 2012;

Cobucci-Ponzano et al., 2011). Recently, a new α-galactosynthase has been obtained for the first time: the TmGalA D327G (Cobucci-Ponzano et al., 2011) mutant derived from the α-galactosidase from the thermophilic bacterium *Thermotoga maritima*. Bayón et al. (2015) explore for the first time the synthesis of α-glycoconjugates using a α-glycosynthase in ILs. Using this biocatalyst, β-Gal-N₃ as donor, pNP-Glc and pNP-Man as acceptors, and ILs, they obtained high yields and excellent selectivities in the synthesis of α-glycoconjugates. In addition, reaction scale-up is feasible and ILs can be recovered and reused, increasing the sustainability of the reaction process (Figure 8.19).

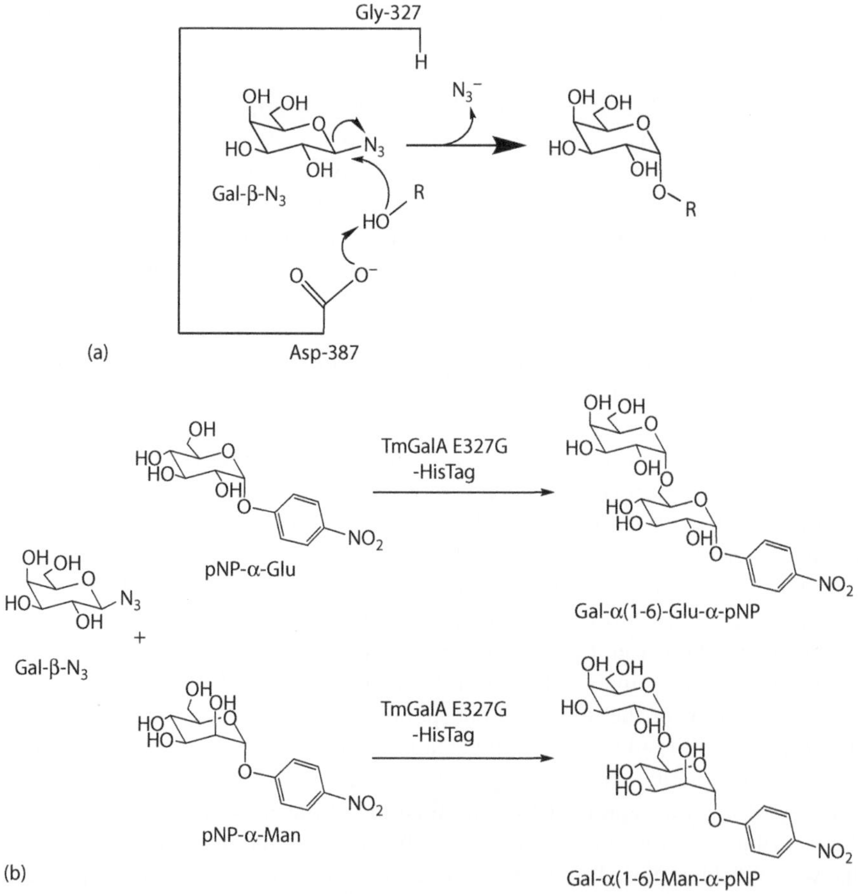

FIGURE 8.19 (a) Mechanism followed by TmGalA D327G α-glycosynthase: the cavity created by the mutation that removed the side chain of Asp327 allowed the access to the active site of the β-Gal-N₃. Then, the galactose moiety is transferred to the acceptor activated by the base catalysis of the Asp387 residue; the disaccharide product, showing the newly formed α-bond, cannot be hydrolyzed by the mutant and thus accumulates in the reaction mixture; (b) Transglycosylation reaction with the α-galactosynthase TmGalA D327G for the synthesis of Gal-α(1-6)-Glc-α-pNP and Gal-α(1-6)-Man-α-pNP using Gal-β-N3 as donor and pNP-α-Glc or pNP-α-Man as acceptors.

REFERENCES

Axelsson, A., Ta, L., and Sundén, H. 2016. Direct highly regioselective functionalization of carbohydrates: A three-component reaction combining the dissolving and catalytic efficiency of ionic liquids. *European Journal of Organic Chemistry*, **2016**(20), 3339–3343.

Badawy, M. E. I. and Rabea, E. I. 2011. A biopolymer chitosan and its derivatives as promising antimicrobial agents against plant pathogens and their applications in crop protection. *International Journal of Carbohydrate Chemistry*, **2011**, 1–29.

Barber, P. S., Griggs, C. S., Bonner, J. R., and Rogers, R. D. 2013. Electrospinning of chitin nanofibers directly from an ionic liquid extract of shrimp shells. *Green Chemistry*, **15**(3), 601–607.

Bayon, C., Cortes, A., Berenguer, J., and Hernaiz, M. J. 2013. Highly efficient enzymatic synthesis of Gal beta-(1 -> 3)-GalNAc and Gal beta-(1 -> 3)-GlcNAc in ionic liquids. *Tetrahedron*, **69**(24), 4973–4978.

Bayon, C., Moracci, M., and Hernaiz, M. J. 2015. A novel, efficient and sustainable strategy for the synthesis of alpha-glycoconjugates by combination of a alpha-galactosynthase and a green solvent. *Royal Society of Chemistry Advances*, **5**(68), 55313–55320.

Bezbradica, D., Corovic, M., Tanaskovic, S. J., Lukovic, N., Carevic, M., Milivojevic, A., and Knezevic-Jugovic, Z. 2017. Enzymatic syntheses of esters—Green chemistry for valuable food, fuel and fine chemicals. *Current Organic Chemistry*, **21**(2), 104–138.

Bodachivskyi, I., Kuzhiumparambil, U., and Williams, D. B. G. 2018. Acid-catalyzed conversion of carbohydrates into value-added small molecules in aqueous media and ionic liquids. *Chemsuschem*, **11**(4), 642–660.

Brakowski, R., Pontius, K., and Franzreb, M. 2016. Investigation of the transglycosylation potential of ß-Galactosidase from Aspergillus oryzae in the presence of the ionic liquid [Bmim][PF$_6$]. *Journal of Molecular Catalysis B: Enzymatic*, **130**, 48–57.

Brode, G. 1991. Polysaccharides: "Naturals" for cosmetics ad pharmaceuticals. In: *Cosmetic and Pharmaceutical Applications of Polymers*, (Eds.) C. Gebelein, T. Cheng, V. Yang, Springer, Newyork. pp. 105–115.

Capello, C., Fischer, U., and Hungerbühler, K. 2007. What is a green solvent? A comprehensive framework for the environmental assessment of solvents. *Green Chemistry*, **9**(9), 927.

Chang, S. S., Lin, C. C., Li, Y. K., and Mong, K. K. 2009. A straightforward alpha-selective aromatic glycosylation and its application for stereospecific synthesis of 4-methylumbelliferyl alpha-T-antigen. *Carbohydrate Research*, **344**(4), 432–438.

Chen, Q. T., Xu, A. R., Li, Z. Y., Wang, J. J., and Zhang, S. J. 2011. Influence of anionic structure on the dissolution of chitosan in 1-butyl-3-methylimidazolium-based ionic liquids. *Green Chemistry*, **13**(12), 3446–3452.

Cobucci-Ponzano, B. and Moracci, M. 2012. Glycosynthases as tools for the production of glycan analogs of natural products. *Natural Products Reports*, **29**(6), 697–709.

Cobucci-Ponzano, B., Conte, F., Bedini, E. et al. 2009. BetaGlycosyl azides as substrates for alpha-glycosynthases: Preparation of efficient alpha-L-fucosynthases. *Chemical Biology*, **16**(10), 1097–1108.

Cobucci-Ponzano, B., Zorzetti, C., Strazzulli, A., Carillo, S., Bedini, E., Corsaro, M. M., Comfort, D. A., Kelly, R. M., Rossi, M., and Moracci, M. 2011. A novel alpha-D-galactosynthase from Thermotoga maritima converts beta-D-galactopyranosyl azide to alpha-galacto-oligosaccharides. *Glycobiology*, **21**(4), 448–456.

Coleman, D. and Gathergood, N. 2010. Biodegradation studies of ionic liquids. *Chemical Society Reviews*, **39**(2), 600–637.

Crich, D. and Li, W. 2007. O-sialylation with N-acetyl-5-n,4-o-carbonyl-protected thiosialoside donors in dichloromethane: Facile and selective cleavage of the oxazolidinone ring. *Journal of Organic Chemistry*, **72**(7), 2387–2391.

Dai, Q., Ren, J., Kong, W., Peng, F., and Meng, L. 2015. Adsorption kinetics and thermody-namics of cellulose dinitrobenzoate prepared in ionic liquid for the removal of creati-nine. *Bioresources*, **10**(2), 3666–3681.

Dasgupta, S. and Nitz, M. 2011. Use of *N,O*-Dimethylhydroxylamine as an anomeric protect-ing group in carbohydrate synthesis. *Journal of Organic Chemistry*, **76**(6), 1918–1921.

Datta, S., Holmes, B., Park, J. I., Chen, Z., Dibble, D. C., Hadi, M., Blanch, H. W., Simmons, B.A., and Sapra, R. 2010. Ionic liquid tolerant hyperthermophilic cellulases for biomass pretreatment and hydrolysis. *Green Chemistry*, **12**(2), 338–345.

De Winter, K., Van Renterghem, L., Wuyts, K., Pelantová, H., Křen, V., Soetaert, W., and Desmet, T. 2015. Chemoenzymatic synthesis of β-D-glucosides using cellobiose phos-phorylase from clostridium thermocellum. *Advanced Synthesis & Catalysis*, **357**(8), 1961–1969.

Díaz, G., Ponzinibbio, A., and Bravo, R. D. 2012. pTSA/[bmim][BF4] Ionic liquid: A pow-erful recyclable catalytic system for the synthesis of α-2-Deoxyglycosides. *Topics in Catalysis*, **55**(7), 644–648.

Edgar, K. J., Buchanan, C. M., Debenham, J. S., Rundquist, P. A., Seiler, B. D., Shelton, M. C., and Tindall, D. 2001. Advances in cellulose ester performance and application. *Progress in Polymer Science*, **26**(9), 1605–1688.

Farran, A., Cai, C., Sandoval, M., Xu, Y. M., Liu, J., Hernaiz, M. J., and Linhardt, R. J. 2015. Green solvents in carbohydrate chemistry: From raw materials to fine chemicals. *Chemical Reviews*, **115**(14), 6811–6853.

Feng, J. X., Zang, H. J., Yan, Q., Li, M. G., Jiang, X. Q., and Cheng, B. W. 2015. Dissolution and utilization of chitosan in a 1-carboxymethyl-3-methylimidazolium hydrochloride ionic salt aqueous solution. *Journal of Applied Polymer Science*, **132**(22).

Ferdjani, S., Ionita, M., Roy, B., Dion, M., Djeghaba, Z., Rabiller, C., and Tellier, C. 2011. Correlation between thermostability and stability of glycosidases in ionic liquid. *Biotechnology Letters*, **33**(6), 1215–1219.

Fischer, F., Happe, M., Emery, J., Fornage, A., and Schutz, R. 2013. Enzymatic synthesis of 6- and 6′-O-linoleyl-alpha-D-maltose: From solvent-free to binary ionic liquid reaction media. *Journal of Molecular Catalysis B-Enzymatic*, **90**, 98–106.

Forsyth, S. A., MacFarlane, D. R., Thomson, R. J., and von Itzstein, M. 2002. Rapid, clean, and mild O-acetylation of alcohols and carbohydrates in an ionic liquid. *Chemical Communications*, **7**(7), 714–715.

Galan, M. C., Brunet, C., and Fuensanta, M. 2009. Bmim OTf: A versatile room temperature glycosylation promoter. *Tetrahedron Letters*, **50**(4), 442–445.

Galan, M. C., Jones, R. A., and Tran, A.-T. 2013. Recent developments of ionic liquids in oligosaccharide synthesis: The sweet side of ionic liquids. *Carbohydrate Research.*, **375**(0), 35–46.

Galan, M. C., Jouvin, K., and Alvarez-Dorta, D. 2010a. Scope and limitations of imidazolium-based ionic liquids as room temperature glycosylation promoters. *Carbohydrate Research*, **345**(1), 45–49.

Galan, M. C., Tran, A. T., and Whitaker, S. 2010b. Bmim OTf as co-solvent/promoter in room temperature reactivity-based one-pot glycosylation reactions. *Chemical Communications*, **46**(12), 2106–2108.

Galonde, N., Nott, K., Debuigne, A., Deleu, M., Jerome, C., Paquot, M., and Wathelet, J. P. 2012. Use of ionic liquids for biocatalytic synthesis of sugar derivatives. *Journal of Chemical Technology and Biotechnology*, **87**(4), 451–471.

Galonde, N., Nott, K., Richard, G., Debuigne, A., Nicks, F., Jerome, C., and Fauconnier, M. L. 2013. Study of the influence of pure ionic liquids on the lipase-catalyzed (trans)esterifi-cation of mannose based on their anion and cation nature. *Current Organic Chemistry*, **17**(7), 763–770.

Ganske, F. and Bornscheuer, U. T. 2005a. Lipase-catalyzed glucose fatty acid ester synthesis in ionic liquids. *Organic Letters*, **7**(14), 3097–3098.

Ganske, F. and Bornscheuer, U. T. 2005b. Optimization of lipase-catalyzed glucose fatty acid ester synthesis in a two-phase system containing ionic liquids and t-BuOH. *Journal of Molecular Catalysis. B: Enzymatic.*, **36**(1–6), 40–42.

García, C., Hoyos, P., and Hernáiz, M. J. 2017. Enzymatic synthesis of carbohydrates and glycoconjugates using lipases and glycosidases in green solvents. *Biocatalysis and Biotransformation*, **80**, 1–10.

Gervaise, C., Daniellou, R., Nugier-Chauvin, C., and Ferrieres, V. 2009. Influencing the regioselectivity of lipase-catalyzed hydrolysis with bmim PF$_6$. *Tetrahedron Letters*, **50**(18), 2083–2085.

Gladden, J. M., Park, J. I., Bergmann, J., Reyes-Ortiz, V., D'haeseleer, P., Quirino, B. F., Sale, K. L., Simmons, B. A., and Singer, S. W. 2014. Discovery and characterization of ionic liquid-tolerant thermophilic cellulases from a switchgrass-adapted microbial community. *Biotechnology for Biofuels*, **7**(1), 15.

Graenacher, C. 1934. Cellulose solution. U.S. Patent 1,943,176.

Ha, S. H., Hiep, N. M., Lee, S. H., and Koo, Y.-M. 2010. Optimization of lipase-catalyzed glucose ester synthesis in ionic liquids. *Bioprocess and Biosystems Engineering*, **33**(1), 63.

Hernaiz, M. J., Alcántara, A. R., García, J. I., and Sinisterra, J. V. 2010. Applied biotransformations in green solvents. *Chemistry: A European. Journal*, **16**, 9422–9437.

Ilmberger, N., Meske, D., Juergensen, J. et al. 2012. Metagenomic cellulases highly tolerant towards the presence of ionic liquids—Linking thermostability and halotolerance. *Applied Microbiology and Biotechnology*, **95**(1), 135–146.

Itoh, T. 2017. Ionic liquids as tool to improve enzymatic organic synthesis. *Chemical Reviews*, **117**(15), 10567–10607.

Jayakumar, R., Prabaharan, M., Nair, S. V., and Tamura, H. 2010. Novel chitin and chitosan nanofibers in biomedical applications. *Biotechnology Advances*, **28**(1), 142–150.

Jiang, J. H., Xiao, Y. F., Huang, W. J., Gong, P. X., Peng, S. H., He, J. P., Fan, M. M., and Wang, K. 2017. An insight into the influence of hydrogen bond acceptors on cellulose/ 1-allyl-3-methyl imidazolium chloride solution. *Carbohydrate Polymers*, **178**, 295–301.

Jing, Y. and Huang, X. 2004. Fluorous thiols in oligosaccharide synthesis. *Tetrahedron Letters*, **45**(24), 4615–4618.

Kaftzik, N., Wasserscheid, P., and Kragl, U. 2002. Use of ionic liquids to increase the yield and enzyme stability in the beta-galactosidase catalysed synthesis of N-acetyllactosamine. *Organic Process Research & Development*, **6**(4), 553–557.

Khan, N. R. and Rathod, V. K. 2015. Enzyme catalyzed synthesis of cosmetic esters and its intensification: A review. *Process Biochemistry*, **50**(11), 1793–1806.

Kim, S.-K. and Rajapakse, N. 2005. Enzymatic production and biological activities of chitosan oligosaccharides (COS): A review. *Carbohydrate Polymers*, **62**(4), 357–368.

King, A. W. T., Asikkala, J., Mutikainen, I., Jarvi, P., and Kilpelainen, I. 2011. Distillable acid-base conjugate ionic liquids for cellulose dissolution and processing. *Angewandte Chemie-International Edition*, **50**(28), 6301–6305.

Lafuente, L., Diaz, G., Bravo, R., and Ponzinibbio, A. 2016. Efficient and selective N-, S- and O-Acetylation in TEAA ionic liquid as green solvent. Applications in synthetic carbohydrate chemistry. *Letters in Organic Chemistry*, **13**(3), 195–200.

Lee, J.-C., Tai, C.-A., and Hung, S.-C. 2002. Sc(OTf)3-catalyzed acetolysis of 1,6-anhydro-β-hexopyranoses and solvent-free per-acetylation of hexoses. *Tetrahedron Letters*, **43**(5), 851–855.

Lee, S. H., Dang, D. T., Ha, S. H., Chang, W. J., and Koo, Y. M. 2008a. Lipase-catalyzed synthesis of fatty acid sugar ester using extremely supersaturated sugar solution in ionic liquids. *Biotechnology and Bioengineering*, **99**(1), 1–8.

Lee, S. H., Ha, S. H., Hiep, N. M., Chang, W.-J., and Koo, Y.-M. 2008b. Lipase-catalyzed synthesis of glucose fatty acid ester using ionic liquids mixtures. *Journal of Biotechnology*, **133**(4), 486–489.

Lee, S. H., Nguyen, H. M., Koo, Y. M., and Ha, S. H. 2008c. Ultrasound-enhanced lipase activity in the synthesis of sugar ester using ionic liquids. *Process Biochemistry*, **43**(9), 1009–1012.

Lehmann, C., Bocola, M., Streit, W. R., Martinez, R., and Schwaneberg, U. 2014. Ionic liquid and deep eutectic solvent-activated CelA2 variants generated by directed evolution. *Applied Microbiology and Biotechnology*, **98**(12), 5775–5785.

Li, H., Kankaanpää, A., Xiong, H., Hummel, M., Sixta, H., Ojamo, H., and Turunen, O. 2013. Thermostabilization of extremophilic Dictyoglomus thermophilum GH11 xylanase by an N-terminal disulfide bridge and the effect of ionic liquid [emim]OAc on the enzymatic performance. *Enzyme and Microbial Technology*, **53**(6–7), 414–419.

Liu, J., Willför, S., and Xu, C. 2015. A review of bioactive plant polysaccharides: Biological activities, functionalization, and biomedical applications. *Bioactive Carbohydrates and Dietary Fibre*, **5**(1), 31–61.

Liu, L., Zhou, S., Wang, B., Xu, F., and Sun, R. 2013. Homogeneous acetylation of chitosan in ionic liquids. *Journal of Applied Polymer Science*, **129**(1), 28–35.

Llevot, A., Dannecker, P. K., von Czapiewski, M., Over, L. C., Soyler, Z., and Meier, M. A. R. 2016. Renewability is not enough: Recent advances in the sustainable synthesis of biomass-derived monomers and polymers. *Chemistry-a European Journal*, **22**(33), 11509–11520.

Lozano, P., Bernal, B., Bernal, J. M., Pucheault, M., and Vaultier, M. 2011. Stabilizing immobilized cellulase by ionic liquids for saccharification of cellulose solutions in 1-butyl-3-methylimidazolium chloride. *Green Chemistry*, **13**(6), 1406–1410.

Luan, Y., Zhang, J., Zhan, M., Wu, J., Zhang, J., and He, J. 2013. Highly efficient propionylation and butyralation of cellulose in an ionic liquid catalyzed by 4-dimethylminopyridine. *Carbohydrate Polymers*, **92**(1), 307–311.

Mackenzie, L. F., Wang, Q., Warren, R. A. J., and Withers, S. G. 1998. Glycosynthases: Mutant glycosidases for oligosaccharide synthesis. *Journal of the American Chemical Society*, **120**(22), 5583–5584.

Malet, C. and Planas, A. 1998. From beta-glucanase to beta-glucansynthase: Glycosyl transfer to alpha-glycosyl fluorides catalyzed by a mutant endoglucanase lacking its catalytic nucleophile. *FEBS Letters*, **440**(1–2), 208–212.

Moracci, M., Trincone, A., Perugino, G., Ciaramella, M., and Rossi, M. 1998. Restoration of the activity of active-site mutants of the hyperthermophilic beta-glycosidase from Sulfolobus solfataricus: Dependence of the mechanism on the action of external nucleophiles. *Biochemistry*, **37**(49), 17262–17270.

Munoz, F. J., Andre, S., Gabius, H. J., Sinisterra, J. V., Hernaiz, M. J., and Linhardt, R. J. 2009. Green glycosylation using ionic liquid to prepare alkyl glycosides for studying carbohydrate-protein interactions by SPR. *Green Chemistry*, **11**(3), 373–379.

Murugesan, S., Karst, N., Islam, T., Wiencek, J. M., and Linhardt, R. J. 2003. Dialkyl imidazolium benzoates—Room temperature ionic liquids useful in the peracetylation and perbenzoylation of simple and sulfated saccharides. *Synlett,* **2003**, (9), 1283–1286.

Muxika, A., Etxabide, A., Uranga, J., Guerrero, P., and de la Caba, K. 2017. Chitosan as a bioactive polymer: Processing, properties and applications. *International Journal of Biological Macromolecules*, **105**, 1358–1368.

Muzzarelli, R. A. A. 2011. Biomedical exploitation of chitin and chitosan via mechanochemical disassembly, electrospinning, dissolution in imidazolium ionic liquids, and supercritical drying. *Marine Drugs*, **9**(9), 1510–1533.

Neta, N. S., Teixeira, J. A., and Rodrigues, L. R. 2015. Sugar ester surfactants: Enzymatic synthesis and applications in food industry. *Critical Reviews in Food Science and Nutrition*, **55**(5), 595–610.

Nicolaou, K. C. and Mitchell, H. J. 2001. Adventures in carbohydrate chemistry: New synthetic technologies, chemical synthesis, molecular design, and chemical biology. *Angewandte Chemie International Edition*, **40**(9), 1576–1624.

No, H. K., Meyers, S. P., Prinyawiwatkul, W., and Xu, Z. 2007. Applications of chitosan for improvement of quality and shelf life of foods: A review. *Journal of Food Science*, **72**(5), R87–R100.

Ohira, K., Abe, Y., Kawatsura, M., Suzuki, K., Mizuno, M., Amano, Y., and Itoh, T. 2012. Design of cellulose dissolving ionic liquids inspired by nature. *Chemsuschem*, **5**(2), 388–391.

Park, S., and Kazlauskas, R. J. 2001. Improved preparation and use of room-temperature ionic liquids in lipase-catalyzed enantio- and regioselective acylations. *Journal of Organic Chemistry*, **66**(25), 8395–8401.

Park, T. J., Weiwer, M., Yuan, X. J., Baytas, S. N., Munoz, E. M., Murugesan, S., and Linhardt, R. J. 2007. Glycoslylation in room temperature ionic liquid using unprotected and unactivated donors. *Carbohydrate Research*, **342**(3–4), 614–620.

Parviainen, A., King, A. W. T., Mutikainen, I., Hummel, M., Selg, C., Hauru, L. K. J., Sixta, H., and Kilpelainen, I. 2013. Predicting cellulose solvating capabilities of acid-base conjugate ionic liquids. *Chemsuschem*, **6**(11), 2161–2169.

Peniche, C., Arguelles-Monal, W., and Goycoolea, F. M. 2008. Chitin and chitosan: Major sources, properties and applications. *Monomers, Polymers and Composites from Renewable Resources*, 517–542.

Prasad, K., Murakami, M., Kaneko, Y., Takada, A., Nakamura, Y., and Kadokawa, J. 2009. Weak gel of chitin with ionic liquid, 1-allyl-3-methylimidazolium bromide. *International Journal of Biological Macromolecules*, **45**(3), 221–225.

Procopiou, P. A., Baugh, S. P. D., Flack, S. S., and Inglis, G. G. A. 1998. An extremely powerful acylation reaction of alcohols with acid anhydrides catalyzed by trimethylsilyl trifluoromethanesulfonate. *The Journal of Organic Chemistry*, **63**(7), 2342–2347.

Rahikainen, J., Anbarasan, S., Wahlström, R., Parviainen, A., King, A. W. T., Puranen, T., Kruus, K., Kilpeläinen, I., Turunen, O., and Suurnäkki, A. 2018. Screening of glycoside hydrolases and ionic liquids for fibre modification. *Journal of Chemical Technology & Biotechnology*, **93**(3), 818–826.

Rinaudo, M. 2006. Chitin and chitosan: Properties and applications. *Progress in Polymer Science*, **31**(7), 603–632.

Rosatella, A. A., Frade, R. F. M., and Afonso, C. A. M. 2011. Dissolution and transformation of carbohydrates in ionic liquids. *Current Organic Synthesis*, **8**(6), 840–860.

Sandoval, M., Cortes, A., Civera, C., Trevino, J., Ferreras, E., Vaultier, M., Berenguer, J., Lozano, P., and Hernaiz, M. J. 2012. Efficient and selective enzymatic synthesis of N-acetyl-lactosamine in ionic liquid: A rational explanation. *RSC Advances*, **2**(15), 6306–6314.

Seddon, K. R. 1997. Ionic liquids for clean technology. *Journal of Chemical Technology & Biotechnology*, **68**(4), 351–356.

Sheldon, R. A. 2016. Biocatalysis and biomass conversion in alternative reaction media. *Chemistry-a European Journal*, **22**(37), 12983–12998.

Silva, S. S., Santos, T. C., Cerqueira, M. T., Marques, A. P., Reys, L. L., Silva, T. H., Caridade, S. G., Mano, J. F., and Reis, R. L. 2012. The use of ionic liquids in the processing of chitosan/silk hydrogels for biomedical applications. *Green Chemistry*, **14**(5), 1463–1470.

Swatloski, R. P., Spear, S. K., Holbrey, J. D., and Rogers, R. D. 2002. Dissolution of cellose with ionic liquids. *Journal of the American Chemical Society*, **124**(18), 4974–4975.

Talisman, I. J., Kumar, V., Razzaghy, J., and Malhotra, S. V. 2011. O-Glycosidation reactions promoted by in situ generated silver N-heterocyclic carbenes in ionic liquids. *Carbohydrate Research*, **346**(7), 883–890.

Uju, Nakamoto, A., Shoda, Y., Goto, M., Tokuhara, W., Noritake, Y., Katahira, S., Ishida, N., Ogino, C., and Kamiya, N. 2013. Low melting point pyridinium ionic liquid pretreatment for enhancing enzymatic saccharification of cellulosic biomass. *Bioresource Technology*, **135**, 103–108.

Ungurean, M., Csanádi, Z., Gubicza, L., and Péter, F. 2014. an integrated process of ionic liquid pretreatment and enzymatic hydrolysis of lignocellulosic biomass with immobilised cellulase. *Bioresources* **9**(6), 6100–6116.

van den Broek, L. A. M. and Boeriu, C. G. 2013. Enzymatic synthesis of oligo- and polysaccharide fatty acid esters. *Carbohydrate Polymers*, **93**(1), 65–72.

Wahlstrom, R. M. and Suurnakki, A. 2015. Enzymatic hydrolysis of lignocellulosic polysaccharides in the presence of ionic liquids. *Green Chemistry*, **17**(2), 694–714.

Wahlstrom, R., King, A., Parviainen, A., Kruus, K., and Suurnakki, A. 2013. Cellulose hydrolysis with thermo- and alkali-tolerant cellulases in cellulose-dissolving superbase ionic liquids. *RSC Advances*, **3**(43), 20001–20009.

Wang, C.-C., Lee, J.-C., Luo, S.-Y., Kulkarni, S. S., Huang, Y.-W., Lee, C.-C., Chang, K.-L., and Hung, S.-C. 2007. Regioselective one-pot protection of carbohydrates. *Nature*, **446**(7138), 896–899.

Wang, F. L., Zhao, G. L., Lang, X. F., Li, J. R., and Li, X. F. 2017. Lipase-catalyzed synthesis of long-chain cellulose esters using ionic liquid mixtures as reaction media. *Journal of Chemical Technology and Biotechnology*, **92**(6), 1203–1210.

Wang, H., Gurau, G., Rogers, R. D. 2012b. Ionic liquid processing of cellulose. *Chemical Society Reviews*, **41**(4), 1519–1537.

Wang, H.-T., Yuan, T.-Q., Meng, L.-J., She, D., Geng, Z.-C., and Sun, R.-C. 2012a. Structural and thermal characterization of lauroylated hemicelluloses synthesized in an ionic liquid. *Polymer Degradation and Stability*, **97**(11), 2323–2330.

Wang, W. T., Zhu, J., Wang, X. L., Huang, Y., and Wang, Y. Z. 2010. Dissolution behavior of chitin in ionic liquids. *Journal of Macromolecular Science Part B-Physics*, **49**(3), 528–541.

Wu, J., Zhang, J., Zhang, H., He, J., Ren, Q., and Guo, M. 2004. Homogeneous acetylation of cellulose in a new ionic liquid. *Biomacromolecules*, **5**(2), 266–268.

Xie, H. B., Zhang, S. B., and Li, S. H. 2006. Chitin and chitosan dissolved in ionic liquids as reversible sorbents of CO_2. *Green Chemistry*, **8**(7), 630–633.

Xiong, X. Q., Yi, C., Han, Q., Shi, L., and Li, S. Z. 2015. I-2/ionic liquid as a highly efficient catalyst for per-O-acetylation of sugar under microwave irradiation. *Chinese Journal of Catalysis*, **36**(2), 237–243.

Yamada, K., Akiba, Y., Shibuya, T., Kashiwada, A., Matsuda, K., and Hirata, M. 2005. Water purification through bioconversion of phenol compounds by tyrosinase and chemical adsorption by chitosan beads. *Biotechnology Progress*, **21**(3), 823–829.

Yang, J. W., Perez, B. C., Anankanbil, S., Li, J. B., Gao, R. J., and Guo, Z. 2017. Enhanced synthesis of alkyl galactopyranoside by thermotoga naphthophila beta-galactosidase catalyzed transglycosylation: Kinetic insight of a functionalized ionic liquid-mediated system. *American Chemical Society Sustainable Chemistry & Engineering*, **5**(2), 2006–2014.

Yu, T., Anbarasan, S., Wang, Y. et al. 2016. Hyperthermostable thermotoga maritima xylanase XYN10B shows high activity at high temperatures in the presence of biomass-dissolving hydrophilic ionic liquids. *Extremophiles*, **20**(4), 515–524.

Zhang, H., Wu, J., Zhang, J., and He, J. S. 2005. 1-Allyl-3-methylimidazolium chloride room temperature ionic liquid: A new and powerful nonderivatizing solvent for cellulose. *Macromolecules*, **38**(20), 8272–8277.

Zhang, J. M., Wu, J., Yu, J., Zhang, X. Y., He, J. S., and Zhang, J. 2017. Application of ionic liquids for dissolving cellulose and fabricating cellulose-based materials: State of the art and future trends. *Materials Chemistry Frontiers*, **1**(7), 1273–1290.

Zhao, G., Lang, X., Wang, F., Li, J., and Li, X. 2017. A one-pot method for lipase-catalyzed synthesis of chitosan palmitate in mixed ionic liquids and its characterization. *Biochemical Engineering Journal*, **126**(Supplement C), 24–29.

9 Sponge-Like Ionic Liquids for Clean Biocatalytic Processes

*Susana Nieto-Cerón, Elena Álvarez-González,
Juana M. Bernal, Antonio Donaire,
and Pedro Lozano*

CONTENTS

9.1 INTRODUCTION

Chemistry and the chemical industry play an important role in our lives, but there is growing demand for chemical processes to follow a policy of environmental care. Ideally, the industry should be non-hazardous, sustainable, and as minimally harmful as possible for the health of people and ecosystems. It was with this aim that the Twelve Principles of Green Chemistry [1] emerged and were developed at the end of the last century. Chemical products and processes must fulfill certain requisites in order to adhere to these principles. Basically, such requirements consist of reducing or eliminating hazardous substances, and, so that the whole process remains economically viable, any new conditions should not increase the price of products. In other words, low cost processes also need to be maintained.

Ionic liquids (ILs) have appeared as new solvents with the potential to ensure the sustainability of an extremely broad range of chemical processes [2–4]: indeed, the use of ILs is a paradigmatic example of green chemistry. ILs, salts which are exclusively composed of ions, are liquid at temperatures below 100°C. Unlike most organic solvents, ILs have very low vapor pressures and, hence, are non-volatile [5–7]. They are also quite inert and non-flammable and have excellent chemical and thermal stability. Consequently, the risk of exposure and the possibility that they will damage the atmosphere are practically non-existent. Moreover, they can be easily reused and separated from their products, reducing the costs of most process in which they are used. Due to these exceptional green features and to their versatility as tunable solvents, they have a vast array of applications [8]. Although they were discovered more than a century ago [9], it is only in the last two decades that ILs have been claiming the attention of those involved in all fields of chemistry, while the number of references related to them has increased exponentially every year in the last decade [3].

ILs are made up of very different cations and anions, some of which are shown in Figure 9.1. Unlike organic solvents, ILs are organized into a network of heterogeneous nanostructures, in which coulombic attractions between ions, hydrogen bond formations, solvation effects, and dispersion forces between hydrophobic chains are responsible for the interactions that occur in the network. This, together with the

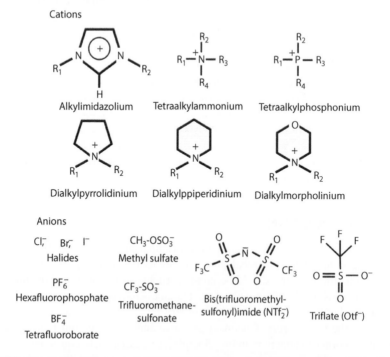

FIGURE 9.1 Some cations and anions found in the most widely used ILs. (From Poole, C.F., *J. Chromatogr. A*, 1037, 49–82, 2004; Wasserscheid, P. and Welton, T., *Ionic Liquids in Synthesis*, Wiley-VCH, Weinheim, Germany, 2007. [10,11])

enormous versatility of the ions, provides ILs with their unique properties. In fact, an extremely wide range of features (such as hydrophilicity/hydrophobicity, viscosity, melting point, electric conductivity, thermal stability, large liquid range, among others) can be obtained just through a combination of specific ions. Hence, a tailored IL can be designed for almost every reaction to obtain an optimal medium. Moreover, ILs can be easily reused, with maximum yields at minimal economic and environmental cost.

The extraordinary physicochemical nature of ILs is a crucial factor for their use as solvents in chemical processes. For example, their high polarity, as can be observed with solvatochromic probe dyes [12], allows solubilization of the different biopolymers used in enzymatic catalysis and their application in separation or extraction processes [10]. Their high density is very useful to set up biphasic systems with other immiscible solvents, favoring processes of liquid-liquid or solid-liquid separation [13]. Their high thermal stability, besides contributing to their harmlessness, opens the door to new applications in conditions that would be difficult for conventional organic solvents [6], such as ultra-high vacuum or extreme temperatures [14]. Yet, one of the most interesting properties is their low (virtually null) vapor pressure, which facilitates their total recovery and reuse, adding to their green character.

Many areas of scientific knowledge have recognized the benefits of ILs. In the field of analysis, new chromatography techniques have been developed by applying ILs [15–20]. Electrochemistry with ILs has permitted the generation of new batteries and fuel cells [21,22], while new different models of photovoltaic solar cells have also been designed [23–25]. ILs have also been used as lubricants because of their excellent tribological behavior [14,26–29] and as extraction solvents in ionic liquid-based aqueous biphasic systems [30,31]. Recently, it has been reported how microcapsules made of magnetic polyatomic liquids are able to absorb and recover oil, showing the great potential of ILs in catalysis and the treatment of wastes [32]. Furthermore, such an ability makes them an efficient tool for tackling environmental disasters, like oil spills. ILs have also been applied in nanoscience, and novel materials, such as films with carbon nanotubes [33,34] or with graphene have been synthesized [35]. They are even useful as chemical sensors [36].

Many applications of ILs have been described in the biology, medicine, and pharmaceutical areas. The ability of ILs to interact with biological systems is related to the hydrophilic features of both the cations and the anions that form them, for example, to their capacity to mix with water molecules. A number of studies demonstrate that some ILs can act as antifungal and antibacterial agents [37], and several articles point to the anticancer properties of ILs through their ability to cause oxidative stress, DNA damage, and apoptosis in cancerous cells [38–40]. ILs have also proved useful as drug carriers due to their ample range of solubilizing chemicals [41]. Among other biological applications, their usefulness in the extraction [42,43] and stabilization of nucleic acids [44] have also been demonstrated.

In this chapter, the most relevant function of ILs is related to their use as solvents in biocatalysis. The tunability of ILs provides them with an extensive range of properties that makes it possible to design solvents able to dissolve a wide spectrum of organic, inorganic, or even polymeric raw materials, opening up the possibility of using new substrates. For instance, the solubilization of cellulose by [C₄mim][Cl], described by

Rogers et al., allowed it to be used as sugar supply for fuel synthesis [45,46]. This discovery is the key to the exploitation of new renewable resources to combat the depletion of finite natural ones [47]. It may also provide an answer to the problem of waste management, with applications in plastics depolymerization (e.g., nylon [48,49] or rubber tires [50]) or for the recovery of metals from industrial wastes [47].

9.2 IONIC LIQUIDS AS NON-AQUEOUS SOLVENTS FOR BIOTRANSFORMATIONS

The reactivity and the range of substrates solvated by water limit their application in many synthetic procedures. Biocatalysis in organic solvents is the common practice for carrying out the chemical synthesis of highly appreciated products in a wide variety of industries. However, among the drawbacks of using organic solvents are the environmental damage they may cause and, in some cases, the high cost of the processes and the effect on the price the final product. For their part, ILs have been revealed as ideal solvents for a wide range of chemical reactions.

Nowadays, many organic synthetic procedures are performed in ionic liquids [11]. Among them, the following can be mentioned: hydroformylation and hydrogenation [51], epoxidation, Friedel-Crafts alkylations, Diels-Alder cycloadditions, [52], and metal-catalyzed reactions [53,54], in all of which ILs have shown their ability to improve the rate, selectivity [55], or enantioselectivity [56–58] of the processes. ILs can even play the role of catalyst, as occurs with [Bmim][N(CN)$_2$], which is able to catalyze mild acetylations of carbohydrates at room temperature [59]. In this respect, it has recently been reported that immobilized ILs act not only as solvents or extractants, but also as highly efficient catalysts in reactions that involve the oxidation of sulfides and alcohols, the epoxidation and oxidation of alkenes, Baeyer-Villiger reactions, etc. [60]. On the other hand, some ILs are able to mimic enzymatic activity. In fact, García-Verdugo et al. [61] described the aldolase activity of the imidazolium ring of certain ionic liquids, which acts as the catalytic histidine residue of some enzymes to produce the synthesis of vitamin esters [62].

More importantly, the ability of ILs as a solvent medium to support biocatalytic reactions should be emphasized. Enzymes are natural catalysts that combine high efficiency and selectivity with mild reaction conditions. These features make it possible to develop processes with high atom and energy economy and limit waste generation in accordance with the principles of sustainability. In agreement with this, ILs are excellent non-aqueous media for enzymatic transformations, not only because both hydrophilic and hydrophobic substrates and enzymes can be solved in the adequate IL mixture, but also due to their flexibility for designing reasonably inexpensive approaches for the green purification and extraction of the products. In fact, ILs can facilitate extraction procedures due to their different degrees of solubility with organic solvents (Figure 9.2). As a consequence, both enzyme and IL can be completely recovered for reuse, maintaining the efficiency over several cycles.

Among enzymes, lipases stand out because of their wide catalytic versatility. They can perform different reactions such as C–C or C–heteroatom bond formation, oxidative processes, and hydrolytic reactions [64]. Lipases are the most common enzymes used in synthetic organic chemistry for hydrolysis, esterification, transesterification

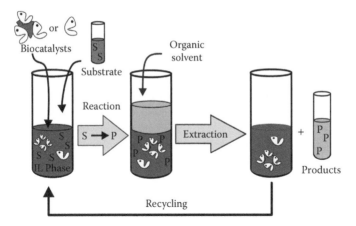

FIGURE 9.2 Cycle of synthesis, product extraction, and recovery of the IL-immobilized enzyme system for reuse. (From Lozano, P., *Green Chem.*, 12, 555–569, 2010. [63])

by acidolysis, transesterification by alcoholysis, inter-esterification, or aminolysis, reactions that are highly valued for the production of pharmaceuticals and drugs [65] and in the biodiesel [66] and food [67] industries.

Of particular note is the improvement in enantioselectivity made possible in enzymatic reactions carried out with ILs. The first attempts in this sense were carried out by Itoh et al. [68], who achieved the enantioselective transesterification of allylic alcohols in an imidazolium-based ionic liquid solvent. Simultaneously, Schöfer et al. [69] obtained the kinetic resolution of 1-phenylethanol. Since then, multiple approaches for enantioselective transformations have been described [70,71]. For example, [Bmim] [NTf$_2$] was shown to be a very efficient medium for the kinetic resolution of *rac*-2-penthanol with vinyl propionate, where it increased the activity and selectivity compared with that possible in hexane [72]. Other ILs were seen to act as excellent solvents for the kinetic resolution of *rac*-menthol [73], *rac*-1-phenylethanol [74], aromatic *sec*-alcohols [75], 4-phenylbut-3-en-2-ol [76], and *rac*-3-phenyllactic acid [77].

Many approaches combine enzymes and ILs to attain high conversion rates and selectivity in biocatalytic transformations. Some examples are:

- Direct addition of the biocatalysts to the IL medium containing substrates. This strategy was used in the laccase oxidation of catechol in [Bmim][Br] or [Bmim][N(CN)$_2$] [78]; the synthesis of flavor esters in [Bmim][PF$_6$] [79,80]; the kinetic resolution of 5-phenyl-1-penten-3-ol with vinyl acetate in [Bmim][PF$_6$] [81]; or the synthesis of biodiesel by alcoholysis of vegetable oils with short chain alcohols [82,83].
- Coating the enzyme with a suitable IL. For instance, in the enantioselective transesterification of vinyl acetate with racemic alcohols [84] or in the encapsulation of enzymes in IL polymers [85]. The direct IL-coating of cellulase carried out by our group provided enhanced enzyme thermal stability [86].

- Supported liquid membranes also improve the product separation step [87,88]. This strategy was used in the in the kinetic resolution of S-ibuprofen [89] and S-ketoprofen [90], as active anti-inflamatory drugs, and in the dynamic kinetic resolution of rac-1-phenylethanol [91].
- ILs may also be immobilized onto the surface of inorganic or organic solid supports, either non-covalently (supported ionic liquid phases, SILPs) [6], or covalently-attached ILs (supported ionic liquid-like phases, SILLPs) [92–94]. These strategies were used in the transesterification of vinyl butyrate with 1-butanol under continuous biphasic liquid-liquid conditions [95], in the asymmetric acylation of 1-phenyethanol [96], and in the dynamic kinetic resolution of alcohols [97].

9.3   SPONGE-LIKE IONIC LIQUIDS: A NEW SUPPORT LOOKING FOR CHEMICAL SUSTAINABILITY

9.3.1   SPONGE-LIKE IONIC LIQUIDS: THE IDEA AND POTENTIAL BENEFITS

Traditional chemical synthesis usually involves the use of organic solvents, the unavoidable production of wastes and by-products, and the need for distillation or analogous purification processes to separate the desired compounds. The organic solvents, needed to carry out the reactions and to extract the products, are contaminant and dangerous, while their elimination is expensive from both economic and environmental points of view. Since wastes and by-products are not reused in conventional processes, which also increases costs. Moreover, purification often involves distillation, which is expensive energetically and, certainly, a non-green process.

As indicated earlier, applying green chemistry to chemical processes is increasingly necessary to reduce both the economic and environmental burden. With this in mind, our group has developed an interesting methodology for achieving green biocatalytic processes in organic synthesis. The idea arose from the well-known fact that cations in ILs containing long alkyl side chains (typically from C_{12} and earlier, Table 9.1) are able to solubilize hydrophobic substrates when ILs are heated to above their melting points, giving rise to a homogeneous liquid phase (Figure 9.3). Cooling maintains the homogeneity, but the mixture is now in solid state, with both the ILs and hydrophobic compounds forming one phase. The novelty of this idea is that the ILs, products, and water can be easily separated by simple centrifugations and changes in temperature. The centrifugation step separates the more dense phase, corresponding to the solid ILs (at the bottom), from the less dense phase containing the product and the remaining fatty acids (as the supernatant), while water, if present, remains in the intermediate phase. At that point, the ILs can be "wrung out" by a combination of decreasing temperatures and centrifugation to release the product. This behavior mimics the action of a sponge, which is why the ILs were dubbed "sponge-like ionic liquids" (SLILs).

The separation can be more efficiently achieved if the process is repeated at different temperatures (the lower the temperature, the better the separation) and/or if a nylon filter is used. A further advantage of the method is that the enzyme is retained in the ILs after centrifugation and can therefore be reused, minimizing wastes. The improvement of this methodology resides in the easy and inexpensive separation

TABLE 9.1

Commonly Used SLILs and Their Melting Points

Sponge-Like Ionic Liquid	Abbreviation	Melting Point (°C)
1-Methyl-3-octadecylimidazolium bis((trifluoromethyl)sulfonyl)imide	[C$_{18}$mim][NTf$_2$]	53
1-Hexadecyl-3-methylimidazolium bis((trifluoromethyl)sulfonyl)imide	[C$_{16}$mim][NTf$_2$]	46
1-Tetradecyl-3-methylimidazolium bis((trifluoromethyl)sulfonyl)imide	[C$_{14}$mim][NTf$_2$]	33
1-Hexadecyl-3-methylimidazolium hexafluorophosphate	[C$_{16}$mim][PF$_6$]	74
1-Tetradecyl-3-methylimidazolium hexafluorophosphate	[C$_{14}$mim][PF$_6$]	67
1-Dodecyl-3-methylimidazolium hexafluorophosphate	[C$_{12}$mim][PF$_6$]	58
1-Methyl-3-octadecylimidazolium tetrafluoroborate	[C$_{18}$mim][BF$_4$]	60
1-Hexadecyl-3-methylimidazolium tetrafluoroborate	[C$_{16}$mim][BF$_4$]	49
1-Tetradecyl-3-methylimidazolium tetrafluoroborate	[C$_{14}$mim][BF$_4$]	36
1-Dodecyl-3-methylimidazolium tetrafluoroborate	[C$_{12}$mim][BF$_4$]	30
Octadecyltrimethylammonium bis((trifluoromethyl)sulfonyl)imide	[C$_{18}$tma][NTf$_2$]	74
Hexadecyltrimethylammonium bis((trifluoromethyl)sulfonyl)imide	[C$_{16}$tma][NTf$_2$]	64
Tetradecyltrimethylammonium bis((trifluoromethyl)sulfonyl)imide	[C$_{14}$tma][NTf$_2$]	52

Source: Lozano, P. et al., *Fuel*, 90, 3461–3467, 2011; Lozano, P., *Green Chem.*, 14, 3026–3033, 2012 [98,99].

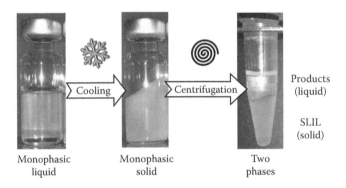

FIGURE 9.3 Schematic behavior of SLILs. (From Lozano, P. et al., *Energy Environ. Sci.*, 6, 1328–1338, 2013. [100])

procedure (cooling, instead of heating), together with a minimal loss of the (solid or liquid, non-gaseous) products.

9.3.2 SLIL STRUCTURE

To understand the mechanism of SLILs acting as biocatalysis, their structural arrangement at a molecular level must be known. Indeed, a large number of works have been undertaken with this aim. Several groups have characterized ILs by X-ray in studies that revealed an ordered organization of the ionic species in ILs in the solid state [101–103]. This order is imposed by a network of hydrogen bonds between the cation and the adjacent anions, on the one hand, and by dispersion (van der Waals)

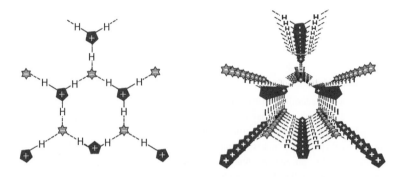

FIGURE 9.4 Arrangement of 1,3-dialkylimidazolium salts. Left: A monolayer view. Right: Three-dimensional array of the layers. Free channels provide polar and non-polar "nano" regions (i.e., hydrophobic and hydrophilic zones) with a high degree of directionality. (From Stassen, H.K. et al., *Chem. Eur. J.*, 21, 8324–8335, 2015. [102])

forces between the hydrophobic alkyl chains, on the other. For ILs containing the imidazolium cation (the most common and most studied cation in ILs), at least three anions interact with the imidazolium moiety through hydrogen bonds, and *vice versa* (Figure 9.4). Additionally, π-π stacking interactions between adjacent heterocyclic rings are responsible for the order within the ILs.

As many studies performed with a wide range of spectroscopies have demonstrated [103–105], this well-arranged nanostructure is basically maintained not only in solid state, but also in liquid state and, partially, in gaseous state. By applying neutron diffraction spectroscopy, Hayes et al. [101,106] discovered the existence of polar and non-polar domains, that depend on the length of the cation. The domains are flexible and have the ability to reorder themselves, in a process whereby holes are continuously created and disappear. They also observed that the anionic nature had little effect on the structure. Metastable atom electron spectroscopy allowed Iwahashi's group to explore the organization of the surfaces in 1-alkyl-3-methyl-imidazolium ILs (C_4, C_8, C_{10}) [107]. Basically, these ILs were seen to consist of a double-layered nanostructure, where the hydrophobic alkyl chains face each other and are flanked by the polar domains (anions and imidazolium rings) protected by domains of a polar nature. Other groups [107–109] described analogous arrangements in alternating polar and non-polar parallel layers that accounted for the tribological properties of ILs containing long-chain cations, such as [C_{16}mim][BF_4]. Interestingly, an increase in the length of the side chains produces stronger hydrophobic bilayer interactions and, in turn, higher packing degrees. Consequently, segregation between the polar and dipolar bilayers can be achieved by lengthening the cation alkyl side chain. Computational studies have indicated that a similar arrangement can also be attained by functionalization of alkyl chains as short as C_2 with an ester group [110].

The plasticity of the arrangement of the ionic species in SLILs has also been studied. The sponge-like structure may be affected in several ways (interaction with different surfaces, interfaces, electrical potentials, mechanical perturbations, etc.) and adopt alternative structures of greater extension and for longer times. Recent computational studies described the ability of symmetric ILs, like 1-methyl-3-octylimidazolium octylsulfate $[C_8mim][C_8-O-SO_3]$, to change from sponge-like to lamellar structures in response to vacuum exposure [111].

9.3.3 Enzymes Trapped in Ionic Liquids: A Close Look to Their Arrangement

Reflecting their versatile structure, polar and non-polar domains are present in SLILs, and so substrates of different solvation degrees can be accommodated inside the holes that are being continuously created in their nanostructure. This property has no size limitations. Macromolecules, such as enzymes, can also be lodged in the polar domains of the ionic network, where the non-polar surroundings enclose the native enzyme within the strong ionic network (Figure 9.5).

Even though ILs are non-aqueous solvents, all of them are hygroscopic to different degrees. The water present in the complex ionic matrix affects the chemical-physical properties of the ionic liquid. A recent study with specially designed imidazolium ILs revealed that water is trapped in the ionic environment, establishing strong hydrogen bonds with the imidazolium rings in solid, liquid, or gas states and forming a guest@host supramolecular structure [113]. However, the low amount of water is the key to ensuring that an essential hydrating shell will allow the enzymes to maintain their active conformation [114,64]. Itoh et al. [68] were the first to demonstrate the enzyme activity of immobilized lipases in different ILs. They carried out the asymmetric transesterification of rac-5-phenyl-1-penten-3-ol with vinyl acetate. Far from deactivating the enzyme, a boost in the activity was observed compared with that observed with hydrophilic solvents, perhaps because the hydrophobicity of the ILs helps to preserve the hydrating coat of the enzymes and, consequently, their tertiary structure. By contrast, hydrophilic solvents tend to strip the essential water molecules, causing enzyme deactivation [115].

The combination of the limited amount of water and the strength of the ionic network is determinant in improving the stabilization of the enzymes. Moreover, some immobilized enzymes are stable and functional at higher temperatures than the optimal [98,116–118], which has been explained as follows: the enzyme unfolding at high temperatures is associated with an increase in the kinetic energy of the water

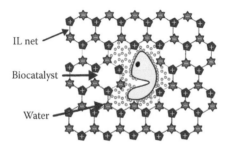

FIGURE 9.5 Native folded enzyme trapped in a wet hole of a water-immiscible IL network. (From De Diego, T. et al., *Biomacromolecules*, 6, 1457–1464, 2005. [112])

molecules when heated. in this case, the non-polar domains in the IL nanostructure create a "mold" that avoids wide-ranging molecular movements [63,116,112].

At the same time, the anchorage/incorporation of the biocatalyst in the strong ionic net serves as a new tool for immobilization in order to facilitate its recovery. Unlike attempts of enzyme immobilization using other supports (Eupergit C, α-alumina, ceramic membranes, or Fe_2O_3 magnetic particles) coated or not with hydrophilic polymers (PEG, PEI, etc.) and/or different techniques (covalent binding, adsorption, cross-linking, entrapment in sol-gel matrices) [119–122], immobilization in ILs provides a series of advantages for the enzyme activity. Among these may be mentioned the stabilization and protection of the enzymes over a wide range of temperatures, the enhancement of their catalytic activities, improved enantioselectivity, and simple protocols, such as ultrafiltration for enzyme recovery [123].

9.3.4 SLIL Separation Mechanism

The key to their sponge-like behavior is the segregation that exists between polar and apolar layers in the IL arrangement. Indeed, due to the plasticity of SLILs, hydrophobic substrates can be accommodated among the apolar layers, making the IL-substrate mix homogenous, whether in solid or liquid state (Figure 9.6). The supramolecular aggregates of imidazolium-based ILs are depicted in Figure 9.6a, where non-polar cation chains are organized in parallel layers flanked by highly cohesive charged groups. This results in the generation of hydrophobic holes able to accommodate the hydrophobic components of a reaction (substrates and products, Figure 9.6b), "soaking" the SLILs, in the manner of a wet sponge. It should be emphasized that the products are included in the ionic matrix, not dissolved by the ILs.

The homogeneous medium, solid at room temperature, can be "wrung out" through controlled centrifugations at decreasing temperatures to obtain the products (Figure 9.6c). In this strategy, the combination of low temperatures and centrifugal force allows gentle compaction of the ionic network, reducing the size of the holes and facilitating the removal of the products. Nevertheless, as in a wet sponge, which remains damp even after it is wrung out, a minimal fraction of the hydrophobic products may still be retained due to the strong interactions with the long alkyl chains of the SLIL cations.

While the usual separation approach in other molecular solvents is based on distillation at high temperatures [124], which involves high costs, the exceptional features of SLILs allow products to be recovered by means of two alternating processes of cooling and centrifugation. This is a great advance for developing simple, clean, and green cyclic protocols for enzymatic synthesis, for product separation, and enzyme-IL recovery and reuse, which can be extrapolated to industrial scale applications. As neither organic solvents nor high consuming purification techniques are needed, the use of SLILs is greener and substantially less expensive than other traditional approaches. A simple series of controlled centrifugations allows the recovery of the products either by decantation or after filtration through a centrifugal nylon filter (0.2 μm pore size). In this last case, the SLIL remains in the filter, while the clean products pass through the filter membrane.

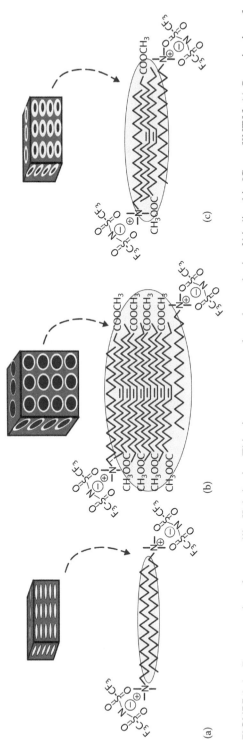

FIGURE 9.6 Representation of sponge-like IL behavior. This picture corresponds to the synthesis of biodiesel in [C$_{18}$tma][NTf$_2$]. (a) Organization of the parallel layers of the IL net, as in a dry sponge. (b) Methyl oleate is incorporated into the hydrophobic holes, causing the sponge to swell. (c) The "damp" sponge after wringing out by centrifugation.

9.4   SPONGE-LIKE IONIC LIQUIDS IN GREEN PROCESSES

9.4.1   Synthesis of Flavor Esters

The sponge-like behavior was first described during the lipase-catalyzed synthesis of natural flavor esters (geranyl acetate, citronellal acetate, neryl acetate, and iso-amyl acetate) in the IL $[C_{16}mim][NTf_2]$ [99]. These products are of great value in the food, cosmetics, and pharmaceutical industries. Nowadays, fragrance compounds are mostly chemically synthesized by non-green processes, in which neither the (organic) solvents nor the by-products are reused. The extraction of fragrance compounds from natural products is a difficult task (low yields and, consequently, expensive). A novel solution presents itself in the form of lipases in ILs. Lipases can convert natural fatty acids and polyols into the desired flavor esters by controlled esterification reactions. The process is highly efficient if it is performed in SLILs.

In the cited work, the highly hydrophobic IL $[C_{16}mim][NTf_2]$ provided a homogeneous liquid medium with the reactants above 50°C [99]. In these conditions, esterification was straightforward (more than 80% in two hours for the four studied carboxylic acids), the longer the cation alkyl side chain, the more efficient the process ($C_{18}tma > C_{16}tma > C_{14}mim > C_{12}mim$). Microwaves even more accelerate the rate of biocatalysis [125]. These yields must be compared with those obtained when solvent-free esterification was carried out (14.9% in four hours reaction for isoamyl acetate). In the absence of ILs, the high concentrations of polyols tended to deactivate the enzyme, and, moreover, a biphasic system was created. The hydrophobic ILs keep the enzyme folded, but also sequester the alcohols, preventing them from reaching high concentrations in the surroundings of the enzyme. These two features are crucial for a high reaction efficiency.

In this sense, the imidazolium cation provides the enzyme with an additional stabilization compared with the reaction performed in conventional organic solvents. Indeed, our group demonstrated that free lipase B from *Candida antarctica* provided substantially higher yields for the synthesis of butyl butyrate when four hydrophobic ILs, comprising dialkylimidazolium cations and perfluorinated or bis(trifluoromethyl) sulfonyl amide anions, were used compared with those obtained in 1-butanol or hexane. The stabilizing effect of the imidazolium cation and the protective effect of the ILs on the enzyme were maintained even after seven operation cycles [126]. Simultaneously, the effect of the nature of both cation and anion was checked in the synthesis of N-acetyl-L-tyrosine propyl ester [126] by using alpha-chymotrypsin as biocatalyst. An extended length of the side chain of the imidazolim cations also contributed to the efficiency of the reaction, as did the anion size. Taking all the aforementioned together, it is concluded that highly apolar ILs present a high degree of protection to the enzyme.

As mentioned in the previous section, the great advantage of the sponge-like behavior of some ILs is the ability to easily separate products from the reaction medium, which, in turn, contains the enzyme that can be reused. Once the reaction has taken place in the liquid state, the homogeneity of the medium is maintained even if the temperature falls sufficiently for it to reach the solid state. Subsequent controlled centrifugations of the homogeneous solid medium leads to its fractionation into a biphasic system, where the solid IL remains at the bottom, while the products are released into the liquid upper phase [99] (Figure 9.7).

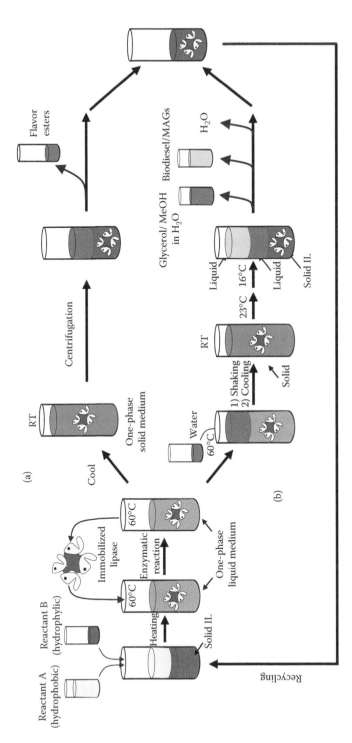

FIGURE 9.7 General scheme of synthesis of products and their purification based on the behavior of sponge-like ionic liquids. Initial step of synthesis is common to the three reactions described in this chapter. (a) Extraction of flavor esters (Section 4.1). (b) Purification of biodiesel (Section 4.2) and monoacylglycerides (Section 4.3).

In order to completely rescue the product, cycles of centrifugation and cooling at progressively lower temperatures (25°C, 21°C, 10°C) were performed with a 50/50 $[C_{16}tma][NTf_2]$/flavor mixture. At these low temperatures, the system continues being monophasic, but solid. Then, centrifugation compacts the ILs, while the product is released, and the system becomes biphasic. Only the accurate application of subsequent centrifugations at decreasing temperatures permitted the extraction of pure products. Indeed, when the reaction medium was rapidly solidified by submerging it in an ice bath, the released product still contained traces of ILs as shown by ^{19}F NMR [99]. This may be explained by the mechanism of compaction. While direct centrifugation at low temperatures leads to a rapid and disorganized packing, resulting in the remaining ILs appearing in the upper phase, successive centrifugations at decreasing temperatures lead to the gradual ordered compaction of the ILs and the release of the pure products.

It is interesting that the same hydrophobic interactions between the products and the long chains in the ILs cation that allow their accommodation in the ionic matrix also prevent their total release. This results in a decrease in the efficiency of product recovery, a decrease that is particularly evident as the number of interactions with long chain products increases. In fact, in the synthesis of flavor esters, the longer the alkyl chain length (for instance in the series isoamyl acetate, neryl acetate, citronellyl acetate, geranyl acetate), the lower the recovery is (82%, 60%, 59%, and 51%, respectively) [99].

A further purification step can be performed using centrifugal filters, which retain the solid SLILs in the nylon membrane, and thus facilitate its separation from the liquid product. Yet, even though these devices greatly improve the recovery, they fail in selectivity. This was observed for the fragrance anisyl acetate: when solid mixtures of 30/70 (w/w) anisyl acetate/$[C_{14/16}tma]$ $[NTf_2]$ were submitted to fractionation with a 0.2 μm pore size nylon filter, the recovery yield reached 95% and 93% of the total fragrance in the mixtures $[C_{14}tma][NTf_2]$ and $[C_{16}tma][NTf_2]$, respectively, however, the purity was affected, and the samples still contained 4.1% and 3.1% (w/w) of residual SLILs [100].

9.4.2 BIOSYNTHESIS OF BIODIESEL

The exponential growth in the use of fossil fuels in the last century has made it necessary to look for alternative and renewable fuels. In this respect, biodiesel has become one of the most widely used propellants since the 1980s. It has been pointed out that biodiesel is an eco-friendly fuel that emits CO_2 levels lower than conventional diesel or other petroleum-based fuels. Biodiesel is composed of fatty acid methyl esters (FAMEs), its main source arising from transesterification of vegetable oils. Although it is considered a renewable sustainable resource, the process of its synthesis is far from being so. Indeed, biodiesel is obtained from the chemical alcoholysis (methanol or ethanol as the typical reactants) of natural triacylglycerides catalyzed by strong acids or, preferentially, in alkaline conditions (NaOH or KOH). Glycerol is the by-product of the reaction. Also, side products (sludge or soap) are obtained in the medium, reducing the quality of the products. As a consequence, the complete process requires many washing steps and subsequent distillation for water elimination. The process is energetically expensive and the wastes are not easy to eliminate. In summary, the current industrial approach does not fulfill the requirements of green chemistry.

Enzymatic catalysis is the preferred approach to fulfill such green chemistry principles in biodiesel synthesis. The product obtained using lipase as biocatalyst is less contaminated than that obtained in alkaline conditions. The immobilized enzymes that are usually employed can be recycled, while their catalytic activities are higher than those of free enzymes. Nevertheless, biocatalysis also has drawbacks that are difficult to avoid. First, as both substrates (triglycerides and methanol) are nonmiscible, the medium is biphasic and, consequently, the efficiency of the reaction is not optimal. Second, the interaction between the lipase and its surrounding alcohol molecules causes the dehydration and deactivation of the enzyme. This is partially prevented by the gradual addition of methanol to the reaction medium, but this does not avoid the inability to reuse the enzyme. Moreover, the by-product glycerol also contributes to the "poisoning" of the lipase, since it can stick to the immobilized particles that form the enzyme. Again, this leads to a reduction in the efficiency of the process. Finally, biodiesel is not easily separated and recovered from the medium. In summary, the process needs to be improved both for economic reasons and to increase sustainability.

In this context, biodiesel production performed in ILs [82,83,127] or in SILLPs [128] have been proposed as platforms to overcome the aforementioned problems. Dupont's group [82] anchored the lipase from *Pseudomonas cepacia* in the ionic liquid 1-*n*-butyl-3-methylimidazolium bis(trifluoromethylsulfonyl)imide ([Bmim] [NTf$_2$]). This hydrophobic IL was seen to be highly efficient not only in terms of stabilization of the enzymatic activity (maintained for at least four operational cycles), but also in the elimination of glycerol from the reaction medium since it was incorporated into the ionic network. Even so, this reaction medium was multiphasic, and due to the low solubilization of triglycerides and to enzyme deactivation (as a consequence of glycerol accumulation), the results only showed moderate efficiency (90% yield for 24 hours at 60°C).

The subsequent use of SILLPs improved the synthesis. A further step forward was the use of continuous flow processes in biphasic ILs/supercriticCO$_2$ reactors, in which ILs long alkyl cations improved the diffusion of long chain fatty acids, providing higher reaction yields [128]. Again, this approach also had drawbacks, the main one being the excessive costs of high-pressure equipment for mass transfer when using SILLPs and supercriticCO$_2$.

The application of SLILs to the synthesis of biodiesel helps to avoid wastes and to optimize the efficiency of the process. By replacing the IL solvent by another one with a longer alkyl chain cation (like [C$_{18}$mim][NTf$_2$]), the medium became monophasic, dissolving both triolein and methanol and increasing the reaction rate and yield (96% biodiesel yield in just six hours of reaction [83]). Our group also demonstrated the suitability of hydrophobic ILs based on the [NTf$_2$] anion and the alkyltrimethylalkylammonium ([C$_{12-18}$tma]) cation to solubilize triolein-methanol mixtures. Monophasic behavior was easily attained using cations with longer alkyl chains at 60°C. The most intriguing aspect in this respect is that although pure C$_{16}$ and C$_{18}$-ILs melt above that temperature (Table 9.1), their melting points decrease in the reaction medium. The explanation lies in the reduction of the hydrophobic interactions between the cation chains as a consequence of the accommodation of hydrophobic compounds (the substrate triolein or the biodiesel product) among them. $[C_{18}tma]$ $[NTf_2]$ was seen to be ideal for biodiesel biocatalysis with 100% yield in eight hours, while protecting the enzyme activity with a half-life up to 1370 days at 60°C [100].

The most interesting novelty of using hydrophobic ILs with the tetramethylammonium cations was the development of an easy protocol for the extraction of the reaction components based on the sponge-like behavior. There is a great interest in the reuse of waste cooking oils for biodiesel synthesis, and many studies have focused on this every-day waste as a supply of free fatty acids, avoiding competition with the use of agricultural land best dedicated to crops for human consumption. The use of SLILs represents a clean method to synthesize and recover biodiesel to obtain the most of this waste, while reducing its accumulation. Bearing in mind that the earlier process renders two products, biodiesel and glycerol, the strategy of purification was slightly modified to incorporate an extra step of water elimination. After shaking and cooling at room temperature, the medium became homogenous and solid. However, successive centrifugations at lower temperatures led to the formation of three clearly separate phases (Figure 9.7): a lower solid IL phase, an intermediate aqueous phase containing glycerol and the excess of methanol, and an upper low-density biodiesel liquid phase [100].

The recovery of glycerol adds another plus to this green protocol because of its high value for other synthesis processes (e.g., the synthesis of chemicals or the production of hydrogen gas). Moreover, glycerol itself may act as a fuel additive, and, as a fermentation substrate for bacterial biomass, it is useful for wastewater treatment, etc [129,130]. Two opposing, but equally useful, behaviors of SLILs for green catalytic processes were demonstrated in a previous study performed by our group [100]: their ability to create homogeneous reaction media that enhance catalytic efficiency, and their use in a protocol of heterogeneous separation, based on temperature and their miscibility with other solvents.

Following this, other strategies for biodiesel synthesis have recently been developed. New formulations are based on the combination of FAMEs and fatty acid glycerol esters, taking advantage of some specific properties (viscosity, cetane number, adiabatic flame temperature, among others) [131]. In particular, solketal (1,2-isopropylideneglycerol) is a glycerol derivative that provides fatty acid solketal esters (FASEs), which are highly appreciated as fuel additives [132].

SLILs and the alcohols methanol and solketal can be used for the synthesis of green oxygenated biofuels, FAMEs and FASEs, using two related approaches: (1) transesterification with vegetable oils (e.g., sunflower, olive, cottonseeds, and waste cooking oil) and (2) direct esterification with different long chain fatty acids (e.g., lauric, myristic, palmitic, and oleic acids) [133].

The IL $[C_{18}tma][NTf_2]$ again revealed its suitability for the biocatalysis of FAMEs and FASEs, irrespectively of the source of the fatty acids or alcohols, and with a yield close to 100% in six hours at 60°C. The innovation here was the demonstration of the ability of this IL to support the combination of both biocatalytic processes (reactions involving mixtures of vegetable oils/free fatty acids and methanol/solketal), resulting in oxygenated biofuels containing *ca.* 80% FAMEs and *ca.* 20% FASEs [133].

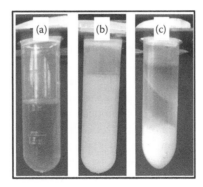

FIGURE 9.8 Extraction of oxygenated fuels from waste cooking oil after water addition and centrifugations. (a) Homogeneous reaction medium. (b) Heterogeneous solid medium after water addition and cooling to room temperature. (c) Iterative centrifugations results in three phases, the low density biofuels appearing in the top one. The color of the SLIL reveals the total release of the products. (From Lozano, P. et al., *ACS Sustain. Chem. Eng.*, 4, 6125–6132, 2016. [133])

Even more importantly, the sponge-like nature of $[C_{18}tma][NTf_2]$ allowed easy extraction of the products and recovery of the SLIL-enzyme system. While direct centrifugation of the solid reaction was ineffective because of the low level release of impure product (Figure 9.8), the addition of water was determinant in achieving the complete separation of pure FAMEs/FASEs from the SLIL and for recovering the hydrophilic components of the mixture (glycerol and non-reacted alcohols). The stability of the SLIL-enzyme system was assessed over six 24-hour reaction cycles at 60°C.

9.4.3 BIOSYNTHESIS OF MONOACYLGLYCERIDES

Monoacylglycerides (MAGs) are amphipathic molecules widely used as surfactants and emulsifiers in many industries. MAGs confer high plasticity to the fibers in textile products. They are widely used as additives and dietary supplements in the food industry. MAGs have antimicrobial action [134] and may act as drug carriers [135]. Some of them have also revealed their protective effect in prostatic hyperplasias [136,137] and cardiovascular diseases [138].

MAGs are obtained from the direct hydrolysis of triacylglycerides or by the esterification of one hydroxyl group of the glycerol molecule with the corresponding fatty acid. Chemical glycerolysis implies high temperatures (210°C–250°C), which often confers inappropriate organoleptic features on the MAGs obtained by this method, such as a dark color and burnt taste, which limits their use in the food and pharmaceutical industries. Moreover, the process provides a very low yield (~35%–60%) [139] and specificity (35%–50% Diacylglycerides [DAGs], 1%–20% Triacylglycerides [TAGs], 1%–10% Free fatty acids [FFAs], and residual metal salts) and is also expensive due to the need for specific inorganic catalysts. The purification step is also expensive since it involves molecular distillation, a high energy consuming process [140].

To obtain greener, more specific, and efficient methods for producing MAGs by glycerolysis that circumvent the aforementioned drawbacks, the use of enzymes in

ILs has been considered. In fact, a highly efficient method (90% yield) for the glycolysis of triacylglyceraldehydes in the IL CPMA-MS with lipase as biocatalyst has been reported [141,142]. The hydrophobicity of CPMA-MS allows the solubilization of TAGs, while the hydrophilic ethoxyl or hydroxyl moieties trap MAGs by means of a hydrogen bond network. This leads to higher yields and selectivities [143]. Nevertheless, the high cost of this IL prevents its use at industrial scale.

Direct esterification of glycerol is the other approach used for MAG synthesis. As mentioned in the previous section, glycerol is the main by-product of biodiesel production and using it as a resource for other applications can be considered a priority for green chemistry. For this purpose, both chemical and enzymatic catalysis have been applied. Mesoporous silicates have been used and have been seen to provide high yields (92%) with acceptable degrees of selectivity (62%). However, the working temperatures continued to be relatively high (>100°C), adding to costs. Direct enzyme-catalysis esterification of glycerol is an interesting alternative for the specific synthesis of MAGs under mild conditions, while immobilized lipases have also been applied for the synthesis of MAGs, with yields and specificities slightly lower than those obtained with silicates (75% yield, 40% selectivity).

The synthesis of MAGs from glycerol presents two major problems. First, DAGs and TAGs are obtained in the process, while a large excess of glycerol is necessary to obtain the monoester. Second, the polar glycerol is immiscible with hydrophobic fatty acids, which significantly reduces the yield. Both surfactants and organic solvents have been applied to the synthesis of MAGs, with high yields and relatively good selectivity. For instance, the use of surfactants, such as cocamidopropyl betaine as additive increases the selectivity toward monolaurin ester catalyzed by Novozym 435 compared with glycerolysis.

In this context, ILs have also been seen to act as efficient media for synthesizing MAGs from glycerol, with their aforementioned advantages that include moderate reaction temperatures, high yields and selectivity, reuse of products—in short, a green process. MAGs made up of long alkyl side chains (like capric, lauric, myristic, palmitic, and oleic acids) show high water solubility and so are of great value in the pharmaceutical and food industries [144–146]. Our group synthesized MAGs from glycerol using hydrophobic SLILs, demonstrating not only the ability of some of these SLILs to achieve the highest possible yields (100% conversion) with total selectivity, but also the greenness of the procedure, which reuses both ILs and the biocatalyst for several cycles [147].

The direct synthesis of long alkyl side chain MAGs was carried out in SLILs containing the methylimidazolium cation with different alkyl chains (C_{10}, C_{12}, C_{14}, C_{15}, and C_{18}) and with two anions of opposing behaviors, the hydrophobic NTf_2 and the hydrophilic BF_4 anions. Although all the ILs tested provided homogeneous monophasic media for biocatalysis, both the conversion and specificity of the reaction were strongly dependent on the nature of the cation and the anion. For both anions, the highest conversions were obtained with the C_{12}mim cation (100% yield). The FFAs were able to accommodate themselves in the ILs when both presented the same length: interchange interactions were maximal, as was the diffusion of this reactant in the hydrophobic surroundings of the ILs, allowing the FFAs to easily reach the enzyme particles.

The selectivity of the reaction was directly dependent on the nature of the anion. The hydrophobic NTf_2 anion provided a mixture of MAGs and DAGs. The selectivity for MAGs varied between 48% (lauric acid in $C_{16}mimNTf_2$) and 85% (lauric acid, $C_{12}mimNTf_2$), whereas ILs containing hydrophylic BF_4 anions provided, in all ILs, 100% of specificity toward MAGs production. The reason for this is related to glycerol diffusion. Indeed, as previously mentioned, the solubility of glycerol is the first limiting step in the direct synthesis of MAGs. In moderately hydrophobic (i.e., partially hydrophilic) ILs, such as those containing the BF_4 anion, glycerol can be efficiently transferred from the hydrophilic surroundings, where it is solved, to the hydrophobic environments, where the lipase is located. Thus, as soon as the first reaction takes place, the generated MAGs will be diffused among the hydrophobic cations and another glycerol molecule can be allocated in the enzyme cavity. When the anion is hydrophobic (the NTf_2 case), the monoacylglycerol cannot diffuse so easily, hence, the same molecule can be esterified twice before a new glycerol molecule can reach the lipase active site.

As in the previous sections, the most interesting aspect of sponge-like behavior is the ease with which products can be extracted and the SLILs reused, in addition it is possible to reuse the enzymes. Based on this possibility, the recovery of monoolein/$C_{18}mimNTf_2$ was tuned. The scheme basically consisted of the same steps as mentioned earlier for biodiesel extraction (see Figure 9.7). As also mentioned, the great advantage of the method resides in the fact that cooling is used rather than heating, for example, the process is energetically and environmentally more efficient.

The same protocol applied to the extraction of monolaurin/$C_{12}mimBF_4$ mixture was not so simple. The addition of water led to a heterogeneous solid gel at room temperature, which was unable to release the MAGs. So, SLIL was precipitated in dodecane at 10°C and the products were isolated after cooling and centrifuging the upper phase at 0°C (Figure 9.9). Even though the use of organic solvents would not

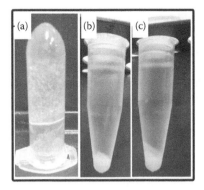

FIGURE 9.9 Phase behavior of monolaurin/[$C_{12}mim$][BF_4] reaction mixtures during the extraction step. (a) Phase behavior of the monolaurin/[$C_{12}mim$][BF_4] reaction mixture after addition of H_2O at 60°C and cooling to room temperature. (b) The addition of dodecane (1 mL) and centrifugation at 10°C lead to the IL precipitating at the bottom. (c) Precipitation of the supernatant with MAGs after centrifugation at 0°C. (From Lozano, P. et al., *Green Chem.*, 19, 390–396, 2017. [147])

normally be considered in a completely green process, dodecane presents lower volatility than other, more widely used organic coadjuvants, such as hexane or octane, indeed, it can be totally recovered for reuse, avoiding its release to the environment.

Finally, in the same work [147], the ability to reuse the enzyme was also studied.

Repeated cycles were performed, and it was confirmed that the lipase was active for up to eight operational phases with 100% of activity.

9.5   PROSPECTS

Society is increasingly aware of problems related with the environment and, consequently, becoming more vociferous in this sense. In this respect, the chemical synthesis industry faces an exciting challenge. A greener philosophy would avoid waste generation, high energy consumption, the depletion of raw materials, and the release of contaminants into the environment.

The discovery of ionic liquids and the increasing development of applications involving them have opened a new door to the design of more environmentally friendly synthesis strategies. Ionic liquids are non-aqueous and innocuous solvents. Due to their singular properties, almost all materials can be solubilized in the appropriate ILs, facilitating the availability of new materials. Moreover, ILs permit reactions that avoid the need for extreme conditions.

In the biocatalysis area, ILs have already been demonstrated to be excellent solvents for improving the efficiency and enantioselectivity of reactions, features that are exclusive to ILs. The parallel advance in genetic engineering points to the possibility of tailored enzymes that can be adapted to the new substrates or conditions. In this respect, ILs and enzymes constitute a good blend for developing biosynthetic processes and clearly have a promising future.

Green chemical processes must always take into consideration the stage of product purification and ensure the recovery of all the elements of the reaction for reuse. In this respect, the sponge-like behavior of certain hydrophobic ILs with long alkyl side chains (SLILs) should be emphasized. The possibility that the product of the reaction can be lodged in hydrophobic holes and then be released in a pure state by simple centrifugation is a fascinating property that permits the easy recovery of the products and the IL-enzyme system. Thus, widening the use of SLILs in chemical manufacturing should lead to the design of clean, inexpensive, and straightforward strategies of production and recovery for a wide range of applications in the food, cosmetics, pharmaceutical, and chemical industries.

Society demands that chemistry and chemists find solutions to environmental problems. We must face these demands and conceive new methodologies to provide possible answers. In the biocatalyst field, sponge-like ionic liquids are, without doubt, one of the most promising responses to satisfy these demands.

ACKNOWLEDGMENTS

We thank the Ministerio de Economía y Competitividad (MINECO), Spain (Ref: CTQ-2015-67927-R, and the Seneca Foundation, Spain (Ref. 19278/PI/14) for supporting the research.

REFERENCES

1. P. T. Anastas and J. C. Warner. *Green Chemistry: Theory and Practice.* Oxford University Press, New York, 1998.
2. K. Faber. *Biotransformations in Organic Chemistry*, Springer, Heidelberg, Germany, 2011.
3. T. Itoh. Ionic liquids as tool to improve enzymatic organic synthesis. *Chem. Rev.*, 2017, 117, 10567–10607.
4. R. Sheldon. Catalytic reactions in ionic liquids. *Chem. Comm.*, 2001, 2399–2407.
5. J. P. Hallett and T. Welton. Room-temperature ionic liquids: Solvents for synthesis and catalysis. 2. *Chem. Rev.*, 2011, 111, 3508–3576.
6. J. Dupont, R. F. de Souza, and P. A. Z. Suarez. Ionic liquid (molten salt) phase organometallic catalysis. *Chem. Rev.*, 2002, 102, 3667–3691.
7. Y. Ding, L. Zhang, J. Xie, and R. Guo. Binding characteristics and molecular mechanism of interaction between ionic liquid and DNA. *J. Phys. Chem. B*, 2010, 114, 2033–2043.
8. K. Egorova, E. Gordeev, and V. Ananikov. Biological activity of ionic liquids and their application in pharmaceutics and medicine. *Chem. Rev.*, 2017, 117, 7132–7189.
9. P. Walden. Molecular weights and electrical conductivity of several fused salts. *Bull Acad. Imper. Sci.*, 1914, 405–422.
10. C. F. Poole. Chromatographic and spectroscopic methods for the determination of solvent properties of room temperature ionic liquids. *J. Chromatogr. A*, 2004, 1037, 49–82.
11. P. Wasserscheid and T. Welton. *Ionic Liquids in Synthesis*, Wiley-VCH, Weinheim, Germany, 2007.
12. C. Reichardt. Pyridinium N-phenoxide betaine dyes and their application to the determination of solvent polarities part 29—Polarity of ionic liquids determined empirically by means of solvatochromic pyridinium N-phenolate betaine dyes. *Green Chem.*, 2005, 7, 339–351.
13. S. Ventura. Ionic-liquid-mediated extraction and separation processes for bioactive compounds: Past, present, and future trends. *Chem. Rev.*, 2017, 117, 6984–7052.
14. M. D. Bermudez, A. E. Jimenez, J. Sanes, and F. J. Carrion. Ionic liquids as advanced lubricant fluids. *Molecules*, 2009, 14, 2888–2908.
15. B. Buszewski and S. Studzinska. A review of ionic liquids in chromatographic and electromigration techniques. *Chromatographia*, 2008, 68, 1–10.
16. C. D. Tran and I. Mejac. Chiral ionic liquids for enantioseparation of pharmaceutical products by capillary electrophoresis. *J. Chromatogr. A*, 2008, 1204, 204–209.
17. Z. S. Breitbach and D. W. Armstrong. Characterization of phosphonium ionic liquids through a linear solvation energy relationship and their use as GLC stationary phases. *Anal. Bioanal. Chem.*, 2008, 390, 1605–1617.
18. C. Yao and J. L. Anderson. Retention characteristics of organic compounds on molten salt and ionic liquid-based gas chromatography stationary phases. *J. Chromatogr. A*, 2009, 1216, 1658–1712.
19. F. H. Liu and Y. Jiang. Room temperature ionic liquid as matrix medium for the determination of residual solvents in pharmaceuticals by static headspace gas chromatography. *J. Chromatogr. A*, 2007, 1167, 116–119.
20. Y. Wang, M. Tian, W. Bi, and K. H. Row. Application of ionic liquids in high performance reversed-phase chromatography. *Int. J. Mol. Sci.*, 2009, 10, 2591–2610.
21. T. Vogl, S. Menne, R.-S. Kühnel, and A. Balducci. The beneficial effect of protic ionic liquids on the lithium environment in electrolytes for battery applications. *J. Mater. Chem. A*, 2014, 2, 8258–8265.
22. E. Sebastiao, C. Cook, A. Hu, and M. Murugesu. Recent developments in the field of energetic ionic liquids. *J. Mater. Chem. A*, 2014, 2, 8153–8173.

23. P. Bonhote, A. P. Dias, N. Papageorgiou, K. Kalyanasundaram, and M. Gratzel. Hydrophobic, highly conductive ambient-temperature molten salts. *Inorg. Chem.*, 1996, 35, 1168–1178.

24. D. Shi, Y. Cao, N. Pootrakulchote, Z. Yi, M. Xu, S. M. Zakeeruddin, M. Graetzel, and P. Wang. New organic sensitizer for stable dye-sensitized solar cells with solvent-free ionic liquid electrolytes. *J. Phys. Chem. C*, 2008, 112, 17478–17485.

25. A. Bharwal, N. Nguyen, C. Iojoiu, C. Henrist, and F. Alloin. New polysiloxane bearing imidazolium iodide side chain as electrolyte for photoelectrochemical cell. *Solid State Ion.*, 2017, 307, 6–13.

26. H. Zhao. Innovative applications of ionic liquids as "green" engineering liquids. *Chem. Eng. Commun.*, 2006, 193, 1660–1677.

27. C. F. Ye, W. M. Liu, Y. X. Chen, and L. G. Yu. Room-temperature ionic liquids: A novel versatile lubricant. *Chem. Comm.*, 2001, 2244–2245.

28. Q. M. Lu, H. Z. Wang, C. F. Ye, W. M. Liu, and Q. J. Xue. Room temperature ionic liquid 1-ethyl-3-hexylimidazolium-bis(trifluoromethylsulfonyl)-imide as lubricant for steel-steel contact. *Tribol. Int.*, 2004, 37, 547–552.

29. D. Blanco, R. Gonzalez, J. Viesca, A. Fernandez-Gonzalez, M. Bartolome, and A. Hernandez Battez. Antifriction and antiwear properties of an ionic liquid with fluorine-containing anion used as lubricant additive. *Tribol. Lett.*, 2017, 65.

30. M. G. Freire. Introduction to ionic-liquid-based aqueous biphasic systems (ABS). *Ionic-Liquid-Based Aqueous Biphasic Systems. Fundamentals and Applications*, Springer-Verlag, Heidelberg, Germany, 2016.

31. L. Sheikhian, M. Akhond, and G. Absalam. Partitioning of reactive red-120, 4-(2-pyridylazo)-resorcinol, and methyl orange in ionic liquid-based aqueous biphasic systems. *J. Environ. Chem. Eng.*, 2014, 2, 137–142.

32. A. M. Fernandes, M. Paulis, J. Yuan, and D. Mecerreyes. Magnetic poly(ionic liquid) microcapsules for oil capture and recovery. *Part. Part. Syst. Char.*, 2016, 33, 734–739.

33. S. Chen, K. Kobayashi, R. Kitaura, Y. Miyata, and H. Shinohara. Direct HRTEM observation of ultrathin freestanding ionic liquid film on carbon nanotube grid. *Acs Nano*, 2011, 5, 4902–4908.

34. S. M. Chen, G. Z. Wu, M. L. Sha, and S. R. Huang. Transition of ionic liquid [bmim] [PF6] from liquid to high-melting-point crystal when confined in multiwalled carbon nanotubes. *J. Amer. Chem. Soc.*, 2007, 129, 2416–2417.

35. A. S. Pensado, F. Malberg, M. F. C. Gomes, A. A. H. Padua, J. Fernandez, and B. Kirchner. Interactions and structure of ionic liquids on graphene and carbon nanotubes surfaces. *Rsc Advances*, 2014, 4, 18017–18024.

36. A. Rehman and X. Q. Zeng. Ionic liquids as green solvents and electrolytes for robust chemical sensor development. *Acc. Chem. Res.*, 2012, 45, 1667–1677.

37. F. Walkiewicz, K. Materna, A. Kropacz, A. Michalczyk, R. Gwiazdowski, T. Praczyk, and J. Pernak. Multifunctional long-alkyl-chain quaternary ammonium azolate based ionic liquids. *New J. Chem.*, 2010, 34, 2281–2289.

38. K. S. Egorova, M. M. Seitkalieva, A. V. Posvyatenko, and V. P. Ananikov. An unexpected increase of toxicity of amino acid-containing ionic liquids. *Toxicol. Res.*, 2015, 4, 152–159.

39. S. V. Malhotra, V. Kumar, C. Velez, and B. Zayas. Imidazolium-derived ionic salts induce inhibition of cancerous cell growth through apoptosis. *Med. Chem. Comm.*, 2014, 5, 1404–1409.

40. X. Y. Li, J. Ma, and J. J. Wang. Cytotoxicity, oxidative stress, and apoptosis in HepG2 cells induced by ionic liquid 1-methyl-3-oct-ylimidazolium bromide. *Ecotoxicol. Environ. Saf.*, 2015, 120, 342–348.

41. X. Yi, D. S. Manickam, A. Brynskikh, and A. V. Kabanov. Agile delivery of protein therapeutics to CNS. *J. Control. Rel.*, 2014, 190, 637–663.

42. J. H. Wang, D. H. Cheng, X. W. Chen, Z. Du, and Z. L. Fang. Direct extraction of double-stranded DNA into ionic liquid 1-butyl-3-methylimidazolium hexafluorophosphate and its quantification. *Anal. Chem.*, 2007, 79, 620–625.

43. S. Satpathi, A. Sengupta, V. Hridya, K. Gavvala, R. K. Koninti, B. Roy, and P. Hazra. A green solvent induced DNA package. *Sci. Rep.*, 2015, 5, 9137–9145.

44. R. Vijayaraghavan, A. Izgorodin, V. Ganesh, M. Surianarayanan, and D. R. MacFarlane. Long-term structural and chemical stability of DNA in hydrated ionic liquids. *Angew. Chem. Int. Ed.*, 2010, 49, 1631–1633.

45. R. P. Swatloski, S. K. Spear, J. D. Holbrey, and R. D. Rogers. Dissolution of cellulose with ionic liquids. *J. Amer. Chem. Soc.*, 2002, 124, 4974–4975.

46. C. Chiappe, M. J. Douton, A. Mezzetta, C. S. Pomelli, G. Assanelli, and A. R. de Angelis. Recycle and extraction: Cornerstones for an efficient conversion of cellulose into 5-hydroxymethylfurfural in ionic liquids. *ACS Sustain. Chem. Eng.*, 2017, 5, 5529–5536.

47. G. Cevasco and C. Chiappe. Are ionic liquids a proper solution to current environmental challenges? *Green Chem.*, 2014, 16, 2375–2385.

48. A. Kamimura and S. Yamamoto. An efficient method to depolymerize polyamide plastics: A new use of ionic liquids. *Org. Lett.*, 2007, 9, 2533–2535.

49. A. Kamimura and S. Yamamoto. A novel depolymerization of nylons in ionic liquids. *Polym. Adv. Technol.*, 2008, 19, 1391–1395.

50. D. Stowe. Recycling of solid waste in ionic liquid media, US 9,303,134 B2, 2016.

51. V. I. Parvulescu and C. Hardacre. Catalysis in ionic liquids. *Chem. Rev.*, 2007, 107, 2615–2665.

52. F. M. Kerton. *Alternative Solvents for Green Chemistry*. RSC Publishing, Cambridge, UK, 2009.

53. J. Kraemer, E. Redel, R. Thomann, and C. Janiak. Use of ionic liquids for the synthesis of iron, ruthenium, and osmium nanoparticles from their metal carbonyl precursors. *Organometallics*, 2008, 27, 1976–1978.

54. E. Redel, R. Thomann, and C. Janiak. Use of ionic liquids (ILs) for the IL-anion size-dependent formation of Cr, Mo and W nanoparticles from metal carbonyl M(CO)(6) precursors. *Chem. Comm.*, 2008, 1789–1791.

55. A. P. Abbott, G. Capper, D. L. Davies, R. H. Rasheed, and V. Tambyrajah. Quaternary ammonium zinc- or tin-containing ionic liquids: Water insensitive, recyclable catalysts for Diels-Alder reactions. *Green Chem.*, 2002, 4, 24–26.

56. M. Schmitkamp, D. Chen, W. Leitner, J. Klankermayer, and G. Francio. Enantioselective catalysis with tropos ligands in chiral ionic liquids. *Chem. Comm.*, 2007, 4012–4014.

57. C. Baudequin, J. Baudoux, J. Levillain, D. Cahard, A. C. Gaumont, and J. C. Plaquevent. Ionic liquids and chirality: Opportunities and challenges. *Tetrahedron: Asymmetry*, 2003, 14, 3081–3093.

58. P. Goodrich, H. Gunaratne, L. Hall et al. Using chiral ionic liquid additives to enhance asymmetric induction in a Diels-Alder reaction. *Dalton Trans.*, 2017, 46, 1704–1713.

59. S. A. Forsyth, D. R. MacFarlane, R. J. Thomson, and M. von Itzstein. Rapid, clean, and mild O-acetylation of alcohols and carbohydrates in an ionic liquid. *Chem. Comm.*, 2002, 714–715.

60. C. Dai, J. Zhang, C. Huang, and Z. Lei. Ionic liquids in selective oxidation: Catalysts and solvents. *Chem. Rev.*, 2017, 117, 6929–6983.

61. E. Karjalainen, D. F. Izquierdo, V. Marti-Centelles, S. V. Luis, H. Tenhu, and E. Garcia-Verdugo. An enzymatic biomimetic system: Enhancement of catalytic efficiency with new polymeric chiral ionic Liquids synthesised by controlled radical polymerisation. *Polym. Chem.*, 2014, 5, 1437–1446.

62. Y. Tao, R. Dong, I. V. Pavlidis, B. Chen, and T. Tan. Using imidazolium-based ionic liquids as dual solvent-catalysts for sustainable synthesis of vitamin esters: Inspiration from bio- and organocatalysis. *Green Chem.*, 2016, 18, 1240–1248.

63. P. Lozano. Enzymes in neoteric solvents: From one-phase to multiphase systems. *Green Chem.*, 2010, 12, 555–569.

64. E. Busto, V. Gotor-Fernandez, and V. Gotor. Hydrolases: Catalytically promiscuous enzymes for non-conventional reactions in organic synthesis. *Chem. Soc. Rev.*, 2010, 39, 4504–4523.

65. R. Kourist, P. D. de Maria, and K. Miyamoto. Biocatalytic strategies for the asymmetric synthesis of profens—recent trends and developments. *Green Chem.*, 2011, 13, 2607–2618.

66. J. M. Bernal, P. Lozano, E. Garcia-Verdugo, M. Isabel Burguete, G. Sanchez-Gomez, G. Lopez-Lopez, M. Pucheault, M. Vaultier, and S. V. Luis. Supercritical synthesis of biodiesel. *Molecules*, 2012, 17, 8696–8719.

67. R. C. Rodrigues and R. Fernandez-Lafuente. Lipase from *Rhizomucor miehei* as a biocatalyst in fats and oils modification. *J. Mol. Catal. B: Enzym.*, 2010, 66, 15–32.

68. T. Itoh, E. Akasaki, K. Kudo, and S. Shirakami. Lipase-catalyzed enantioselective acylation in the ionic liquid solvent system: Reaction of enzyme anchored to the solvent. *Chem. Lett.*, 2001, 262–263.

69. S. H. Schofer, N. Kaftzik, P. Wasserscheid, and U. Kragl. Enzyme catalysis in ionic liquids: Lipase catalysed kinetic resolution of 1-phenylethanol with improved enantioselectivity. *Chem. Comm.*, 2001, 425–426.

70. F. van Rantwijk and R. A. Sheldon. Biocatalysis in ionic liquids. *Chem. Rev.*, 2007, 107, 2757–2785.

71. B. Sandig and M. R. Buchmeiser. Highly productive and enantioselective enzyme catalysis under continuous supported liquid-liquid conditions using a hybrid monolithic bioreactor. *Chemsuschem*, 2016, 9, 2917–2921.

72. M. Noel, P. Lozano, M. Vaultier, and J. L. Iborra. Kinetic resolution of rac-2-pentanol catalyzed by Candida antarctica lipase B in the ionic liquid, 1-butyl-3-methylimidazolium bis[(trifluoromethyl)sulfonyl]amide. *Biotechnol. Lett.*, 2004, 26, 301–306.

73. D. H. Zhang, S. Bai, M. Y. Ren, and Y. Sun. Optimization of lipase-catalyzed enantio selective esterification of (+/−)-menthol in ionic liquid. *Food Chem.*, 2008, 109, 72–80.

74. M. Habulin and Z. Knez. Optimization of (R, S)-1-phenylethanol kinetic resolution over Candida antarctica lipase B in ionic liquids. *J. Mol. Catal. B: Enzym.*, 2009, 58, 24–28.

75. P. Hara, U. Hanefeld, and L. T. Kanerv. Immobilised *Burkholderia cepacia* lipase in dry organic solvents and ionic liquids: A comparison. *Green Chem.*, 2009, 11, 250–256.

76. K. Yoshiyama, Y. Abe, K. Kude, A. Ishioka, S. Hayase, M. Kawatsura, and T. Itoh. Design of phosphonium ionic liquids appropriate for lipase-catalyzed transesterification. *J. Mol. Catal. B:Enzym.*, 2010, 62, 121–121.

77. L. Banoth, M. Singh, A. Tekewe, and U. C. Banerjee. Increased enantioselectivity of lipase in the transesterification of dl-(+/−)-3-phenyllactic acid in ionic liquids. *Biocatal. Biotransform.*, 2009, 27, 263–270.

78. S. Shipovskov, H. Gunaratne, K. R. Seddon, and G. Stephens. Catalytic activity of laccases in aqueous solutions of ionic liquids. *Green Chem.*, 2008, 10, 806–810.

79. L. Gubicza, K. Belafi-Bako, E. Feher, and T. Frater. Waste-free process for continuous flow enzymatic esterification using a double pervaporation system. *Green Chem.*, 2008, 10, 1284–1287.

80. D. Barahona, P. H. Pfromm, and M. E. Rezac. Effect of water activity on the lipase catalyzed esterification of geraniol in ionic liquid [bmim]PF6. *Biotechnol. Bioeng.*, 2006, 93, 318–324.

81. T. Itoh, E. Akasaki, and Y. Nishimura. Efficient lipase-catalyzed enantioselective acylation under reduced pressure conditions in an ionic liquid solvent system. *Chem. Lett.*, 2002, 154–155.

82. M. Gamba, A. A. Lapis, and J. Dupont. Supported ionic liquid enzymatic catalysis for the production of biodiesel. *Adv. Synth. Catal.*, 2008, 350, 160–164.

83. P. Lozano, J. Maria Bernal, R. Piamtongkam, D. Fetzer, and M. Vaultier. One-phase ionic liquid reaction medium for biocatalytic production of biodiesel. *Chemsuschem*, 2010, 3, 1359–1363.

84. J. K. Lee and M. J. Kim. Ionic liquid-coated enzyme for biocatalysis in organic solvent. *J. Org. Chem.*, 2002, 67, 6845–6847.

85. K. Nakashima, N. Kamiya, D. Koda, T. Maruyama, and M. Goto. Enzyme encapsulation in microparticles composed of polymerized ionic liquids for highly active and reusable biocatalysts. *Org. Biomol. Chem.*, 2009, 7, 2353–2358.

86. P. Lozano, B. Bernal, J. M. Bernal, M. Pucheault, and M. Vaultier. Stabilizing immobilized cellulase by ionic liquids for saccharification of cellulose solutions in 1-butyl-3-methylimidazolium chloride. *Green Chem.*, 2011, 13, 1406–1410.

87. L. C. Branco, J. G. Crespo, and C. A. M. Afonso. Highly selective transport of organic compounds by using supported liquid membranes based on ionic liquids. *Angew. Chem. Int. Ed.*, 2002, 41, 2771–2773.

88. M. Mori, R. G. Garcia, M. P. Belleville, D. Paolucci-Jeanjean, J. Sanchez, P. Lozano, M. Vaultier, and G. Rios. A new way to conduct enzymatic synthesis in an active membrane using ionic liquids as catalyst support. *Catal. Today*, 2005, 104, 313–317.

89. E. Miyako, T. Maruyama, N. Kamiya, and M. Goto. Enzyme-facilitated enantioselective transport of (S)-ibuprofen through a supported liquid membrane based on ionic liquids. *Chem. Comm.*, 2003, 2926–2927.

90. P. Lozano, T. de Diego, A. Manjon, M. A. Abad, M. Vaultier, and J. L. Iborra. Enzymatic membrane reactor for resolution of ketoprofen in ionic liquids and supercritical carbon dioxide. In *Ionic Liquids Applications: Pharmaceuticals, Therapeutics, and Biotechnology.* (Ed. S.V. Malhotra), pp. 25–34. ACS Symposium Series Vol. 1038. 2010.

91. P. Lozano, T. De Diego, S. Gmouh, M. Vaultier, and J. L. Iborra. A continuous reactor for the (chemo)enzymatic dynamic kinetic resolution of rac-1-phenylethanol in ionic liquid/supercritical carbon dioxide biphasic systems. *Int. J. Chem. React. Eng.*, 2007, 5, A53, 12.

92. V. Sans, N. Karbass, M. I. Burguete, V. Compan, E. Garcia-Verdugo, S. V. Luis, and M. Pawlak. Polymer-supported ionic-liquid-like phases (SILLPs): Transferring ionic liquid properties to polymeric matrices. *Chem. Eur. J.*, 2011, 17, 1894–1906.

93. V. Sans, F. Gelat, N. Karbass, M. I. Burguete, E. Garcia-Verdugo, and S. V. Luis. Polymer cocktail: A multitask supported ionic liquid-like species to acilitate multiple and consecutive C-C coupling reactions. *Adv. Synth. Catal.*, 2010, 352, 3013–3021.

94. B. Altava, V. Compan, A. Andrio, L. del Castillo, S. Molla, M. Burguete, I. E. Garcia-Verdugo, and S. V. Luis. Conductive films based on composite polymers containing ionic liquids absorbed on crosslinked polymeric ionic-like liquids (SILLPs). *Polymer*, 2015, 72, 69–81.

95. B. Sandig, L. Michalek, S. Vlahovic, M. Antonovici, B. Hauer, and M. R. Buchmeiser. A monolithic hybrid cellulose-2.5-acetate/polymer bioreactor for biocatalysis under continuous liquid-liquid conditions using a supported ionic liquid phase. *Chem. Eur. J.*, 2015, 21, 15835–15842.

96. P. Hara, J. P. Mikkola, D. Y. Murzin, and L. T. Kanerva. Supported ionic liquids in *Burkholderia cepacia* lipase-catalyzed asymmetric acylation. *J. Mol. Catal. B: Enzym.*, 2010, 67, 129–134.

97. D. F. Izquierdo, J. M. Bernal, M. Burguete, I. E. Garcia-Verdugo, P. Lozano, and S. V. Luis. An efficient microwave-assisted enzymatic resolution of alcohols using a lipase immobilised on supported ionic liquid-like phases (SILLPs). *Rsc Advances*, 2013, 3, 13123–13126.

98. P. Lozano, J. M. Bernal, and M. Vaultier. Towards continuous sustainable processes for enzymatic synthesis of biodiesel in hydrophobic ionic liquids/supercritical carbon dioxide biphasic systems. *Fuel*, 2011, 90, 3461–3467.

99. P. Lozano, J. M. Bernal, and A. Navarro. A clean enzymatic process for producing flavour esters by direct esterification in switchable ionic liquid/solid phases. *Green Chem.*, 2012, 14, 3026–3033.

100. P. Lozano, J. M. Bernal, G. Sanchez-Gomez, G. Lopez-Lopez, and M. Vaultier. How to produce biodiesel easily using a green biocatalytic approach in sponge-like ionic liquids. *Energy Environ. Sci.*, 2013, 6, 1328–1338.

101. R. Hayes, G. G. Warr, and R. Atkin. Structure and nanostructure in ionic liquids. *Chem. Rev.*, 2015, 115, 6357–6426.

102. H. K. Stassen, R. Ludwig, A. Wulf, and J. Dupont. Imidazolium salt ion pairs in solution. *Chem. Eur. J.*, 2015, 21, 8324–8335.

103. K. Iwata, H. Okajima, S. Saha, and H. O. Hamaguchi. Local structure formation in alkyl-imidazolium-based ionic liquids as revealed by linear and nonlinear Raman spectroscopy. *Acc. Chem. Res.*, 2007, 40, 1174–1181.

104. B. A. Marekha, M. Bria, M. Moreau, I. De Waele, F. A. Miannay, Y. Smortsova, T. Takamuku, O. N. Kalugin, M. Kiselev, and A. Idrissi. Intermolecular interactions in mixtures of 1-n-butyl-3-methylimidazolium acetate and water: Insights from IR, Raman, NMR spectroscopy and quantum chemistry calculations. *J. Mol. Liq.*, 2015, 210, 227–237.

105. M. N. Garaga, M. Nayeri, and A. Martinelli. Effect of the alkyl chain length in 1-alkyl-3-methylimidazolium ionic liquids on inter-molecular interactions and rotational dynamics A combined vibrational and NMR spectroscopic study. *J. Mol. Liq.*, 2015, 210, 169–177.

106. R. Hayes, S. Imberti, G. G. Warr, and R. Atkin. Pronounced sponge-like nanostructure in propylammonium nitrate. *Phys. Chem. Chem. Phys.*, 2011, 13, 13544–13551.

107. T. Iwahashi, T. Nishi, H. Yamane, T. Miyamae, K. Kanai, K. Seki, D. Kim, and Y. Ouchi. Surface structural study on ionic liquids using metastable atom electron spectroscopy. *J. Phys. Chem. C*, 2009, 113, 19237–19243.

108. C. Zhang, S. Zhang, L. Yu, P. Zhang, Z. Zhang, and Z. Wu. Tribological behavior of 1-Methyl-3-Hexadecylimidazolium tetrafluoroborate ionic liquid crystal as a neat lubricant and as an additive of liquid paraffin. *Tribol. Lett.*, 2012, 46, 49–54.

109. D. Pontoni, J. Haddad, M. Di Michiel, and M. Deutsch. Self-segregated nanostructure in room temperature ionic liquids. *Soft Matter*, 2017, 13, 6947–6955.

110. M. Fakhraee and M. R. Gholami. Probing the effects of the ester functional group, alkyl side chain length and anions on the bulk nanostructure of ionic liquids: A computational study. *Phys. Chem. Chem. Phys.*, 2016, 18, 9734–9751.

111. W. D. Amith, J. J. Hettige, E. W. Castner, and C. J. Margulis. Structures of ionic liquids having both anionic and cationic octyl tails: Lamellar vacuum interface vs sponge-like bulk order. *J. Phys. Chem. Lett.*, 2016, 7, 3785–3790.

112. T. De Diego, P. Lozano, S. Gmouh, M. Vaultier, and J. L. Iborra. Understanding structure—Stability relationships of Candida antartica lipase B in ionic liquids. *Biomacromolecules*, 2005, 6, 1457–1464.

113. M. Zanatta, A. L. Girard, G. Marin, G. Ebeling, F. P. dos Santos, C. Valsecchi, H. Stassen, P. R. Livotto, W. Lewis, and J. Dupont. Confined water in imidazolium based ionic liquids: A supramolecular guest@host complex case. *Phys. Chem. Chem. Phys.*, 2016, 18, 18297–18304.

114. A. M. Klibanov. Improving enzymes by using them in organic solvents. *Nature*, 2001, 409, 241–246.

115. T. Arakawa, Y. Kita, and S. N. Timasheff. Protein precipitation and denaturation by dimethyl sulfoxide. *Biophys. Chem.*, 2007, 131, 62–70.

116. P. Lozano, T. De Diego, D. Carrie, M. Vaultier, and J. L. Iborra. Over-stabilization of Candida antarctica lipase B by ionic liquids in ester synthesis. *Biotechnol. Lett.*, 2001, 23, 1529–1533.

117. P. Lozano, T. De Diego, C. Mira, K. Montague, M. Vaultier, and J. L. Iborra. Long term continuous chemoenzymatic dynamic kinetic resolution of rac-1-phenylethanol using ionic liquids and supercritical carbon dioxide. *Green Chem.*, 2009, 11, 538–542.

118. P. Lozano, S. Nieto, J. L. Serrano, J. Perez, G. Sanchez-Gomez, E. Garcia-Verdugo, and S. V. Luis. Flow biocatalytic processes in ionic liquids and supercritical fluids. *Mini Rev. Org. Chem.*, 2017, 14, 65–74.

119. C. Garcia-Galan, A. Berenguer-Murcia, R. Fernandez-Lafuente, and R. C. Rodrigues. Potential of different enzyme immobilization strategies to improve enzyme performance. *Adv. Synth. Catal.*, 2011, 353, 2885–2904.

120. O. Barbosa, R. Torres, C. Ortiz, A. Berenguer-Murcia, R. C. Rodrigues, and R. Fernandez-Lafuente. Heterofunctional supports in enzyme immobilization: From traditional immobilization protocols to opportunities in tuning enzyme properties. *Biomacromolecules*, 2013, 14, 2433–2462.

121. J. K. Poppe, R. Fernandez-Lafuente, R. C. Rodrigues, and M. A. Z. Ayub. Enzymatic reactors for biodiesel synthesis: Present status and future prospects. *Biotechnol. Adv.*, 2015, 33, 511–525.

122. R. A. Sheldon. Characteristic features and biotechnological applications of cross-linked enzyme aggregates (CLEAs). *Appl. Microbiol. Biotechnol.*, 2011, 92, 467–477.

123. N. Kaftzik, P. Wasserscheid, and U. Kragl. Use of ionic liquids to increase the yield and enzyme stability in the beta-galactosidase catalysed synthesis of N-acetyllactosamine. *Org. Process Res. Dev.*, 2002, 6, 553–557.

124. P. Lozano, J. M. Bernal, E. Garcia-Verdugo, G. Sanchez-Gomez, M. Vaultier, M. Isabel Burguete, and S. V. Luis. Sponge-like ionic liquids: A new platform for green biocatalytic chemical processes. *Green Chem.*, 2015, 17, 3706–3717.

125. P. Lozano, J. M. L. A. Bernal, D. Romera, E. Garcia-Verdugo, G. Sanchez-Gomez, M. Pucheault, M. Vaultier, M. I. Burguete, and S. V. Luis. A green approach for producing solvent-free anisyl acetate by enzyme-catalyzed direct esterification in sponge-like ionic liquids under conventional and microwave heating. *Curr. Green Chem.*, 2014, 1, 145–154.

126. P. Lozano, T. De Diego, J. P. Guegan, M. Vaultier, and J. L. Iborra. Stabilization of alpha-chymotrypsin by ionic liquids in transesterification reactions. *Biotechnol. Bioeng.*, 2001, 75, 563–569.

127. T. De Diego, A. Manjon, P. Lozano, M. Vaultier, and J. L. Iborra. An efficient activity ionic liquid-enzyme system for biodiesel production. *Green Chem.*, 2011, 13, 444–451.

128. P. Lozano, E. Garcia-Verdugo, J. M. Bernal, D. F. Izquierdo, M. Isabel Burguete, G. Sanchez-Gomez, and S. V. Luis. Immobilised lipase on structured supports containing covalently attached ionic liquids for the continuous synthesis of biodiesel in scCO$_2$. *Chemsuschem*, 2012, 5, 790–798.

129. M. Anitha, S. Kamarudin, and N. Kofli. The potential of glycerol as a value-added commodity. *Chem. Eng. J.*, 2016, 295, 119–130.

130. J. D. Rivaldi, L. C. Duarte, R. D. C. Rodrigues, H. J. Izario Filho, M. D. G. de Almeida Felipe, and I. M. de Mancilha. Valorization of glycerol from biodiesel industries as a renewable substrate for co-producing probiotic bacteria biomass and acetic acid. *Biomass Convers. Biorefin.*, 2017, 7, 81–90.

131. M. Lapuerta, J. Rodriguez-Fernandez, C. Estevez, and N. Bayarri. Properties of fatty acid glycerol formal ester (FAGE) for use as a component in blends for diesel engines. *Biomass Bioenergy*, 2015, 76, 130–140.

132. M. R. Nanda, Y. Zhang, Z. Yuan, W. Qin, H. S. Ghaziaskar, and C. Xu. Catalytic conversion of glycerol for sustainable production of solketal as a fuel additive: A review. *Renew. Sustain. Energy Rev.*, 2016, 56, 1022–1031.

133. P. Lozano, C. Gomez, A. Nicolas, R. Polo, S. Nieto, J. M. Bernal, E. Garcia-Verdugo, and S. V. Luis. Clean enzymatic preparation of oxygenated biofuels from vegetable and waste cooking oils by using spongelike ionic liquids technology. *ACS Sust. Chem. Eng.*, 2016, 4, 6125–6132.

134. H. Zhang, Y. Cui, S. Zhu, F. Feng, and X. Zheng. Characterization and antimicrobial activity of a pharmaceutical microemulsion. *Int. J. Pharm.*, 2010, 395, 154–160.

135. D. Vollhardt and G. Brezesinski. Mono layer characteristics of 1-Monostearoyl-rac-glycerol at the air-water interface. *J. Phys. Chem. C*, 2015, 119, 9934–9946.

136. H. Shimada, V. E. Tyler, and J. L. McLaughlin. Biologically active acylglycerides from the berries of saw-palmetto (Serenoa repens). *J. Nat. Prod.*, 1997, 60, 417–418.

137. J. B. Monteiro, M. G. Nascimento, and J. L. Ninow. Lipase-catalyzed synthesis of monoacylglycerol in a homogeneous system. *Biotechnol. Lett.*, 2003, 25, 641–644.

138. U. T. Bornscheuer. Lipase-catalyzed syntheses of monoacylglycerols. *Enzyme Microb. Technol.*, 1995, 17, 578–586.

139. N. O. V. Sonntag. Glycerolysis of fats and methyl-esters—Status, review and critique. *J. Amer. Oil Chem. Soc.*, 1982, 59, A795–A802.

140. L. V. Fregolente, P. B. L. Fregolente, A. M. Chicuta, C. B. Batistella, R. M. Filho, and M. R. Wolf-Maciel. Effect of operating conditions on the concentration of monoglycerides using molecular distillation. *Chem. Eng. Res. Des.*, 2007, 85, 11524–1528.

141. Z. Guo and X. B. Xu. New opportunity for enzymatic modification of fats and oils with industrial potentials. *Org. Biomol. Chem.*, 2005, 3, 2615–2619.

142. Z. Guo and X. B. Xu. Lipase-catalyzed glycerolysis of fats and oils in ionic liquids: A further study on the reaction system. *Green Chem.*, 2006, 8, 54–62.

143. N. Zhong, L. Z. Cheong, and X. Xu. Strategies to obtain high content of monoacylglycerols. *Eur. J. Lipid Sci. Technol.*, 2014, 116, 97–107.

144. R. DiCosimo, J. McAuliffe, A. J. Poulose, and G. Bohlmann. Industrial use of immobilized enzymes. *Chem. Soc. Rev.*, 2013, 42, 6437–6474.

145. M. B. Ansorge-Schumacher and O. Thum. Immobilised lipases in the cosmetics industry. *Chem. Soc. Rev.*, 2013, 42, 6475–6490.

146. A. Sorrenti, O. Illa, and R. Ortuno. Amphiphiles in aqueous solution: Well beyond a soap bubble. *Chem. Soc. Rev.*, 2013, 42, 8200–8219.

147. P. Lozano, C. Gomez, S. Nieto, G. Sanchez-Gomez, E. Garcia-Verdugo, and S. V. Luis. Highly selective biocatalytic synthesis of monoacylglycerides in sponge-like ionic liquids. *Green Chem.*, 2017, 19, 390–396.

10 Ionic Liquids for Biomass Processing

Wei-Chien Tu and Jason P. Hallett

CONTENTS

10.1 INTRODUCTION TO LIGNOCELLULOSIC BIOMASS

Biomass and specifically lignocellulosic biomass has become an increasingly attractive alternative to fossil fuels to mitigate climate change due to ever-increasing consumption of energy and materials. There are two major categories of biomass used to produce bioethanol. First generation biomasses are derived from food crops, such as starch, sugar, and vegetable oils. Bioethanol is produced from the fermentation of these starches and sugars. Second generation biofuels are derived from biomass feedstocks that are composed of cellulose, hemicellulose, and lignin; the organic biopolymers that make up the cell walls of woody trees, shrubs, and grasses. These renewable and sustainable resources do not compete with food crops and have the potential to be converted into fuels, polymers, such as plastics, and can be used in the production of other chemical feedstocks.[1] However, unlike first-generation biomass feedstocks, an additional pretreatment step is required to fractionate the lignocellulosic biomass into its components for biofuel and or materials production.

The process of biorefining lignocellulosic biomass for biofuels involves the fermentation of sugars, mainly glucose from depolymerized cellulose, to produce ethanol. However, other elements in lignocellulosic biomass, notably lignin, inhibit the fermentation and significantly reduce the efficiency of the biofuel production process. A pretreatment process is necessary to separate the cellulose from hemicellulose and lignin for efficient fuel production. This process can be one of the costliest and energy-intensive aspects of biofuel production from lignocellulosic biomass, with estimates of 8¢/liter. For lignocellulosic biomass to be competitive with fossil fuels and ultimately commercially viable, improvements in pretreatment efficiency and processing costs are necessary. The production of bioethanol fuels is not currently competitive with traditional fossil fuels without government subsidies. However, research over time has focused on improving the viability of lignocellulosic biomass as a fuel source. An early feasibility study in 2000 pegged the cost of a liter of ethanol produced from lignocellulose to be 40% higher than from corn.[2] Estimates in 2000 forecast that the cost per liter of lignocellulosic ethanol would reach $0.20–$0.29 in the United States, where similar studies in Europe forecasted the average cost to be between $0.35–$0.48.[3,4] However, lignocellulosic biofuels were seen as having the potential to out yield fuels derived from first generation feedstocks and have additional climate benefits. In 2000, the theoretical yield of corn stover was estimated to be 91 gallons of ethanol per dry ton vs. 69 gallons per ton for corn.[2] With improved advances in technology and efficiency, a 2013 estimate of the minimum selling price for lignocellulosic bioethanol was at $0.58 per liter (gasoline was sold at $0.61 per liter in the United States in 2012 adjusted for taxes).[5] Later integrated studies in 2015 have begun to expand from the simple evaluation of raw cost by taking a more environmentally holistic approach to cost by comparing various biomasses. The study examined production of biofuels from different feedstocks, including *Miscanthus*, switchgrass, and corn stover over various regions and soil types and found that production costs for bioethanol can range between $0.88 and $1.66 per liter.[6] Perhaps more importantly, the study has modelled the greenhouse gas cost of producing bioethanol and compared it to its impact with as much as $48/Mg of savings.[6] Other studies that examined the efficiency of producing ethanol from sugarcane bagasse measured the minimum ethanol selling price at $0.55/L–$0.63/L. This implied that while advancements have been made since the earlier 2000s, further improvements in the efficiency, and particularly enhancing the yield of bioethanol production without increasing costs, is required to be competitive with traditional fossil fuels.[7] Avenues for increasing efficiency not only include reducing the cost of producing bioethanol, but also in deriving value from by-products of the production process. A 2017 study explores the use of lignin residue in the production of jet fuel grade hydrocarbon which resulted in the increase to the 2011 National Renewable Energy Laboratory (NREL) baseline case by 60 million liters of fuel while only increasing costs by 4%.[8] These types of studies have motivated researchers to focus on lignocellulosic biomass as a source of energy for bioethanol production.

10.2   SOURCES OF LIGNOCELLULOSIC BIOMASS

McKendry, in his 2002 work, established that sources of lignocellulose for biorefinery production can be obtained from by-products of other industries.[9] Lignocellulosic biomass could include forestry and agricultural residues or dedicated crops grown for bioenergy purposes. Food waste from food processing, consumer food items, municipal, or industrial waste are also potential streams of biomass components.

Lignocellulosic biomass is typically classified as hardwoods, softwoods, and grasses. Each present different potential for biofuel and chemical feedstock production, as well as challenges due to their varying compositions. An advantage of lignocellulosic biomass, compared to other renewable resources, is that production is possible in diverse climates, potentially offering energy solutions to multiple world regions. Many regions already possess the means to harvest hardwood and adapt waste from the fuel-wood, paper industries, and municipal solid waste collection facilities.[10] Kraft mills in the U.S. already produce over 50 million tons of degraded lignin from 130 million tons of kraft pulp every year.[11] Energy production employing lignocellulosic biomass energy is well underway with annual U.S. production in excess of 125 million liters of bioethanol from lignocellulosic sources, and this will increase to 780 million by 2016.[11] The challenge presented by the variety of lignocellulosic biomass is that different biomasses have varying compositions and require different conditions and procedures for optimal processing. In addition, addressing logistical challenges, such as harvesting and transporting of the feedstocks is essential.

10.3   COMPOSITION OF BIOMASS

Cellulose, hemicellulose, and lignin are the primary components of lignocellulosic biosmass.[12] These components make up *ca.* 90% of dry lignocellulosic matter. Pectins, inorganic compounds, proteins, and extractives comprise the remaining portion (Figure 10.1).

Ratios of these three components vary among plant types, species, and even within species due to differences in growing conditions, genetics, and environments.[14] Hardwood species generally contain a range between 40%–50% cellulose, 24%–40% hemicellulose, and 18%–25% lignin, while softwoods typically range between 45%–50% cellulose,

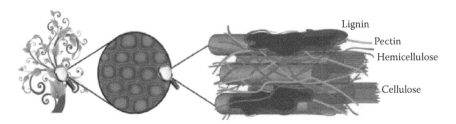

FIGURE 10.1   Sample plant cell wall representation. The composition of the cell wall can vary between plant species, as well as in the same species depending on environmental factors. The development stages of a plant will also impact the ratio of cellulose, hemicellulose, and lignin. In general, pectin and hemicellulose surround the cellulose fibrils, while lignin is covalently bound to the hemicellulose via glyosidic bonds. (Reprinted with permission from Turumtay, H., *BioEnergy Res.*, 8, 1574–1588, 2015. Copyright 2015 BioEnergy Research.[13])

25%–35% hemicellulose, and 25%–35% lignin. Grasses range from 25%–40% cellulose, 35%–50% hemicellulose, and 10%–30% lignin.[14] The variation of these components have direct impact on the optimum procedure for the pretreatment of the biomass.

Cellulose is a polysaccharide of D-glucose sugars linked via β(1–4) glycosidic bonds forming long chains with a degree of polymerization above 10,000 units. Though six different polymorphs of cellulose exist, cellulose Iβ is the most predominant conformation in terrestrial plants. These β linkages create a rigid inflexible flat polymer which can easily stack one atop another via hydrogen bonding and van der Waals forces forming cellulose fibrils and giving cellulose its crystalline structure (Figure 10.2).[15] Glucose as a monomer and its oligomers (up to four to five units) are water soluble. The cellulose fibrils are not soluble in water or most organic solvents due to the high molecular weight, low flexibility, as well as the intimate ordered packing of the polymer strands.[2] There are few solvents that can dissolve cellulose due to its highly crystalline structure and strong intermolecular and intramolecular bonds.

Hemicellulose is composed of five major linked carbohydrate monomers: xylose, arabinose (pentose sugars), mannose, galactose, and glucose (hexose sugars).[14] It consists of combinations of glyosidic linkages, forming a heteropolymer between these sugars in groups of 200–3000 units. Branching hydrophilic and hydrophobic areas make hemicellulose a much less rigid and amorphous structure.[17] These branched areas often contain various functional groups including; methyl, acetyl, cinnamic, glucuronic, and galacturonic acid moieties. The predominant hemicellulose carbohydrate in grasses and hardwoods plants is xylose, while the most common hemicellulose for softwoods is mannose.[18] Hemicellulose sheaths and provides support to cellulose via hydrogen bonds while covalently bound to the surrounding lignin.[19] Due to its amorphous structure, hemicellulose is more readily

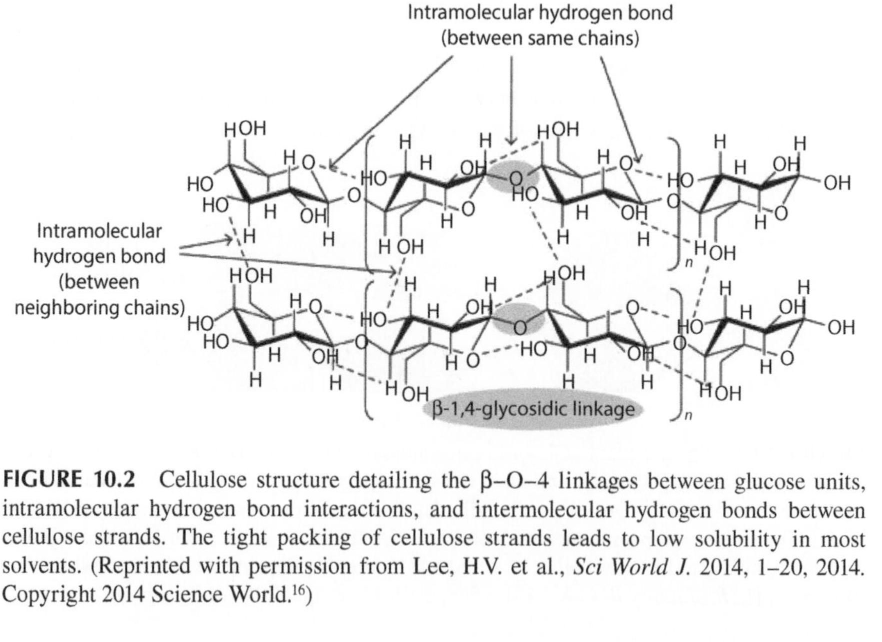

FIGURE 10.2 Cellulose structure detailing the β–O–4 linkages between glucose units, intramolecular hydrogen bond interactions, and intermolecular hydrogen bonds between cellulose strands. The tight packing of cellulose strands leads to low solubility in most solvents. (Reprinted with permission from Lee, H.V. et al., *Sci World J.* 2014, 1–20, 2014. Copyright 2014 Science World.[16])

depolymerized than cellulose. Hemicellulose can be dissolved or depolymerized in alkali or acidic aqueous solutions, as well as many organic solvents.

Lignin is an aromatic biopolymer that provides plants with waterproofing and structural reinforcement and acts as a barrier against microbial attack. It accounts for approximately 1/3 of the vascular tissues in the lignocellulose composite. Lignin is the result of oxidative coupling of 4-hydroxyphenylpropanoids subunits.[21] These subunits: p-coumaryl alcohol, coniferyl alcohol, and sinapyl alcohol combine during lignification to form various common motifs (Figure 10.3).[21]

Lignification occurs as a free-radical mechanism involving multiple steps and enzymes. The ratio of these subunits and how they are linked vary by plant species and the growing environment. The random cross-linking of subunits provides protection for the plant, but also contributes to its recalcitrance. Guaiacyl subunits (Figure 10.3b) can condense to form carbon-carbon crosslinks, making them difficult to decouple. Generally, grasses contain guaiacyl, syringyl, and small amounts of p-hydroxyphenyl units, while hardwood lignin contains a mix of guaiacyl and syringyl units requiring more severe pretreatments to fractionate the lignins. Softwood lignin is normally composed almost exclusively of guaiacyl units, and these linkages make softwood lignin the most difficult to depolymerize because of the potential C–C bonds by coupling at the free C-5 position of the aromatic ring (Figure 10.3 [5–5 linkage]). Delignification is therefore more difficult with softwoods than hardwoods or grasses because neither acid or base can hydrolyze these C-C bonds. Deconstruction of these carbon-carbon linkages (C–C) is one of the major obstacles associated with an energy efficient pretreatment process with most depolymerizations taking place at the β–O–4 linkage.[16] β–O–4 bonds (Figure 10.3c), which results in the linear elongation of the biopolymer, are the

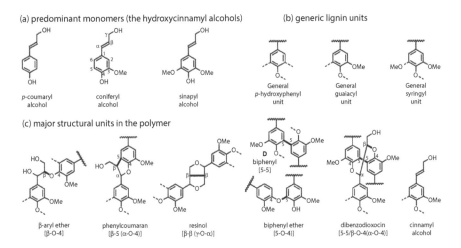

FIGURE 10.3 Amino acids are converted into three major subunits of lignin. (a) p-coumaryl, coniferyl, and sinapyl alcohols; (b) their representation in lignin; and (c) the common condensation motifs in lignin structure. (Reprinted with permission from Ralph, J. et al., *Phytochem Rev.*, 3, 29–60, 2004. Copyright 2004 Phytochemistry Reviews.[20])

easiest to cleave.[16] Hardwood lignin contains approximately 50% β–O–4 intersub-unit, softwoods 60%, and grasses 74%–84%.[22]

10.4   PRETREATMENT OF BIOMASS

While pretreatment is a costly process, the impact of removing it from the bio-fuel production process renders the prospect of lignocellulosic fuel unfeasible. The resulting reduction in yield drives costs far beyond the amount saved by eliminating pretreatment. The efficacy of pretreatment is complicated, as it is impacted by both the collection/harvest of the biomass and the downstream pro-cesses including hydrolysis and fermentation. Commercial feasibility depends more specifically on energy costs and solvent requirements for separation, depolymeriza-tion, and conversion of lignocellulose to the product(s) of choice. Cellulose fibrils are inaccessible to enzymes and chemical treatment due to protective lignin and hemicellulose components. Therefore, lignocellulosic biomass requires pretreatment for effective fractionation and subsequent conversion into materials, chemicals, and fuels.[23] Pretreatment aims to separate the surrounding hemicellulose and lignin from cellulose. This fractionation provides access to enzymes for saccharification (hydro-lysis of sugars to their monomers for fermentation) so that the process may proceed under relatively mild conditions (Figure 10.4).

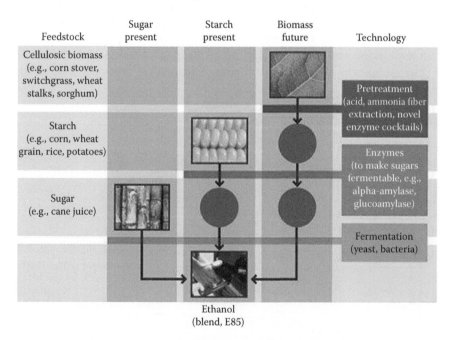

FIGURE 10.4   Processing steps of biomass to bioethanol production. Hydrolysis of sugar monomers after pretreatment of lignocellulosic biomass is required prior to fermentation by micro-organisms. First generation starches and sugars do not require the pretreatment step. (Reprinted with permission from WordPress.com, Ethanol|Bio-energy, https://biowesleyan.wordpress.com/first-generation-biofuels/ethanol/, Accessed April 6, 2018. Copyright 2018 WordPress.com.[24])

10.4.1 Methods of Pretreatment

Several different pretreatment processes have developed as a reaction to the variety of biomass types, each requires different conditions for maximum efficiency and produces different yields in the energy conversion process.[25] Any pretreatment process should be as effective as possible, with low capital and operational costs. Physical pretreatments may include mechanical extrusion, milling, and pyrolysis. Hybrid pretreatments, such as steam explosion, combine physical and chemical processing. In order to fractionate lignocellulosic biomass, some of the most common chemical pretreatments include dilute acid, kraft alkali, organosolv processing, and the use of ionic liquids.[23] Most pretreatments focus on producing pulp for paper making or cellulose for fermentation into bioethanol, though some processes such as organosolv and ionic liquids offer a potentially more diverse product output, such as platform chemical and materials production.

10.4.2 Physical and Physiochemical Pretreatments

Pretreatment via mechanical extrusion utilizes rotating screw blades to mix the biomass while simultaneously forcing it through a die. The cellulose in the biomass is disrupted by the mechanical shearing and elevated temperatures, which can reach an excess of 300°C.[23] Due to exposure to high temperatures and significant mechanical pressure, the pulp exiting the die experiences a sudden expansion which aids in fractionating the lignocellulosic structure. Saccharification (enzymatic hydrolysis), of the cellulose-rich pulp produced after extrusion, is performed to hydrolyze the sugars into monomers for fermentation to produce bioethanol. Mechanical extrusion can also be used in conjunction with alkali and acid pretreatments to improve fractionation of the biomass feedstock.[26] A high energy requirement is one of the disadvantages of this process, though reactor design and optimization can help to reduce the costs.[23]

Milling uses a grinder to reduce particle size of the lignocellulosic biomass, usually below 1 mm. Similar to mechanical extrusion, shearing disrupts the crystallinity of the cellulose, making it easier to hydrolyze and convert to ethanol. Various milling methods are available to reduce biomass size, such as planetary ball, hammer, and vibratory milling.[27] Milling results vary by preparation method and feedstock and can be used in conjunction with chemical processes to improve access to cellulose for further processing. Improvements to the milling process have been made to include alkali or acid, however, energy requirements are still one of the main obstacles to its commercialization.[28]

Pyrolysis is the thermal decomposition of biomass in the absence of oxygen, yielding bio-oils and several gaseous by-products including methane, hydrogen, carbon monoxide, and carbon dioxide. The lignocellulosic constituents degrade at different temperatures. Controlled heating between 400°C and 800°C is applied to the biomass allowing regulation of the product yield. This pretreatment technique can be applied to any type of lignocellulosic biomass.[29] Two major forms of pyrolysis exist, conventional or slow pyrolysis and fast pyrolysis, though no exact definition of "fast" or "slow" are given for the rate and temperature of heating. In general, slow pyrolysis heats biomass to approximately 500°C, forming charcoal, whereas fast pyrolysis rapidly heats the biomass to temperatures in excess of 600°C, producing gaseous products and bio-oils.[29]

Steam explosion demonstrates a combination of physical and chemical processes for the pretreatment of biomass.[23] It uses steam, usually between 180°C and 250°C, at an elevated pressure to expose the cellulose fibers for recovery. Similar to extrusion, sudden decompression when exiting the reaction chamber causes the rupture of the fibers due to the expansion of the steam trapped in the biomass.[30] Industrial application using steam explosion first started in the 1980s and can be run via batch[31] or continuous flow[32] reactors.[31] Enzymatic hydrolysis of the pulp is used to hydrolyze the cellulose to glucose for fermentation. Steam explosion is attractive because it has limited solvent use and low capital expenditure, however, improvement to reactor design and energy consumption need to be addressed.[31]

10.4.3 Chemical Pretreatments

Dilute acid pretreatment disrupts the ligocellulosic biomass by breaking ester bonds of lignin and by hydrolyzing the carbohydrates, making the sugars easier to ferment. Dilute acid pretreatment uses acids, such as sulfuric, oxalic, and maleic acids, between 1% w/v and 4% w/v, with reactions taking places between 100°C and 180°C. One of the main challenges of this pretreatment is the formation of sugar degradation products or humins, which reduce yield and inhibit the enzymes used for saccharification.[33] Further reductions in efficiency come from the common practice of using sulfuric acid in dilute acid pretreatment. It is toxic to yeast and other microorganisms and research into other acids is ongoing.[14] Though the acids used are relatively inexpensive, the corrosion of the reaction vessels over time has proven to be an issue in the long-term efficiency of dilute acid pretreatment.[34]

The kraft process is the most common method for paper production. Wood chips are placed in a mixture of sodium hydroxide and sodium sulfide and heated at elevated temperatures of around 170°C under high pressure.[10] The alkaline reactions improve the solubility of lignin via hydrolysis of ester bonds, breaking the polymer down into smaller subunits and creating anions via deprotonation of hydroxyl alcohol groups. The recovered pulps are washed, bleached, and pressed into paper.[35] The pulps can also be fermented into bioethanol. Though this pretreatment itself uses inexpensive chemicals for the process, it requires copious amounts of water to wash the pulps, producing additional cost and energy requirement for water remediation.[36]

The organosolv process uses organic solvents, such as ethanol, acetone, ethylene glycol, formic, and acetic acid between 160°C and 200°C and pressure from 5 bar–30 bar to break down lignin and hemicellulose.[37] This pretreatment method is versatile and has been commercially developed for papermaking, as well as bioethanol production.[38] The solvent is recyclable and reusable, however, the pulps require cumbersome wash steps because most of the solvents are toxic to microorganisms or inhibitory to enzymes.[37,39] In addition, processing the large quantities of highly flammable solvents at elevated temperatures and pressures presents a safety concern.[40] Organosolv does have some advantages over kraft pulping in that it produces less air and water pollution.[41]

Ionic liquids are a relatively new pretreatment option when compared to the previously mentioned methods. In addition to pretreatment, ionic liquids also present other processing options, such as modifications and transformations, which are not available by other methods and solvents. The possibilities presented by ionic liquids warrant further examination, and the chapter will focus on some of the specifics of their applications to lignocellulosic biomass. The deconstruction of lignocellulosic biomass using ionic liquids can be placed into two broad categories. The most prevalent technique is the dissolution process, whereby the entire biomass feedstock is dissolved, with the main focus on the regeneration and recovery of cellulose.[21] Dissolution with ionic liquids expands the potential of lignocellulosic biomass as a feedstock for many additional platform chemicals or materials. The alternative method involves the disruption of the lignocellulose composite by solubilization of the lignin and hemicellulose biopolymers without complete biomass dissolution, i.e. the cellulose is not dissolved.[42] Ionic liquids allows for selectivity of processing to extract, dissolve, or fractionate the components of biomass depending on the application. Conditions for optimum pretreatment vary with the ionic liquid and feedstock used.

10.5 IONIC LIQUIDS

Ionic liquids are salts that have weakly coordinating anion-cation pairs. Their structures prevent the formation of a crystalline lattice, resulting in melting points below 100°C.[43] Because of the substantial number of potential anion and cation pairings, ionic liquids possess flexible applications with a wide range of properties to fit specific tasks. They have been studied for catalysis, chemical synthesis, electrochemistry, as well as several industrial processes.[44–49] Ionic liquids possess low vapor pressures and have been considered as a replacement for reactions that involve the use of volatile organic solvents.[50] However, one of the major obstacles to the application of bulk ionic liquids as alternative solvents is the high cost of these solvents. Both the starting materials and the preparation of ionic liquids must be cost effective, non-toxic, and effective to compete with conventional solvents.

10.5.1 SOLUBILITY OF LIGNOCELLULOSIC BIOMASS IN IONIC LIQUIDS

Ionic liquids show promise in biorefining due to their customizability and multiple applications, but this property significantly increases the complexity of employing them for the chemical processing of lignocellulose. It is important to understand the mechanisms that drive dissolution of lignocellulose in ionic liquids to optimize its processing. Several reviews provide details into the chemical interactions of the biopolymers (cellulose, hemicellulose, and lignin) with select ionic liquids.[21,51] Characterization and correlations of ionic liquid properties have been used to predict their effectiveness for pretreatments. However, the results of the dissolution of the components of lignocellulosic biomass are different from the dissolution of the composite as a whole and cannot be used to predict the outcome of the pretreatment on the lignocellulosic biomass itself. Particle size, moisture content, and feedstock are all variables that will affect the final product. Further, the method used to produce the particle size may cause changes to the structure of the biomass.[52] Total dissolution of the biomass is usually not achieved, as hazy solutions or small amounts of solid residue remain.[53] Ball milling for extensive periods of time to reduce particle size can aid in achieving complete dissolution, however, such techniques are not economically feasible for industrial processing.[54] The extensive customizability across multiple variables prompts further research to optimize chemical processes with ionic liquids.

10.5.2 CELLULOSE SOLUBILITY IN IONIC LIQUIDS

Cellulose solubility can be important for efficient pretreatment, as well as for modification and production of platform chemicals. However, cellulose is not soluble in water or in most organic solvents with the exception of N-methylmorpholine-N-oxide and concentrated phosphoric acid.[55–57] Neither of these solvents are feasible for lignocellulosic biomass pretreatment, as the solvent is not compatible with downstream processes. Other pretreatment methods include the modification of cellulose to increase solubility, as in the production of viscose (a synthetic fiber also known as rayon).[58] The process for the modifications would be an added cost, and in the case of viscose, one of the main reagents used is toxic and flammable (carbon disulphide). Unlike other solvents, several ionic liquids can dissolve cellulose, with 1-butyl-3-methylimidazolium chloride ([BMIM]Cl) dissolving up to 25 wt%, potentially increasing the efficiency and flexibility of cellulose processing.[59] Ionic liquids would have the potential to improve the safety and cost for viscose production and other existing industrial processes.

Ionic liquids that dissolve or swell the whole lignocellulosic biomass are followed by a regeneration step, usually via the addition of an anti-solvent.[61] These cellulose I fibrils are then converted into the thermodynamically more stable cellulose II (Figure 10.5). There is a significant loss of crystallinity because of an increased abundance of amorphous regions.[62] This transformation is attractive to lignocellulosic bioethanol processing because the hydrolysis of cellulose II is significantly more efficient than native cellulose. Imidazolium-based ionic liquids with strong hydrogen bonding and highly mobile anions are the most commonly used for these types of reactions. Halides, acetates, dialkylphosphates, trialkylphosphonates, and amino acids anions improve cellulose solubility compared to non-coordinating anions.[59,63] It is believed that these anions disrupt the hydrogen bonding of the equatorial hydroxyl groups of cellulose, increasing solubility.[63]

Specific parameters and computer modelling techniques systematically determine the ability of ionic liquids to dissolve cellulose. The most common experimental technique uses Kamlet-Taft polarity parameters, while quantum mechanical conductor-like screening model (COSMO) for realistic solvents have been able to provide computer predictions and aid in screening of ionic liquids. Kamlet-Taft parameters include: the hydrogen bond donor/acidity measurement denoted by (α), the hydrogen bond acceptor/basicity measurement determined by (β), and polarity measurement (π^*). The β is determined by the anion and is (mostly) independent of the cation of the ionic liquid.[12] Ionic liquids that dissolve

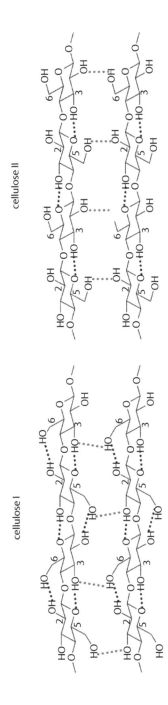

FIGURE 10.5 The structure native cellulose I (left) compared with that of cellulose II which is more thermodynamically stable. (Reprinted with permission from Credou, J. and Berthelot, T., *J. Mater. Chem. B.*, 2, 4767–4788, 2014. Copyright 2014 Journal of Materials Chemistry B.[60])

cellulose are usually characterized by high β values. Conductor-like screening model for realistic solvents predictive modelling has suggested that the difference between the α and β parameters can also be used to predict if an ionic liquid will dissolve cellulose.[64] The size and geometry of the anion also impacts dissolution of cellulose, for example, [BMIM]Cl with a β = 0.80 dissolved 10 wt% cellulose compared with 1-butyl-3-methylimidazolium O,Se-dimethyl phosphoroselenoate ([BMIM][dmpSe]) β = 0.82, which does not dissolve cellulose.

Unlike the anion, the cation of the ionic liquid interacts via dispersion forces and hydrogen bonds with the glycosidic and hydroxyl oxygens located in the axial positions of the cellulose fibers.[65] The impact of the cation on the solubility of cellulose has also been studied by holding the anion constant and varying the cation. In these studies, the functionalization of cations has shown a direct correlation between the natural logarithm of cellulose solubility and α.[66] Functional groups, such as hydroxyl or ether groups, decrease the solubility of cellulose.[67] Hydrogen bonding of the cation to the hydroxyl and ether oxygens of cellulose may be responsible for solubility where the polarizability of the cation interrupts the hydrogen bonds between the anion and cellulose. High polarity and weak ionic association of the cation leads to improvements.[68] Though the cation does effect solubility, the β value (anion) is a far greater contributor to hydrogen bond disruption, and therefore cellulose solubility.[68]

While the use of ionic liquids to dissolve lignocellulosic biomass/cellulose is attractive, converting native cellulose to the more hydrolyzable cellulose II presents several challenges that need to be addressed. Many of these ionic liquids require dry conditions and have lower thermal stability than the optimum processing temperature.[69] In addition to these technical challenges, these ionic liquids are expensive, making large-scale applications economically challenging.[70,71]

10.5.3 Hemicellulose Solubility in Ionic Liquids

Existing literature focuses on the dissolution of cellulose with little discussion on the dissolution of hemicellulose in ionic liquids. Cellulose strands connect via strong hydrogen bonds, and ionic liquids with high hydrogen bond basicity are necessary for dissolution. Hemicellulose is non-crystalline, with lower molecular weights and fewer hydrogen bonds. This means that a high hydrogen bond basicity (β value) which is necessary to dissolve cellulose is not necessary for the dissolution of hemicellulose. Ionic liquids that dissolve cellulose will also dissolve hemicellulose, though different ionic liquids are more suited (hydrophilic are generally better than hydrophobic).[72] The number of hydroxyl groups and hydrophobicity can impact the solubility of the hemicellulose sugars. For example, xylan with two hydroxyl groups will be less soluble than dextrin, which has three. Solubility of 3 wt%–25 wt%, dependent on the hemicellulose sugars, has been achieved with 1-allyl-3-methylimidazolium chloride ([AMIM]Cl) at a temperature range between 30°C and 95°C.[73] Selective hemicellulose extraction/dissolution has been shown to be possible by modifying the ionic liquid anion to have Kamlet-Taft parameters β below 0.85 on delignified pulp.[74] The reduction of high β ionic liquids can be achieved in some cases by the addition of water. It has been demonstrated that these ionic liquids will selectively dissolve and extract hemicellulose from a processed pulp.[75]

10.5.4   LIGNIN SOLUBILITY IN IONIC LIQUIDS

Studying the dissolution of lignin is more complicated than analyzing the dissolution of the complete cellulose composite. There are a greater number of ionic liquid choices for lignin dissolution than there are for cellulose.[76] Once extracted, chemically produced lignin is structurally different from native lignin and solubility tests (mostly on lignin produced via kraft or organosolv), though valuable, do no truly represent the solubility of native lignin.[21,22] Yet the use of the entire lignocellulosic composite means that the other biomass components will impact the results, true solubility of lignin is difficult to characterize. Ultimately, the dissolution of native lignin from lignocellulosic biomass can vary depending on the ionic liquid, feedstock, as well as pretreatment conditions and solubilities of 70% wt have been achieved.[21]

Lignin dissolution is also dependent on the anion, the higher the β value, the better the dissolution for lignin. The β threshold does not need to be as high as that for cellulose, with 55 wt% achieved with 1-ethyl-3-methylimidazolium trifluoromethanesulfonate ([EMIM][OTf]), which has a $\beta = 0.50$.[12] Protic ionic liquids have also demonstrated successful extraction of native lignin via acid-catalyzed depolymerization and cleaving of β-O-4 bonds.[77] Several protic ionic liquids are also attractive for pretreatment due to their relative low costs.[78] However, protic ionic liquids have a tendency to degrade lignocellulosic biomass, forming enzyme inhibiting by-products which reduce yield and efficiency of saccharification.[79] The addition of functional groups, such as a methyl group onto the imidazole ring localizes the electrons causing anions to congregate above and below the cation. This impacts the ability of the cation to solubilize lignin due to reduced access of the lignin to the cation because of increased competition by the anion.[80] Dispersion-corrected density functional theory modeling suggests that imidazolium ionic liquids are especially effective at lignin dissolution due to π-stacking and hydrogen bonding interactions with lignin.

10.6   POTENTIAL FOR BIOREFINING

Lignocellulosic components have the potential to replace the fuels, platform chemicals, and materials derived from petroleum production.[81] Ionic liquids can also be utilized for chemical modification and functionalization of lignocellulosic components for materials production, thus adding value beyond bioethanol production for some of the potential platform chemicals (Figure 10.6). Galactose, mannose, and glucose (hexose sugars) potentially feed the production of 5-hydroxymethylfurfural (HMF). HMF is a gateway to other platform chemicals, such as levulinic acid, 2,5-dimethylfuran, sorbitol, and 2,5-furan dicarboxylic acid, all of which have been identified by the United States Department of Energy (DOE) as the "top value-added chemicals from biomass."[82] Furfural and xylitol can be produced from pentose sugars xylose and arabinose. Each of these chemicals can be produced via ionic liquids.

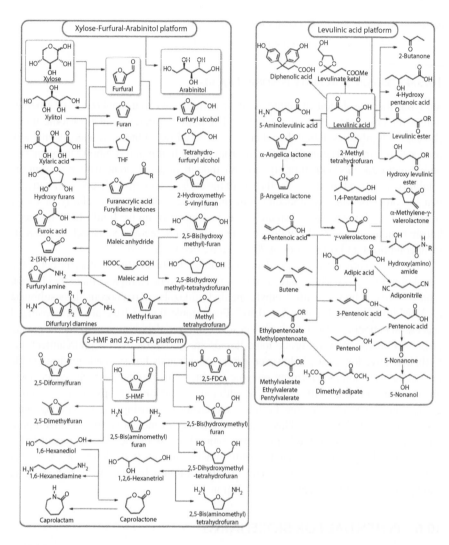

FIGURE 10.6 Potential platform chemicals and derivatives from pentose and hexose sugars. Lignocellulosic carbohydrates can be dehydrated to furans. Derivates of these furans can be used for fuel additives, chemical solvents, as well as polymeric building blocks. (Reprinted with permission from Isikgor, F.H. and Becer, C.R., *Polym. Chem.*, 6, 4497–4559, 2015. Copyright 2015 Polymer Chemistry.[83])

10.6.1 CELLULOSIC MATERIALS FROM IONIC LIQUIDS PROCESSING

Cellulose has various properties that make it an attractive polymer for materials production, such as biocompatibility, non-toxicity, and biodegradability.[84] Unlike aqueous or organic solvents, ionic liquids have the potential to aid in the modification and preparation of cellulosic materials due to their ability to dissolve cellulose.

The most commonly utilized ionic liquids are dialkylimidazolium chloride/acetate for dissolution, chemical modification, and platform chemical production.[85]

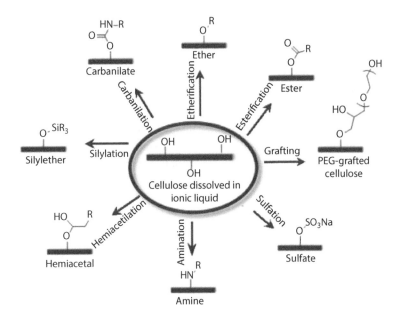

FIGURE 10.7 Potential surface modifications using ionic liquids for cellulose/nano-cellulose. (Reprinted with permission from Isik, M. et al., *Int. J. Mol. Sci.*, 15, 11922–11940, 2014. Copyright 2014 International Journal of Molecular Sciences.[89])

Chemical modifications include grafting, esterification, etherification, and silylation of the hydroxyl groups (Figure 10.7). These cellulose-derived materials impact a broad range of industries including medicine, construction, optics, food, and textiles. For example, chemical modification of cellulose has been functionalized with [BMIM]Cl to form composites that can be used for thermal stimuli response-related applications.[86] Ionic Liquids have also been used for the modification and spinning of cellulose fibers for textile and pharmaceuticals providing higher thermal stability, mechanical strength, and chemical resistance.[85] Thin cellulose films produced with [AMIM]Cl not only have an even, smooth surface, but with the addition of plasticizer, better thermal stability, tensile strength, and hydrophobicity than unmodified films.[87] Cellulose and mixtures of cellulose, soda lignin, and xylan from spruce wood using [BMIM][Cl] have also successfully produced aerogels.[88] These gels have open pore network structures and the surface area can be varied depending on the ratio of the components of the mixture.

Ionic liquids can be used in the production and modification of nanocrystalline cellulose, crystalline cellulose fibers with high aspect ratios, widths ranging from 10 nm–20 nm, and lengths under 1000 nm.[90] This includes the use of acidic ionic liquids to produce nanocellulose from microcrystalline cellulose.[91] Surface modifications can change the properties of these nanocrystalline celluloses, for example, increasing or decreasing hydrophobicity.[92] One-pot processing with acidic ionic liquids for hydrolysis with the addition of acetic anhydride for acetylation altering hydrophobicity has been successfully demonstrated.[93]

[AMIM]Cl has been used to graft polyisoprene to cellulose to demonstrate the ability to use renewable resources on a process previously manufactured via petroleum feedstocks.[94]

10.6.2 LIGNIN MATERIALS AND COMPOSITES

Lignin has traditionally been viewed as a waste product of lignocellulosic processing.[95] Currently, the most common use for lignin is combustion for energy recovery, however, being the most abundant naturally occurring aromatic polymer, lignin has generated interest for other possibilities. Lignin has potential for use as an adhesive, anti-coagulant, emulsifier, pyrolysis, and gasification.[96–98] Materials, composite production, and chemical modification have also become a focus for the upgrading of lignin. Ionic liquids can be used for chemical modification of lignin or just dissolution of the lignin for further processing.

Production of high-strength carbon fibers from lignin has been demonstrated from the combination of pulp and lignin, produced using the ionic liquid [DBNH] [OAc], 1,5-diazabicyclo[4.3.0]non-5-enium acetate.[99] The production method and concentration of lignin impacts tensile strength of the fibers, crystallinity, elasticity, as well as thermal stability. Ionic liquids have been combined with lignin to produce lubricants, where it was demonstrated that amino acid-based ionic liquids are capable of withstanding shearing and friction forces comparable to conventional lubricants with significantly less corrosive effects.[100] Ionic liquid and lignin mixtures have also demonstrated the ability to improve mechanical and water vapor permeability of poly(butylene-adipate-co-terephtalate)-polylactide polymer blends.[101] Lignin, in this case, is used as a reinforcing agent enhancing the interaction between the poly(butylene-adipate-co-terephtalate)-polylactide, while the ionic liquid is a multifunctional additive, resulting in improved tensile strength and flexibility. These examples not only demonstrate the potential for lignin materials and composites, but the important role that ionic liquids have for the development of these applications.

10.7 LIGNOCELLULOSIC BIOMASS TO PRODUCE PLATFORM CHEMICALS

The processing of lignocellulosic biomass with ionic liquids may provide efficient paths to the production of necessary chemicals and materials for other chemical reactions. Lignocellulosic biomass has the potential to provide "top-value-added" platform chemicals identified in DOE reports.[82] Specifically, cellulose and hemicellulose have been investigated as potential sources for furfural and HMF, both of which have been identified as the top 30 chemicals that can be derived from biomass.[82] There has been some success in acid-catalyzed conversions of these monomeric sugars to HMF or furfural, however, these reactions usually suffer from low yield due to humin formation and sugar degradation.[102] Feasible production of renewable platform chemicals from sugars is highly dependent on the improvement of these yields, while maintaining a green process. Research has focused on dialkylimidazolium-based cations paired with the chloride anion, due to their ability to solubilize cellulose and hemicellulose. Chromium catalysts and other metal catalysts have been found to facilitate the conversion of these sugars.[103] Water content also aids in the hydrolysis of the polymeric sugars to their monomers. Varying water content can lead to the formation of humins and other platform chemicals.[104] Temperature generally increases conversion rates, but also leads to the formation of undesirable side products.[105] Degradation of hexose sugars differ from pentose sugars, with pentose sugars more likely to be involved in secondary reactions.[106]

10.7.1 HMF Production

According to the DOE reports *Top Value Added Chemicals from Biomass*, HMF is a building block for various polymers and plastics. Hydrolysis of cellulose to glucose is a necessary pre-step to the production of HMF from lignocellulosic biomass. The precise mechanism of conversion of glucose to HMF is not certain, but the isomerization of glucose to fructose is likely necessary (Figure 10.8). This is demonstrated by the high yield of HMF produced directly from fructose, in comparison to the much reduced amount of HMF produced from glucose.

Several methods employ biphasic extraction or resins to separate HMF and drive the reaction.[107,108]

Conversions as high as 96% have been achieved with fructose using catalysts with no humin or side products detected.[110] Despite the high yields achieved, this is currently not a viable means for commercial production of HMF due to the prohibitive cost of fructose.

[EMIM]Cl serves as a solvent for the conversion of glucose to HMF. Initial experiments using [EMIM]Cl with a chromium catalyst demonstrated that glucose conversion to HMF with promising yields was possible.[111] However, without the use of a catalyst, conversion to HMF was slow and accompanied by various degradation products. Furans (HMF and furfural) demonstrated high stability in the ionic liquid solvent, however, in the presence of other compounds, they quickly polymerize and form humins.[106] This means that the limitation of secondary reaction products is necessary to maximizing yield. Conversion of sugar oligomers with [EMIM]Cl and [BMIM]Cl demonstrated the difficulty of conversion even with catalyst.[112] Various acidic catalysts (mineral acids, such as hydrochloric acid and organic acids, such as acetic acids) and metallic catalysts have been attempted while varying the cation of the ionic liquid and water content.[113] For example, in the presence of chromium (III)

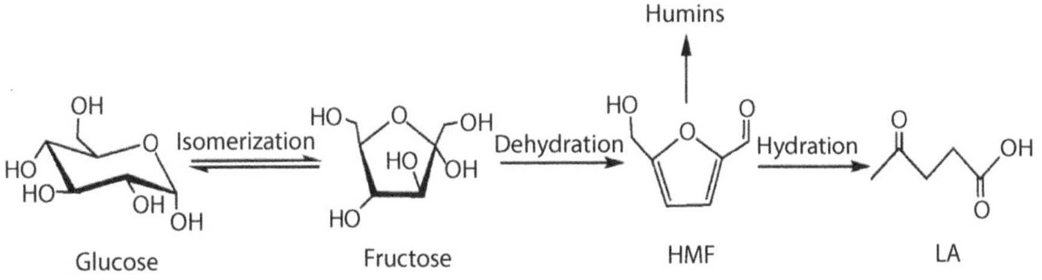

FIGURE 10.8 A proposed mechanism for the conversion of glucose to HMF and levulinic acid. (Reprinted with permission from Li, X. et al., *Catal. Sci. Technol.*, 6, 7586–7596, 2016. Copyright 2016 Catalysis Science & Technology.[109])

chloride, 90% conversion of glucose was achieved, however, the yield of conversion dropped to 50% when using cellobiose.[112]

Conversion of cellulose directly to HMF is difficult without first hydrolyzing by monomeric glucose, making a successful one-pot synthesis unlikely.[114] Water can potentially be used for the hydrolysis, though the moisture content must be carefully monitored.[115] A combination of copper(II) chloride and chromium (II) chloride catalyst with [EMIM] Cl has demonstrated success in disrupting the cellulose structure and converting glucose to HMF.[116] Acidic ionic liquids have also been used for hydrolysis of cellulose and HMF conversion because they can act as both solvent and catalyst.[117] Degradation and side products persist as one of the fundamental issues in production of HMF from cellulose.

Recycling of the ionic liquid (IL) and catalyst is possible though purification and extraction can be a challenge. Addressing these considerations is critical for economic valorization. Separation of HMF with supercritical carbon dioxide or biphasic extraction from ionic liquids is possible.[118] Solvents, such as methyl isobutyl ketone, have been used for extraction from ionic liquids.[118] Once extracted, HMF is best preserved in its crystalline form as opposed to oil (liquid) HMF, which is not stable and may rapidly degrade at room temperature.[119]

10.7.2 FURFURAL PRODUCTION

Furfural shows strong potential as a renewable feedstock for resins, polymers, and solvents, but self-polymerization and humin production make efficient production challenging.[120,121] The most abundant and most researched pentose sugar in lignocellulosic biomass is xylose. Dehydration of xylose results in several useful platform chemicals: acids, ketones, and aldehydes (such as furfural).[122,123]

Due to the similar structure to glucose, chloride ILs and metal catalysts are also potential agents for the dehydration of xylose to furfural. Similar to glucose, xylose may also require an isomerization step, in this case to xylulose (Figure 10.9). This process can be aided via the addition of a catalyst.[125] Metal catalysts have also demonstrated successful conversion of xylans to furfural and microwave-assisted reactions can aid in reducing reaction times.[126]

Acidic ionic liquids, such as 1-butyl-3-methylimidazolium hydrogen sulfate [BMIM][HSO$_4$], have been used as both solvent and catalyst for the conversion of xylose to furfural. Though high levels of conversion have been achieved, yields of furfural remain low due to production of side products.[123] Ionic liquids can also be

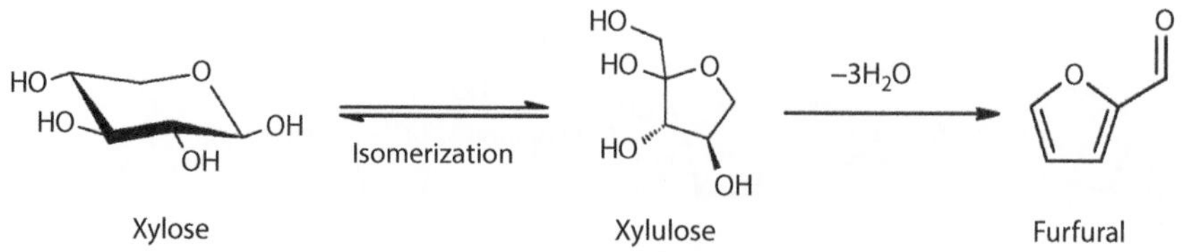

FIGURE 10.9 Potential mechanism for the conversion of xylose to furfural. Isomerization to xylulose is believed to be necessary to facilitate final conversion to furfural. (Reprinted with permission from Shirotori, M. et al., *Catal. Sci. Technol.*, 4, 971–978, 2014. Copyright 2014 Catalysis Science & Technology.[124])

functionalized for different purposes. For example, the addition of sulfonic acid to the cation can improve the thermal stability of the ionic liquid and catalyzes the conversion of xylose to furfural.[127] Biphasic extractions with methyl isobutyl ketones during reaction have been used to improve furfural yields and prevent humin formation.[128]

HMF and furfural production directly from lignocellulosic biomass is not currently feasible. Interactions and interference of other components of the biomass reduce yields and create difficult separations. Fractionation and pretreatment of the biomass are necessary and require multiple steps and process considerations.[129] Purification of pentose sugars from biomass would be a necessary consideration in addition to other scale-up matters, such as IL corrosion and energy consumption.[130]

10.7.3 LEVULINIC ACID PRODUCTION

Levulinic acid is a hydration product of HMF, though direct conversion from cellulose via catalysis is possible.[131] Levulinic acid has potential applications as a fuel additive, solvent, uses in pharmaceuticals, building blocks for polymers, as well as generation of other platform chemicals.[132] There are several major challenges in producing levulinic acid. One issue is the lack of selectivity, side reactions, and humin formation detracts from yields. In addition, recovery and separation of catalyst are necessary for production economic realization.

The formation of levulinic acid has been demonstrated to be temperature dependent, forming at a range between 120°C and 160°C. Below these temperatures, the glucose remained unchanged and above 160°C side products dominate.[133] Catalysts are necessary for selectivity, and modified ionic liquids (such as SO_3H-functionalized) may potentially be utilized for solvent and catalyst. Levulinic acid and derivatives, such as ethyl levulinate have been produced from cellulose with the use of acidic ionic liquids with ethanol and water mixtures.[134] Acidic ionic liquids facilitate the hydrolysis of cellulose and its conversion to HMF, and water is necessary for the hydration of HMF to levulinic acid and formic acid.[134] In the presence of ethanol, the levulinic acid can be esterified to ethyl levulinate.

Many have suggested the possibility of one-pot synthesis of levulinic acid, most of them using mineral acid-based ionic liquids or even functionalized ionic liquids to facilitate catalysis.[134–137] Water and temperature play a large role in the optimization of the reaction, impacting the viscosity and kinetics of the reaction.[132] However, these reaction yields are often compromised, with glucose, HMF, furfural, lactic acid, formic acid, acetic acid, and humins. The addition of dimethyl sulfoxide (DMSO) has been used to reduce the viscosity of the ionic liquid, as well as act as a potential stabilizer for HMF and to prevent the formation of humins.[138] However, this addition seems to impact levulinic acid production yield.

10.7.4 PLATFORM CHEMICALS FROM LIGNIN

The production of platform chemicals from lignin have mostly been performed in organic solvents like dioxane and pyridine with the aid of a catalyst.[139,140] The solubility of lignin in ionic liquids provides a potentially greener method for the valorization and production of chemicals and materials from lignin. Bioengineering of the feedstock can aid in lignin structural determination and depending on the pretreatment method for lignin production, structural motifs may differ.[22] Most reactions were conducted in imidazolium-based ionic liquids due to their ability to solubilize lignin, with most lignin produced via kraft or organosolv processing.

Lignin has the potential to produce aromatic platform chemicals via the use of catalysts in combination with ionic liquids as the solvent (Figure 10.10). Depolymerization and subsequent transformation of the lignin monomers has been the focus of most studies.[141] Ionic liquids can aid in stabilizing the catalysts, demonstrated during the oxidative cleavage of lignin. This may facilitate the recycling of the catalyst.[142] Cleavage products vary depending on reaction conditions. Syringaldehyde, vanillin, and p-hydroxybenzaldehyde, syringol, sinapinic acid, syringic acid, catechols, and 2,6-dimethoxy-1,4-benzoquinone in addition to other aromatic compounds have been produced (Figure 10.10).[83] Ionic liquids have also been used in the production of low molecular weight aromatic oils, catalyzed by methyltrioxorhenium, from lignin.[143]

Choice of anion and or cation pairing for lignin depolymerization varies the compounds produced. There is evidence that varying the anion can influence the oxidative depolymerization of lignin, and that cations impact extraction and depolymerization

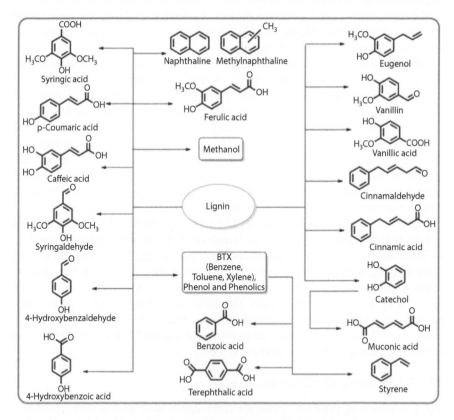

FIGURE 10.10 Potential aromatic platform chemicals from the depolymerization of lignin including the potential production of benzene, toluene, and xylene (BTX), which employ fossil fuels in production. (Reprinted with permission from Isikgor, F.H. and Becer, C.R., *Polym. Chem.*, 6, 4497–4559, 2015. Copyright 2014 Polymer Chemistry.[83])

of lignin from the biomass.[144] Non-acidic imidazolium ionic liquids can interact with the various motifs of lignin aiding in the cleavage of β–O–4 bonds, with some being able to generate free radicals to aid in depolymerization and conversion.[145] Varying ionic liquid catalyst systems also impact polydispersity index and reaction rate of the depolymerized lignin.[146] In addition, the catalyst used to depolymerize lignin can also impact the molecular weight.[147]

10.8 ENVIRONMENTAL CONCERNS

In addition to the techno-economic challenges of making ionic liquids commercially feasible, environmental and safety concerns also need to be addressed. The properties, such as low vapor pressure and high thermal stability, that make ionic liquids extremely attractive for various chemical reactions present some challenges when disposal is necessary. Ionic liquids can have serious and persistent consequences for the environment. Many ionic liquids are soluble in water, which has implications for aquatic ecosystems.[148] Clear guidelines and protocols for assessing the toxicology and environmental impacts for ionic liquids have not yet been determined, though a various battery of tests are often employed (Figure 10.11).[149]

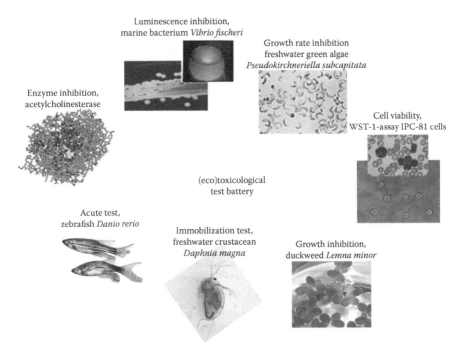

FIGURE 10.11 Common toxicology tests that are used to determine ecotoxicity. Changes in behavior due to ionic liquid exposure have been observed in *Danio rerioas*, *Vibrrio fischeri*, and *Daphnia magna*. (Reprinted with permission from Thuy Pham, T.P., *Water Res.*, 44, 352–372, 2010. Copyright 2010 Water Research; Stolte, S. et al., (Eco)Toxicology and biodegradation of ionic liquids, in *Ionic Liquids Completely UnCOILed*, John Wiley & Sons, Hoboken, NJ, 189–208, 2015.[148,149])

Ionic liquids are often marketed as being "tailor made," making their ecological impact very difficult to generalize.[150] Attempts have been made using characteristics like: hydrophobicity/hydrophilicity and anion/cation functionalities.[151,152] Some protic ionic liquids have been found to be less toxic and potentially more biodegradable than their aprotic counterparts.[153] Hydrophobic ionic liquids can have higher cytotoxicity due to interactions with the lipid bilayers of cell membranes.[150] Computer modelling of some ionic liquids suggests that the cation structure has a large impact on cytotoxicity, with larger, more branched cations being more hazardous. While polar functional groups reduce toxicity and the anion seems to have little effect.[154]

Though generalizations are useful to give a broad toxicology picture, they do not necessarily address specific effects of an ionic liquid. Comparative studies between ionic liquids that have similar physical and chemical properties, but different structures, revealed the difficulties in relying on generalizations.[155] These ionic liquids disrupted the same biological systems in very different ways. Disruptions to living organisms are not necessarily seen only at the cellular level, but can also impact other systems, including enzymes and hormones, leading to behavioral changes.[148]

Due to the robust nature of most ionic liquids, bioaccumulation and persistence in the environment are other factors to consider in addition to toxicity. Long-term effects are not well understood, though trials have been conducted using in vitro assays and use of various biological organisms such as *V. fisheri*, *D. rerio*, and *D. magna*. These studies conducted are not necessarily representative of real-life conditions.[156–158] Using biodegradable starting materials, such as metabolite-based ionic liquids, was considered a potential solution.[159] These ionic liquids were thought to be inherently nontoxic and easily broken down by microorganisms. However, when a selection of ionic liquids were tested, all failed to meet the Organization for Economic Cooperation and Development standards for ready biodegradability.[160] Environmental conditions also impacted the degradation of the ionic liquids. In addition, depending on the starting material, as well as the bacterial metabolic pathways, the microorganisms produce various by-products, not all of which were benign.[161] Research is ongoing into genetically modifying bacteria to control the products of these transformations.

10.9   TECHNO-ECONOMIC CONCERNS

There are a number of technical and economic factors that underpin the viability of using a solvent in a given application, particularly an ionic liquid. Most academic studies have naturally focused on technical considerations, as these are rightly viewed as go/no go criteria for research purposes. One example is the high viscosity of most ionic liquids, which can limit transport in applications involving multiphase processing (especially gas/liquid transport). Other potential issues (stability, toxicity, disposal, corrosion) can also be important and are usually examined late in a development cycle.

Economic considerations are, however, vital if translation of IL-based technologies are to become a reality. In the context of bioethanol, it is important to note that if the ionic liquid itself is 100 times more valuable than the proposed product, solvent recovery will quickly dominate the economics of the process. Other important objectives include minimizing solvent cost, minimizing solvent losses, and maximizing biomass loading (to minimize solvent use), all of which contribute to the cost of using a solvent in a process. Energy and sustainability considerations also must be explored, and the attending energy costs regarding solvent regeneration are included. Only a full, specific process techno-economic model can identify these key issues, and these unfortunately require much technical development (and data!) to complete.

We can make some general conclusions about IL cost and use. The dissolution pretreatment relies on a relatively small number of ionic liquids containing highly hydrogen-bond basic anions (required for cellulose solubility), the most widely used solvent being [Emim][OAc]. These ILs suffer from a number of drawbacks, such as high cost (price estimates for [Emim][OAc] range from $20/kg–$101/kg), low thermal stability, and low tolerance to moisture.[162–164] To be effective at dissolving cellulose, these ionic liquids also require low water contents that are at odds with the high moisture content of freshly harvested biomass (up to 50%) and the ionic liquids' high affinity for water, thereby increasing energy usage in the process.[165,166]

The ionoSolv pretreatment approach (lignin/hemicellulose dissolution) avoids some of these drawbacks.[167] This approach requires 10%–40% water in the IL for effective fractionation to occur. Also, the use of a protic ionic liquid in the pretreatment of lignocellulose was a major boost to economic viability of ionoSolv pretreatment, as these ionic liquids are inevitably cheaper than their peralkylated analogues. This is due to replacing the conversion of the base into the cation by an alkylation reaction and a potential subsequent ion exchange step with the simple mixing of an acid and a base. A recent techno-economic analysis of the bulk-scale synthesis of [TEA][HSO4], an effective solvent for fractionation, would cost as little as $1.24/kg.[78,167,168] This price is similar to that of common organic solvents, such as acetone and toluene, addressing one of the greatest concerns raised in conjunction with applying ionic liquids in large volumes, their (normally) high cost.

Ionic liquid cost is not the consideration, their recovery and recycling is also important.[70,71,167] For example: if 99% of the ionic liquid was recycled, a solvent priced at $50/kg (price selected based on the study by Klein-Marcuschamer et al.) would incur solvent replacement cost of $5.56 per liter of ethanol (assuming 300 L/dry ton of biomass), while an IL at $1.24/kg would incur $0.13 per liter.[70] These values represent 1400% and 32% of the current ethanol selling price, respectively ($0.77/L, National Association of Securities Dealers Automated Quotations (NASDAQ)). Despite this, only a limited number of studies have investigated the reuse of ionic liquids after pretreatment to date, often recycling at low biomass loadings (5% or less) and at recovery rates below 95%. A shift in focus to these economic considerations would greatly aid the determination of IL pretreatment feasibility.

10.10 CONCLUSIONS AND OUTLOOK

Ionic liquids have many interesting properties that have attracted researchers to their vast potential, including applications as solvents for the processing of lignocellulose. In particular, ILs can dissolve, fractionate, convert, or modify the biopolymers comprising the lignocellulosic matrix, often under mild conditions, which opens up possibilities for improving the processing of lignocellulose for applications beyond biofuel production.

These biopolymers (especially lignin and, unusually, cellulose) are highly soluble in ILs. However, the relatively more limited range of solvents available for cellulose raises questions about the potential cost of these applications compared to the more flexible solvent design allowed for lignin dissolution. Conversely, the unusual solubility of cellulose (compared to limited, and undesirable, organic solvent options) presents exciting opportunities for cellulose modification and derivitization in homogeneous solution using ionic liquids.

Most of the solubility control in ILs has been linked to the choice of anion, with a relatively smaller cation effect, due to the need for strong hydrogen bond acceptance from cellulose for dissolution to occur. The low toxicity and relatively high thermal and chemical stability of these ionic liquids present excellent opportunities for commercial exploitation and a contrast to toxic organic solvent alternatives. However, there is a trade-off between chemical and physical properties (e.g., high viscosity) and high cost of solvent production (compared to organic solvents), which could hinder the development of an ionic liquid biorefinery. A more extensive analysis of IL price, biomass loading, and IL recycling is therefore needed.

Several technical challenges still remain for full biopolymer valorization, as we transition to a bio-based economy, where cellulosic and hemicellulosic sugars present a renewable alternative source for fuels, chemicals, and materials, replacing fossil fuels. Ionic liquids will have a crucial role to play in this transition, as they possess the versatility to fractionate the biomass composite, catalyze polysaccharide transformations to chemicals, and produce useful derivatives and materials. However, several processing challenges still remain, including the prominent challenge of monomeric sugar and sugar derivative (HMF, furfural) recovery from the involatile IL. Finally, lignin valorization presents the next great challenge in economical biorefining. The wide range of ILs with extremely high lignin solubilities, and the potential of these solvents for chemical transformation, provide an exciting alternative medium to organic solvents for lignin transformation. Though product separation also remains a challenge, ionic liquids have shown exceptional promise in this final frontier of biorefining.

REFERENCES

1. Field, C., Campbell, J., and Lobell, D. Biomass energy: The scale of the potential resource. *Trends Ecol Evol.* 2008;23(2):65–72. doi:10.1016/j.tree.2007.12.001.
2. McAloon, A., Taylor, F., Yee, W., Ibsen, K., and Wooley, R. *Determining the Cost of Producing Ethanol from Corn Starch and Lignocellulosic Feedstock.* National Renewable Energy Laboratory, Golden, CO. 2000. https://scholar.google.co.uk/scholar?hl=en&q=A.+McAloon%2C+F.+Taylor%2C+W.+Yee%2C+K.+Ibsen+and+R.+Wooley%2C+National+Renewable+Energy+Laboratory+Report%2C+2000&btnG=&as_sdt=1%2C5&as_sdtp=. Accessed August 29, 2017.
3. Mosier, N., Wyman, C., Dale, B. et al. Features of promising technologies for pretreatment of lignocellulosic biomass. *Bioresour Technol.* 2005;96(6):673–686. doi:10.1016/j.biortech.2004.06.025.
4. Viikari, L., Vehmaanperä, J., and Koivula, A. *Biomass and Bioenergy.* Vol. 46. Pergamon, Turkey. 2012. https://www.cabdirect.org/cabdirect/abstract/20133013277. Accessed March 17, 2018.

5. Chovau, S., Degrauwe, D., and Van Der Bruggen, B. Critical analysis of techno-economic estimates for the production cost of lignocellulosic bio-ethanol. *Renew Sustain Energy Rev.* 2013;26:307–321. doi:10.1016/J.RSER.2013.05.064.

6. Dwivedi, P., Wang, W., Hudiburg, T. et al. Cost of abating greenhouse gas emissions with cellulosic ethanol. *Environ Sci Technol.* 2015;49(4):2512–2522. doi:10.1021/es5052588.

7. Gubicza, K., Nieves, I. U., Sagues, W. J., Barta, Z., Shanmugam, K. T., and Ingram, L. O. Techno-economic analysis of ethanol production from sugarcane bagasse using a liquefaction plus simultaneous saccharification and co-fermentation process. *Bioresour Technol.* 2016;208:42–48. doi:10.1016/j.biortech.2016.01.093.

8. Ge, Y., Dababneh, F., and Li, L. Economic evaluation of lignocellulosic biofuel manufacturing considering integrated lignin waste conversion to hydrocarbon fuels. *Procedia Manuf.* 2017;10:112–122. doi:10.1016/J.PROMFG.2017.07.037.

9. McKendry, P. Energy production from biomass (part 1): Overview of biomass. *Bioresour Technol.* 2002;83(1):37–46. doi:10.1016/S0960-8524(01)00118-3.

10. Chakar, F. and Ragauskas, A. Review of current and future softwood kraft lignin process chemistry. *Ind Crops Prod.* 2004. http://www.sciencedirect.com/science/article/pii/S0926669004000664. Accessed August 29, 2017.

11. Li, H.-Q., Li, C.-L., Sang, T., and Xu, J. Pretreatment on Miscanthus lutarioriparious by liquid hot water for efficient ethanol production. *Biotechnol Biofuels.* 2013;6(1):76. doi:10.1186/1754-6834-6-76.

12. Rani, M. A., Brandt, A., and Crowhurst, L. Understanding the polarity of ionic liquids. (*Phys. Chem. Chem. Phys.* (2011) doi: 10.1039/c1cp21262a). *Phys Chem.* 2011. https://ukm.pure.elsevier.com/en/publications/erratum-understanding-the-polarity-of-ionic-liquids-physical-chem. Accessed August 30, 2017.

13. Turumtay, H. Cell wall engineering by heterologous expression of cell wall-degrading enzymes for better conversion of lignocellulosic biomass into biofuels. *BioEnergy Res.* 2015;8(4):1574–1588. doi:10.1007/s12155-015-9624-z.

14. Kumar, P., Barrett, D. M., Delwiche, M. J., and Stroeve, P. Methods for pretreatment of lignocellulosic biomass for efficient hydrolysis and biofuel production. *Ind Eng Chem Res.* 2009;48(8):3713–3729. doi:10.1021/ie801542g.

15. McMillan, J. Pretreatment of lignocellulosic biomass. 1994. http://pubs.acs.org/doi/abs/10.1021/bk-1994-0566.ch015. Accessed August 29, 2017.

16. Lee, H. V., Hamid, S. B. A., and Zain, S. K. Conversion of lignocellulosic biomass to nanocellulose: Structure and chemical process. *Sci World J.* 2014;2014:1–20. doi:10.1155/2014/631013.

17. Laureano-Perez, L., Teymouri, F., and Alizadeh, H. Understanding factors that limit enzymatic hydrolysis of biomass. *Biotechnol ....* 2005. http://www.springerlink.com/index/V68258V8R53M8570.pdf. Accessed August 29, 2017.

18. Timell, T. E. Recent progress in the chemistry of wood hemicelluloses. *Wood Sci Technol.* 1967;1(1):45–70. doi:10.1007/BF00592255.

19. Vanholme, R., Demedts, B., Morreel, K., and Ralph, J. Lignin biosynthesis and structure. *Plant Physiol.* 2010. http://www.plantphysiol.org/content/153/3/895.short. Accessed August 29, 2017.

20. Ralph, J., Lundquist, K., Brunow, G. et al. Lignins: Natural polymers from oxidative coupling of 4-hydroxyphenyl- propanoids. *Phytochem Rev.* 2004;3(1–2):29–60. doi:10.1023/B:PHYT.0000047809.65444.a4.

21. Brandt, A., Gräsvik, J., Hallett, J., and Welton, T. Deconstruction of lignocellulosic biomass with ionic liquids. *Green Chem.* 2013. http://pubs.rsc.org/en/content/articlehtml/2013/gc/c2gc36364j. Accessed April 16, 2017.

22. Rinaldi, R., Jastrzebski, R., Clough, M. T. et al. Paving the way for lignin valorisation: Recent advances in bioengineering, biorefining and catalysis. *Angew Chemie Int Ed.* 2016;55(29):8164–8215. doi:10.1002/anie.201510351.

23. Kumar, A. K. and Sharma, S. Recent updates on different methods of pretreatment of lignocellulosic feedstocks: A review. *Bioresour Bioprocess.* 2017;4(1):7. doi:10.1186/s40643-017-0137-9.

24. WordPress.com. Ethanol|Bio-energy. https://biowesleyan.wordpress.com/first-generation-biofuels/ethanol/. Accessed April 6, 2018.

25. Shen, H., He, X., Poovaiah, C., and Wuddineh, W. Functional characterization of the switchgrass (Panicum virgatum) R2R3-MYB transcription factor PvMYB4 for improvement of lignocellulosic feedstocks. *Phytologist.* 2012. http://onlinelibrary.wiley.com/doi/10.1111/j.1469-8137.2011.03922.x/full. Accessed August 29, 2017.

26. Zheng, J. and Rehmann, L. Extrusion pretreatment of lignocellulosic biomass: A review. *Int J Mol Sci.* 2014. http://www.mdpi.com/1422-0067/15/10/18967/htm. Accessed August 29, 2017.

27. Kim, H., Chang, J., Jeong, B., and Lee, J. Comparison of milling modes as a pretreatment method for cellulosic biofuel production. *J Clean Energy Technol.* 2013. http://jocet.org/papers/011-J036.pdf. Accessed August 29, 2017.

28. Schell, D. and Harwood, C. Milling of lignocellulosic biomass. *Appl Biochem Biotechnol.* 1994. http://www.springerlink.com/index/5137683403KQ7H64.pdf. Accessed August 29, 2017.

29. Mohan, D., Pittman, C. U., and Steele, P. H. Pyrolysis of wood/biomass for bio-oil: A critical review. *Energy & Fuels.* 2006;20(3):848–889. doi:10.1021/ef0502397.

30. Wang, C., Li, H., Li, M., Bian, J., and Sun, R. Revealing the structure and distribution changes of Eucalyptus lignin during the hydrothermal and alkaline pretreatments. *Sci Rep.* 2017;7(1):593. doi:10.1038/s41598-017-00711-w.

31. Jacquet, N., Maniet, G., and Vanderghem, C. Application of steam explosion as pretreatment on lignocellulosic material: A review. *Ind.* 2015. http://pubs.acs.org/doi/abs/10.1021/ie503151g. Accessed August 29, 2017.

32. Stelte, W. Steam explosion for biomass pre-treatment. *Danish Technol Inst.* 2013. https://www.teknologisk.dk/_/media/52681_RK report steam explosion.pdf. Accessed September 20, 2017.

33. Lenihan, P., Orozco, A., O'Neill, E., Ahmad, M. N. M., Rooney, D. W., and Walker, G. M. Dilute acid hydrolysis of lignocellulosic biomass. *Chem Eng J.* 2010;156(2):395–403. doi:10.1016/j.cej.2009.10.061.

34. Vázquez, B., Roa-Morales, G., and Rangel, R. Thermal hydrolysis of orange peel and its fermentation with alginate beads to produce ethanol. 2017. https://ojs.cnr.ncsu.edu/index.php/BioRes/article/view/BioRes_12_2_2955_Corona_Vazquez_Thermal_Hydrolysis_Orange_Peel. Accessed August 30, 2017.

35. Dence, C. and Lin, S. Introduction. *Methods Lignin Chem.* 1992. https://link.springer.com/chapter/10.1007/978-3-642-74065-7_1. Accessed August 30, 2017.

36. Rudie, A. W. and Hart, P. W. Catalysis: A potential alternative to kraft pulping. A synthesis of the literature. 2014. https://www.fpl.fs.fed.us/documnts/pdf2014/fpl_2014_rudie002.pdf. Accessed April 6, 2018.

37. Zhao, X., Cheng, K., and Liu, D. Organosolv pretreatment of lignocellulosic biomass for enzymatic hydrolysis. *Appl Microbiol Biotechnol.* 2009;82(5):815–827. doi:10.1007/s00253-009-1883-1.

38. Johansson, A., Aaltonen, O., and Ylinen, P. Organosolv pulping—Methods and pulp properties. *Biomass.* 1987;13(1):45–65. doi:10.1016/0144-4565(87)90071-0.

39. Audu, I., Brosse, N., Desharnais, L., and Rakshit, S. Ethanol organosolv pretreatment of typha capensis for bioethanol production and co-products. *BioResources.* 2012. https://ojs.cnr.ncsu.edu/index.php/BioRes/article/view/BioRes_07_4_5917_Audu_Ethanol_Organosolv_Typha. Accessed August 30, 2017.

40. Mussatto, S. *Biomass Fractionation Technologies for a Lignocellulosic Feedstock Based Biorefinery.*; 2016. https://books.google.co.uk/books?hl=en&lr=&id=LrZ7Bg AAQBAJ&oi=fnd&pg=PP1&dq=ISBN:+978-0-12-802323-5&ots=2aSduMus4F&sig =T0u3yYVgxH3PkiQAPFqzXuYRR7g. Accessed August 30, 2017.

41. Mutjé, P., Pèlach, M. A., Vilaseca, F., García, J. C., and Jiménez, L. A comparative study of the effect of refining on organosolv pulp from olive trimmings and kraft pulp from eucalyptus wood. *Bioresour Technol.* 2005;96(10):1125–1129. doi:10.1016/j. biortech.2004.10.001.

42. Weigand, L., Mostame, S., and Brandt-Talbot, A. Effect of pretreatment severity on the cellulose and lignin isolated from Salix using ionoSolv pretreatment. *Faraday.* 2017. https://www.ncbi.nlm.nih.gov/pubmed/28718847. Accessed August 30, 2017.

43. Lui, M. Y., Crowhurst, L., Hallett, J. P., Hunt, P. A., Niedermeyer, H., and Welton, T. Salts dissolved in salts: Ionic liquid mixtures. *Chem Sci.* 2011;2(8):1491. doi:10.1039/ c1sc00227a.

44. Welton, T. Room-temperature ionic liquids. Solvents for synthesis and catalysis. *Chem Rev.* 1999. http://pubs.acs.org/doi/abs/10.1021/cr980032t. Accessed April 16, 2017.

45. Pârvulescu, V. and Hardacre, C. Catalysis in ionic liquids. *Chem Rev.* 2007. http://pubs. acs.org/doi/abs/10.1021/cr050948h. Accessed April 16, 2017.

46. Dinarès, I., de Miguel, C., Ibáñez, A., and Mesquida, N. Imidazolium ionic liquids: A simple anion exchange protocol. *Chemistry (Easton).* 2009. http://pubs.rsc.org/en/ content/articlehtml/2009/gc/b915743n. Accessed August 30, 2017.

47. Dupont, J., de Souza, R., Suarez P. Room-temperature ionic liquids. Solvents for synthesis and catalysis. *Chem Rev.* 2002. https://scholar.google.co.uk/scholar?q=J.+Dupont %2C+R.+F.+de+Souza+and+P.+A.+Z.+Suarez%2C+Chem.+Rev.%2C+2002&btnG=& hl=en&as_sdt=0%2C5. Accessed August 30, 2017.

48. Wasserscheid, P. and Keim, W. Ionic liquids—New "solutions" for transition metal catalysis. *Angew Chemie Int.* 2000. http://onlinelibrary.wiley.com/doi/10.1002/1521-3773(20001103)39:21%3C3772::AID-ANIE3772%3E3.0.CO;2-5/full. Accessed August 30, 2017.

49. Plechkova, N. and Seddon, K. Applications of ionic liquids in the chemical industry. *Chem Soc Rev.* 2008. http://pubs.rsc.org/en/content/articlehtml/2008/cs/b006677j. Accessed August 30, 2017.

50. van Rantwijk, F. and Sheldon, R. Biocatalysis in ionic liquids. *Chem Rev.* 2007. https:// pubs.acs.org/doi/full/10.1021/cr050946x. Accessed August 30, 2017.

51. Elhi, F., Aid, T., and Koel, M. Ionic liquids as solvents for making composite materials from cellulose. *Proc Est Acad.* 2016. http://search.proquest.com/openview/e4fff63ec4339ba040 92a53a6ea2a236/1?pq-origsite=gscholar&cbl=106016. Accessed August 30, 2017.

52. Lopicic, Z., R., Stojanovic, M. D., Markovic, S. B. et al. Effects of different mechanical treatments on structural changes of lignocellulosic waste biomass and subsequent Cu(II) removal kinetics. *Arab J Chem.* April 2016. doi:10.1016/J.ARABJC.2016.04.005.

53. Fort, D. A., Remsing, R. C., Swatloski, R. P., Moyna, P., Moyna, G., and Rogers, R. D. Can ionic liquids dissolve wood? Processing and analysis of lignocellulosic materials with 1-n-butyl-3-methylimidazolium chloride. *Green Chem.* 2007;9(1):63–69. doi:10.1039/B607614A.

54. Zoia, L., King, A. W. T., and Argyropoulos, D. S. Molecular weight distributions and linkages in lignocellulosic materials derivatized from ionic liquid media. *J Agric Food Chem.* 2011;59(3):829–838. doi:10.1021/jf103615e.

55. Kuo, C. and Lee, C. Enhancement of enzymatic saccharification of cellulose by cellulose dissolution pretreatments. *Carbohydr Polym.* 2009. http://www.sciencedirect.com/ science/article/pii/S0144861708005535. Accessed April 16, 2017.

56. Zhang, Y.-H. P., Ding, S.-Y., Mielenz, J. R. et al. Fractionating recalcitrant lignocellulose at modest reaction conditions. *Biotechnol Bioeng.* 2007;97(2):214–223. doi:10.1002/bit.21386.

57. Fink, H., Weigel, P., Purz, H., and Ganster, J. Structure formation of regenerated cellulose materials from NMMO-solutions. *Prog Polym Sci.* 2001. http://www.science direct.com/science/article/pii/S0079670001000259. Accessed August 30, 2017.

58. Liebert, T. Cellulose solvents—Remarkable history, bright future. *Cellul solvents Anal Shap Chem.* 2010. http://pubs.acs.org/doi/abs/10.1021/bk-2010-1033.ch001. Accessed August 30, 2017.

59. Swatloski, R., Spear, S., and Holbrey, J. Dissolution of cellose with ionic liquids. *J Am.* 2002. http://pubs.acs.org/doi/abs/10.1021/ja025790m. Accessed April 16, 2017.

60. Credou, J. and Berthelot, T. Cellulose: From biocompatible to bioactive material. *J Mater Chem B.* 2014;2(30):4767–4788. doi:10.1039/C4TB00431K.

61. Gupta, Krishna, M., and Jiang, J. Cellulose dissolution and regeneration in ionic liquids: A computational perspective. *Chem Eng Sci.* 2015;121:180–189. doi:10.1016/J.CES.2014.07.025.

62. Huang, Q., Huang, J., and Chang, P. Polycaprolactone grafting of cellulose nanocrystals in ionic liquid Cl. *Wuhan Univ J Nat.* 2014. http://link.springer.com/article/10.1007/s11859-014-0987-3. Accessed April 16, 2017.

63. Minnick, D. L., Flores, R. A., DeStefano, M. R., and Scurto, A. M. Cellulose solubility in ionic liquid mixtures: Temperature, cosolvent, and antisolvent effects. *J Phys Chem B.* 2016;120(32):7906–7919. doi:10.1021/acs.jpcb.6b04309.

64. Kahlen, J., Masuch, K., and Leonhard, K. Modelling cellulose solubilities in ionic liquids using COSMO-RS. *Green Chem.* 2010;12(12):2172. doi:10.1039/c0gc00200c.

65. Youngs, T. G. A., Hardacre, C., and Holbrey J. D. Glucose solvation by the ionic liquid 1,3-Dimethylimidazolium chloride: A simulation study. *J Phys Chem B.* 2007;111(49):13765–13774. doi:10.1021/jp076728k.

66. Lu, B., Xu, A., and Wang, J. Cation does matter: How cationic structure affects the dissolution of cellulose in ionic liquids. *Green Chem.* 2014. http://pubs.rsc.org/-/content/articlehtml/2014/gc/c3gc41733f. Accessed August 30, 2017.

67. Zhao, H., Baker, G. A., Song, Z. et al. Designing enzyme-compatible ionic liquids that can dissolve carbohydrates. *Green Chem.* 2008;10(6):696. doi:10.1039/b801489b.

68. Zhang, J., Xu, L., Yu, J. et al. Understanding cellulose dissolution: Effect of the cation and anion structure of ionic liquids on the solubility of cellulose. *Sci China Chem.* 2016;59(11):1421–1429. doi:10.1007/s11426-016-0269-5.

69. Groff, D., George, A., Sun, N., and Sathitsuksanoh, N. Acid enhanced ionic liquid pretreatment of biomass. *Green.* 2013. http://pubs.rsc.org/-/content/articlehtml/2013/gc/c3gc37086k. Accessed April 16, 2017.

70. Klein-Marcuschamer, D., Simmons, B. A., and Blanch, H. W. Techno-economic analysis of a lignocellulosic ethanol biorefinery with ionic liquid pre-treatment. *Biofuels, Bioprod Biorefining.* 2011;5(5):562–569. doi:10.1002/bbb.303.

71. Baral, N. R. and Shah, A. Techno-economic analysis of cellulose dissolving ionic liquid pretreatment of lignocellulosic biomass for fermentable sugars production. *Biofuels, Bioprod Biorefining.* 2016;10(5):664–664. doi:10.1002/bbb.1673.

72. Rogers, R. D. and Seddon, K. R. (Eds.) *Ionic Liquids.* Vol 818. Washington, DC: American Chemical Society; 2002. doi:10.1021/bk-2002-0818.

73. Fukaya, Y., Sugimoto, A., and Ohno, H. Superior solubility of polysaccharides in low viscosity, polar, and halogen-free 1,3-dialkylimidazolium formates. *Biomacromolecules.* 2006;7(12):3295–3297. doi:10.1021/bm060327d.

74. Froschauer, C., Hummel, M., Laus, G. et al. Dialkyl phosphate-related ionic liquids as selective solvents for xylan. *Biomacromolecules.* 2012;13(6):1973–1980. doi:10.1021/bm300582s.

75. Froschauer, C., Hummel, M., Iakovlev, M., Roselli, A., Schottenberger, H., and Sixta, H. Separation of hemicellulose and cellulose from wood pulp by means of ionic liquid/cosolvent systems. *Biomacromolecules*. 2013;14(6):1741–1750. doi:10.1021/bm400106h.

76. Liu, Z., Wang, L., Jenkins, B., Li, Y., Yi, W., and Li, Z. Influence of alkali and alkaline earth metallic species on the phenolic species of pyrolysis oil. *BioResources*. 2017. http://ojs.cnr.ncsu.edu/index.php/BioRes/article/view/BioRes_12_1_1611_Liu_Alkali_Earth_Metallic_Species_Pyrolysis_Oil. Accessed September 20, 2017.

77. da Costa Lopes, A. M. and Bogel-Łukasik, R. Acidic ionic liquids as sustainable approach of cellulose and lignocellulosic biomass conversion without additional catalysts. *ChemSusChem*. 2015;8(6):947–965. doi:10.1002/cssc.201402950.

78. George, A., Brandt, A., Tran, K. et al. Design of low-cost ionic liquids for lignocellulosic biomass pretreatment. *Green Chem*. 2015;17(3):1728–1734. doi:10.1039/C4GC01208A.

79. Yu, H., Hu, J., Fan, J., and Chang, J. One-pot conversion of sugars and lignin in ionic liquid and recycling of ionic liquid. *Ind Eng Chem Res*. 2012;51(8):3452–3457. doi:10.1021/ie2025807.

80. Hart, W. E. S., Harper, J. B., and Aldous, L. The effect of changing the components of an ionic liquid upon the solubility of lignin. *Green Chem*. 2015;17(1):214–218. doi:10.1039/C4GC01888E.

81. Bai, Y., Clark, J., Farmer, T., Ingram, I., and North, M. Wholly biomass derivable sustainable polymers by ring-opening metathesis polymerisation of monomers obtained from furfuryl alcohol and itaconic anhydride. *Polym Chem*. 2017. http://pubs.rsc.org/en/content/articlehtml/2017/py/c7py00486a. Accessed September 22, 2017.

82. Werpy, T., Pacific Northwest National Laboratory, Petersen, G., National Renewable Energy Laboratory. Top value added chemicals from biomass volume I—Results of screening for potential candidates from sugars and synthesis gas energy efficiency and renewable energy. https://www.nrel.gov/docs/fy04osti/35523.pdf. Accessed September 22, 2017.

83. Isikgor, F. H. and Becer, C. R. Lignocellulosic biomass: A sustainable platform for the production of bio-based chemicals and polymers. *Polym Chem*. 2015;6(25):4497–4559. doi:10.1039/C5PY00263J.

84. Kalia, S., Dufresne, A., Cherian, B. M. et al. Cellulose-based bio- and nanocomposites: A review. *Int J Polym Sci*. 2011;2011:1–35. doi:10.1155/2011/837875.

85. Wendler, F., Kosan, B., Krieg, M., and Meister, F. Possibilities for the physical modification of cellulose shapes using ionic liquids. *Macromol Symp*. 2009;280(1):112–122. doi:10.1002/masy.200950613.

86. Hufendiek, A., Trouillet, V., Meier, M. A. R., and Barner-Kowollik, C. Temperature responsive cellulose-*graft*-copolymers via cellulose functionalization in an ionic liquid and RAFT polymerization. *Biomacromolecules*. 2014;15(7):2563–2572. doi:10.1021/bm500416m.

87. Zailuddin, N., Husseinsyah, S., Hahary, F., and Ismail, H. Treatment of oil palm empty fruit bunch regenerated cellulose biocomposite films using methacrylic acid. *BioResources*. 2015. http://152.1.0.246/index.php/BioRes/article/view/BioRes_11_1_873_Zailuddin_Oil_Palm_Empty_Fruit_Bunch. Accessed September 22, 2017.

88. Aaltonen, O. and Jauhiainen, O. The preparation of lignocellulosic aerogels from ionic liquid solutions. *Carbohydr Polym*. 2009;75(1):125–129. doi:10.1016/j.carbpol.2008.07.008.

89. Isik, M., Sardon, H., and Mecerreyes, D. Ionic liquids and cellulose: Dissolution, chemical modification and preparation of new cellulosic materials. *Int J Mol Sci*. 2014;15(7):11922–11940. doi:10.3390/ijms150711922.

90. Peng, B. L., Dhar, N., Liu, H. L., and Tam, K. C. Chemistry and applications of nanocrystalline cellulose and its derivatives: A nanotechnology perspective. *Can J Chem Eng*. 2011;89(5):1191–1206. doi:10.1002/cjce.20554.

91. Man, Z., Muhammad, N., Sarwono, A., Bustam, M. A., Vignesh, Kumar, M., and Rafiq, S. Preparation of cellulose nanocrystals using an ionic liquid. *J Polym Environ.* 2011;19(3):726–731. doi:10.1007/s10924-011-0323-3.

92. Habibi, Y. Key advances in the chemical modification of nanocelluloses. *Chem Soc Rev.* 2014;43(5):1519–1542. doi:10.1039/C3CS60204D.

93. Miao, J., Yu, Y., Jiang, Z., and Zhang, L. One-pot preparation of hydrophobic cellulose nanocrystals in an ionic liquid. *Cellulose.* 2016;23(2):1209–1219. doi:10.1007/s10570-016-0864-7.

94. Wang, Z., Zhang, Y., Jiang, F., Fang, H., and Wang, Z. Synthesis and characterization of designed cellulose-graft-polyisoprene copolymers. *Polym Chem.* 2014;5(10):3379. doi:10.1039/c3py01574b.

95. Stewart, D. Lignin as a base material for materials applications: Chemistry, application and economics. *Ind Crops Prod.* 2008. http://www.sciencedirect.com/science/article/pii/S0926669007001094. Accessed August 29, 2017.

96. Kawamoto, H. Lignin pyrolysis reactions. *J Wood Sci.* 2017;63(2):117–132. doi:10.1007/s10086-016-1606-z.

97. Pu, Y., Jiang, N., and Ragauskas, A. J. Ionic liquid as a green solvent for lignin. *J Wood Chem Technol.* 2007;27(1):23–33. doi:10.1080/02773810701282330.

98. Pure Lignin Environmental Technology. http://purelignin.com/. Accessed September 22, 2017.

99. Ma, Y., Asaadi, S., Johansson, L.-S. et al. High-strength composite fibers from cellulose-lignin blends regenerated from ionic liquid solution. *ChemSusChem.* 2015;8(23):4030–4039. doi:10.1002/cssc.201501094.

100. Mu, L., Shi, Y., Guo, X. et al. Enriching heteroelements in lignin as lubricating additives for bioionic liquids. *ACS Sustain Chem Eng.* 2016;4(7):3877–3887. doi:10.1021/acssuschemeng.6b00669.

101. Livi, S., Bugatti, V., Marechal, M. et al. Ionic liquids–lignin combination: An innovative way to improve mechanical behaviour and water vapour permeability of eco-designed biodegradable polymer blends. *RSC Adv.* 2015;5(3):1989–1998. doi:10.1039/C4RA11919C.

102. Pagán-Torres, Y. J., Wang, T., Gallo, J. M. R., Shanks, B. H., and Dumesic, J. A. Production of 5-Hydroxymethylfurfural from glucose using a combination of Lewis and Brønsted acid catalysts in water in a biphasic reactor with an alkylphenol solvent. *ACS Catal.* 2012;2(6):930–934. doi:10.1021/cs300192z.

103. Zhou, C., Zhao, J., Yagoub, A. E. A., Ma, H., Yu, X., Bao, X., and Liu, S. Conversion of glucose into 5-hydroxymethylfurfural in different solvents and catalysts: Reaction kinetics and mechanism. *Egypt J Pet.* 2017;26(2):477–487. doi:10.1016/J.EJPE.2016.07.005.

104. Dee, S. and Bell, A. A study of the acid-catalyzed hydrolysis of cellulose dissolved in ionic liquids and the factors influencing the dehydration of glucose and the formation of humins. *ChemSusChem.* 2011. http://onlinelibrary.wiley.com/doi/10.1002/cssc.201000426/full. Accessed April 16, 2017.

105. Sievers, C., Musin, I., Marzialetti, T., Valenzuela-Olarte, M. B., Agrawal, P. K., and Jones, C. W. Acid-catalyzed conversion of sugars and furfurals in an ionic-liquid phase. *ChemSusChem.* 2009;2(7):665–671. doi:10.1002/cssc.200900092.

106. Sievers, C., Valenzuela-Olarte, M. B., Marzialetti, T., Musin, I., Agrawal, P. K., and Jones, C. W. Ionic-liquid-phase hydrolysis of pine wood. *Ind Eng Chem Res.* 2009;48(3):1277–1286. doi:10.1021/ie801174x.

107. Hu, S., Zhang, Z., Zhou, Y. et al. Conversion of fructose to 5-hydroxymethylfurfural using ionic liquids prepared from renewable materials. *Green Chem.* 2008;10(12):1280. doi:10.1039/b810392e.

108. Qi, X., Watanabe, M., Aida, T. M., and Smith, R. L. Efficient process for conversion of fructose to 5-hydroxymethylfurfural with ionic liquids. *Green Chem.* 2009;11(9):1327. doi:10.1039/b905975j.

109. Li, X., Xia, Q., Nguyen, V. C. et al. High yield production of HMF from carbohydrates over silica–alumina composite catalysts. *Catal Sci Technol.* 2016;6(20):7586–7596. doi:10.1039/C6CY01628F.

110. Eminov, S., Filippousi, P., Brandt, A., Wilton-Ely, J., and Hallett, J. Direct catalytic conversion of cellulose to 5-Hydroxymethylfurfural using ionic liquids. *Inorganics.* 2016;4(4):32. doi:10.3390/inorganics4040032.

111. Zhao, H., Holladay, J. E., Brown, H., and Zhang, Z. C. Metal chlorides in ionic liquid solvents convert sugars to 5-hydroxymethylfurfural. *Science.* 2007;316(5831):1597–1600. doi:10.1126/science.1141199.

112. Lima, S., Neves, P., Antunes, M. M., Pillinger, M., Ignatyev, N., and Valente, A. A. Conversion of mono/di/polysaccharides into furan compounds using 1-alkyl-3-methylimidazolium ionic liquids. *Appl Catal A Gen.* 2009;363(1–2):93–99. doi:10.1016/j.apcata.2009.04.049.

113. Hu, L., Sun, Y., and Lin, L. Efficient conversion of glucose into 5-Hydroxymethylfurfural by chromium(III) chloride in inexpensive ionic liquid. *Ind Eng Chem Res.* 2012;51(3):1099–1104. doi:10.1021/ie202174f.

114. Guo, X., Zhu, C., and Guo, F. Direct transformation of fructose and glucose to 5-hydroxymethylfurfural in ionic liquids under mild conditions. *BioResources.* 2016. http://ojs.cnr.ncsu.edu/index.php/BioRes/article/view/BioRes_11_1_2457_Guo_Transformation_Fructose_Glucose_Ionic_Liquids. Accessed September 22, 2017.

115. Wang, Y., Song, H., Peng, L., Zhang, Q., and Yao, S. Recent developments in the catalytic conversion of cellulose. *Biotechnol Biotechnol Equip.* 2014;28(6):981–988. doi:10.1080/13102818.2014.980049.

116. Su, Y., Brown, H. M., Huang, X., Zhou, X.-D., Amonette, J. E., and Zhang, C. Single-step conversion of cellulose to 5-hydroxymethylfurfural (HMF), a versatile platform chemical. *Appl Catal A Gen.* 2009;361(1–2):117–122. doi:10.1016/J.APCATA.2009.04.002.

117. Feng, J., Zhu, Q., Ma, D., Liu, X., and Han, X. Direct conversion and NMR observation of cellulose to glucose and 5-hydroxymethylfurfural (HMF) catalyzed by the acidic ionic liquids. *J Mol Catal A Chem.* 2011;334(1–2):8–12. doi:10.1016/J.MOLCATA.2010.10.006.

118. Wen, Y., Jiang, M., Kitchens, C. L., and Chumanov, G. Synthesis of carbon nanofibers via hydrothermal conversion of cellulose nanocrystals. *Cellulose.* August 2017:1–6. doi:10.1007/s10570-017-1464-x.

119. Ryabukhin, D. and Zakusilo, D. Superelectrophilic activation of 5-hydroxymethylfurfural and 2, 5-diformylfuran: Organic synthesis based on biomass-derived products. *Beilstein J.* 2016. https://www.ncbi.nlm.nih.gov/pmc/articles/PMC5082471/. Accessed September 22, 2017.

120. Lange, J.-P., van der Heide, E., van Buijtenen, J., and Price, R. Furfural—A promising platform for lignocellulosic biofuels. *ChemSusChem.* 2012;5(1):150–166. doi:10.1002/cssc.201100648.

121. Hoang, T. M. C., Lefferts, L., and Seshan, K. Valorization of humin-based byproducts from biomass processing-A route to sustainable hydrogen. *ChemSusChem.* 2013;6(9):1651–1658. doi:10.1002/cssc.201300446.

122. Antal, M., Leesomboon, T., Mok, W., and Richards, G. Mechanism of formation of 2-furaldehyde from D-xylose. *Carbohydr Res.* 1991. http://www.sciencedirect.com/science/article/pii/000862159184118X. Accessed September 22, 2017.

123. Peleteiro, S., da Costa Lopes, A. M., Garrote, G., Parajó, J. C., and Bogel-Łukasik, R. Simple and efficient furfural production from xylose in media containing

1-Butyl-3-Methylimidazolium hydrogen sulfate. *Ind Eng Chem Res*. 2015;54(33):8368–8373. doi:10.1021/acs.iecr.5b01771.

124. Shirotori, M., Nishimura, S., and Ebitani, K. One-pot synthesis of furfural derivatives from pentoses using solid acid and base catalysts. *Catal Sci Technol*. 2014;4(4):971–978. doi:10.1039/C3CY00980G.

125. Binder, J. B., Blank, J. J., Cefali, A. V., and Raines, R. T. Synthesis of furfural from xylose and xylan. *ChemSusChem*. 2010;3(11):1268–1272. doi:10.1002/cssc.201000181.

126. Zhang, L., Yu, H., Wang, P., Dong, H., and Peng, X. Conversion of xylan, d-xylose and lignocellulosic biomass into furfural using AlCl3 as catalyst in ionic liquid. *Bioresour Technol*. 2013;130:110–116. doi:10.1016/j.biortech.2012.12.018.

127. Serrano-Ruiz, J. C., Campelo, J. M., Francavilla, M. et al. Efficient microwave-assisted production of furfural from C5 sugars in aqueous media catalysed by Brönsted acidic ionic liquids. *Catal Sci Technol*. 2012;2(9):1828. doi:10.1039/c2cy20217d.

128. Tao, F., Song, H., and Chou, L. Efficient process for the conversion of xylose to furfural with acidic ionic liquid. *Can J Chem*. 2011;89(1):83–87. doi:10.1139/V10-153.

129. Peleteiro, S., Santos, V., Garrote, G., and Parajó, J. C. Furfural production from Eucalyptus wood using an acidic ionic liquid. *Carbohydr Polym*. 2016;146:20–25. doi:10.1016/j.carbpol.2016.03.049.

130. Ferrández-García, A. and Ferrández-Villena, M. Potential use of phoenix canariensis biomass in binderless particleboards at low temperature and pressure. 2017. https://ojs.cnr.ncsu.edu/index.php/BioRes/article/view/BioRes_12_3_6698_Ferrandez_Garcia_Phoenix_Biomass_Binderless_Particleboards. Accessed September 22, 2017.

131. Szabolcs, Á., Molnár, M., Dibó, G., and Mika, L. T. Microwave-assisted conversion of carbohydrates to levulinic acid: An essential step in biomass conversion. *Green Chem*. 2013;15(2):439–445. doi:10.1039/C2GC36682G.

132. Ren, H., Girisuta, B., Zhou, Y., and Liu, L. Selective and recyclable depolymerization of cellulose to levulinic acid catalyzed by acidic ionic liquid. *Carbohydr Polym*. 2015;117:569–576. doi:10.1016/j.carbpol.2014.09.091.

133. Ren, H., Zhou, Y., and Liu, L. Selective conversion of cellulose to levulinic acid via microwave-assisted synthesis in ionic liquids. *Bioresour Technol*. 2013;129:616–619. doi:10.1016/j.biortech.2012.12.132.

134. Amarasekara, A. S. and Wiredu, B. Acidic ionic liquid catalyzed one-pot conversion of cellulose to ethyl levulinate and levulinic acid in ethanol-water solvent system. *BioEnergy Res*. 2014;7(4):1237–1243. doi:10.1007/s12155-014-9459-z.

135. Shen, Y., Sun, J.-K., Yi, Y.-X., Wang, B., Xu, F., and Sun, R.-C. One-pot synthesis of levulinic acid from cellulose in ionic liquids. *Bioresour Technol*. 2015;192:812–816. doi:10.1016/j.biortech.2015.05.080.

136. Qu, Y., Li, L., Wei, Q., Huang, C., Oleskowicz-Popiel, P., and Xu, J. One-pot conversion of disaccharide into 5-hydroxymethylfurfural catalyzed by imidazole ionic liquid. *Sci Rep*. 2016;6(1):26067. doi:10.1038/srep26067.

137. Li, K., Bai, L., Amaniampong, P. N., Jia, X., Lee, J.-M., and Yang, Y. One-pot transformation of cellobiose to formic acid and levulinic acid over ionic-liquid-based polyoxometalate hybrids. *ChemSusChem*. 2014;7(9):2670–2677. doi:10.1002/cssc.201402157.

138. Hu, F. and Ragauskas, A. Suppression of pseudo-lignin formation under dilute acid pretreatment conditions. *RSC Adv*. 2014;4(9):4317–4323. doi:10.1039/C3RA42841A.

139. Taverna, M. E., Ollearo, R., Moran, J. I., Nicolau, V. V., Estenoz, D. A., and Frontini, P. M. *Bioresources*. Department of Wood and Paper Science, College of Natural Resources, North Carolina State University; 2016. http://ri.conicet.gov.ar/handle/11336/10043. Accessed October 4, 2017.

140. Palazzolo, M. A. and Kurina-Sanz, M. Microbial utilization of lignin: Available biotechnologies for its degradation and valorization. *World J Microbiol Biotechnol*. 2016;32(10):173. doi:10.1007/s11274-016-2128-y.

141. Zakzeski, J., Jongerius, A. L., and Weckhuysen, B. M. Transition metal catalyzed oxidation of Alcell lignin, soda lignin, and lignin model compounds in ionic liquids. *Green Chem.* 2010;12(7):1225. doi:10.1039/c001389g.

142. Zhu, Y., Chuanzhao, L., Sudarmadji, M. et al. An efficient and recyclable catalytic system comprising nanopalladium(0) and a pyridinium salt of iron bis(dicarbollide) for oxidation of substituted benzyl alcohol and lignin. *ChemistryOpen.* 2012;1(2):67–70. doi:10.1002/open.201100014.

143. Zhang, B., Li, C., Dai, T., Huber, G. W., Wang, A., and Zhang, T. Microwave-assisted fast conversion of lignin model compounds and organosolv lignin over methyl-trioxorhenium in ionic liquids. *RSC Adv.* 2015;5(103):84967–84973. doi:10.1039/C5RA18738A.

144. Prado, R., Brandt, A., Erdocia, X., Hallet, J., and Welton, T. Lignin oxidation and depolymerisation in ionic liquids. *Green.* 2016. http://pubs.rsc.org/-/content/articlehtml/2016/gc/c5gc01950h. Accessed April 16, 2017.

145. Yang, Y., Fan, H., Song, J. et al. Free radical reaction promoted by ionic liquid: a route for metal-free oxidation depolymerization of lignin model compound and lignin. *Chem Commun.* 2015;51(19):4028–4031. doi:10.1039/C4CC10394G.

146. Nanayakkara, S., Patti, A. F., and Saito, K. Lignin depolymerization with phenol via redistribution mechanism in ionic liquids. *ACS Sustain Chem Eng.* 2014;2(9):2159–2164. doi:10.1021/sc5003424.

147. Dai, J., Nanayakkara, S., Lamb, T. C. et al. Effect of the N-based ligands in copper complexes for depolymerisation of lignin. *New J Chem.* 2016;40(4):3511–3519. doi:10.1039/C5NJ03152D.

148. Thuy Pham, T. P., Cho, C.-W., and Yun, Y.-S. Environmental fate and toxicity of ionic liquids: A review. *Water Res.* 2010;44(2):352–372. doi:10.1016/j.watres.2009.09.030.

149. Stolte, S., Matzke, M., and Arning, J. (Eco)Toxicology and biodegradation of ionic liquids. In: *Ionic Liquids Completely UnCOILed.* Hoboken, NJ: John Wiley & Sons; 2015:189–208. doi:10.1002/9781118840061.ch9.

150. Stolte, S., Arning, J., Bottin-Weber, U. et al. Effects of different head groups and functionalised side chains on the cytotoxicity of ionic liquids. *Green Chem.* 2007;9(7):760–767. doi:10.1039/B615326G.

151. Chatel, G. and MacFarlane, D. R. Ionic liquids and ultrasound in combination: Synergies and challenges. *Chem Soc Rev.* 2014;43(23):8132–8149. doi:10.1039/C4CS00193A.

152. Ninomiya, K., Ohta, A., Omote, S., Ogino, C., Takahashi, K., and Shimizu, N. Combined use of completely bio-derived cholinium ionic liquids and ultrasound irradiation for the pretreatment of lignocellulosic material to enhance enzymatic saccharification. *Chem Eng J.* 2013;215–216:811–818. doi:10.1016/J.CEJ.2012.11.020.

153. Peric, B., Sierra, J., Martí, E. et al. (Eco)toxicity and biodegradability of selected protic and aprotic ionic liquids. *J Hazard Mater.* 2013;261:99–105. doi:10.1016/j.jhazmat.2013.06.070.

154. Grzonkowska, M., Sosnowska, A., Barycki, M., Rybinska, A., and Puzyn, T. How the structure of ionic liquid affects its toxicity to Vibrio fischeri? *Chemosphere.* 2016;159:199–207. doi:10.1016/j.chemosphere.2016.06.004.

155. Perales, E., García, C. B., Lomba, L., Aldea, L., García, J. I., and Giner, B. Comparative ecotoxicology study of two neoteric solvents: Imidazolium ionic liquid versus glycerol derivative. *Ecotoxicol Environ Saf.* 2016;132:429–434. doi:10.1016/j.ecoenv.2016.05.021.

156. Shao, Y., Du, Z., Zhang, C., Zhu, L., Wang, J., and Wang, J. Acute toxicity of imidazole nitrate ionic liquids with varying chain lengths to earthworms (Eisenia foetida). *Bull Environ Contam Toxicol.* 2017;99(2):213–217. doi:10.1007/s00128-017-2082-x.

157. Eva, U. Vibrio fischeri bioluminescence inhibition test|Enfo. http://enfo.agt.bme.hu/drupal/en/gallery/12739. Accessed October 4, 2017.

158. Vaughan, M. and van Egmond, R. The use of the zebrafish (Danio rerio) embryo for the acute toxicity testing of surfactants, as a possible alternative to the acute fish test. *Altern Lab Anim.* 2010;38(3):231–238. http://www.ncbi.nlm.nih.gov/pubmed/20602539. Accessed October 4, 2017.

159. Haiß, A., Jordan, A., Westphal, J., Logunova, E., Gathergood, N., and Kümmerer, K. On the way to greener ionic liquids: Identification of a fully mineralizable phenylalanine-based ionic liquid. *Green Chem.* 2016;18(16):4361–4373. doi:10.1039/C6GC00417B.

160. van Ginkel, C. G. and Geerts, R. Biodegradation of *N,N*-bis(carboxymethyl)-L-glutamate and its utilization as sole source of carbon, nitrogen, and energy by a *Rhizobium radiobacter* strain in seawater. *Toxicol Environ Chem.* 2016;98(1):26–35. doi:10.1080/02772248.2015.1113287.

161. Deng, Y., Beadham, I., Ghavre, M. et al. When can ionic liquids be considered readily biodegradable? Biodegradation pathways of pyridinium, pyrrolidinium and ammonium-based ionic liquids. *Green Chem.* 2015;17(3):1479–1491. doi:10.1039/C4GC01904K.

162. Viell, J. *A Pretreatment Process for Wood Based on Ionic Liquids.* Düsseldorf, Germany: VDI-Verl; 2014.

163. Clough, M. T., Geyer, K., Hunt, P. A., Mertes, J., and Welton, T. Thermal decomposition of carboxylate ionic liquids: Trends and mechanisms. *Phys Chem Chem Phys.* 2013;15(47):20480. doi:10.1039/c3cp53648c.

164. Doherty, T. V., Mora-Pale, M., Foley, S. E., Linhardt, R. J., and Dordick, J. S. Ionic liquid solvent properties as predictors of lignocellulose pretreatment efficacy. *Green Chem.* 2010;12(11):1967. doi:10.1039/c0gc00206b.

165. Hall, C. A., Le, K. A., Rudaz, C. et al. Macroscopic and microscopic study of 1-Ethyl-3-methyl-imidazolium acetate–water mixtures. *J Phys Chem B.* 2012;116(42):12810–12818. doi:10.1021/jp306829c.

166. Brandt, A. Ionic liquid pretreatment of lignocellulosic biomass. 2012. https://spiral.imperial.ac.uk/handle/10044/1/9166. Accessed April 6, 2018.

167. Brandt-Talbot, A., Gschwend, F. J. V., Fennell, P. S. et al. An economically viable ionic liquid for the fractionation of lignocellulosic biomass. *Green Chem.* 2017. doi:10.1039/C7GC00705A.

168. Chen, L., Sharifzadeh, M., Mac Dowell, N., Welton, T., Shah, N., and Hallett, J. P. Inexpensive ionic liquids: [HSO$_4$]$^-$-based solvent production at bulk scale. *Green Chem.* 2014;16(6):3098–3106. doi:10.1039/C4GC00016A.

11 Membrane Technology for Catalytic Processes in Ionic Liquids

Julio Romero, René Cabezas,
Claudio Araya-López, and Gastón Merlet

CONTENTS

11.1 INTRODUCTION: CATALYTIC PROCESSES COUPLED TO MEMBRANE TECHNOLOGY

Membrane processes involve a wide range of configurations, which can be considered to enhance catalytic conversions through the design of operations that combine reaction steps with simultaneous separation or contactor steps, as well as with the use of membranes as catalyst supports. Membranes can be integrated in catalyzed or non-catalyzed reactive processes. If the membranes are understood as separation elements, these can be used in the in-situ separation of product or by-products from a reaction phase. Thus, the conversion in a conventional reactor could be significantly increased by the modification of the thermodynamic equilibrium, increasing the concentration of the reagents.

Figure 11.1 shows only some of the operational configurations of catalytic membrane reactors. In this figure, membranes are represented as units separated from the reactors. However, specific applications, such as membrane bioreactors, commonly used in wastewater treatment, consider the use of immersed membrane modules.

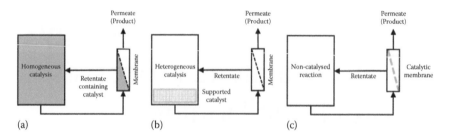

FIGURE 11.1 Some operational configurations of catalytic membrane reactors: (a) homogeneous catalyzed reactor coupled to a membrane separation unit; (b) supported heterogeneous catalyst in a membrane reactor; and (c) membrane reactor with a catalytic membrane.

Membranes can be used to separate products from reaction systems with homogeneous catalysis (Figure 11.1a) retaining the non-reacted reagents, as well as the catalyst. Nevertheless, heterogeneous catalytic systems can also be integrated to membrane operations to increase the conversion of reactions, either with the catalyst supported into the reactor (Figure 11.1b) or in the membrane itself (Figure 11.1c) as a catalytic membrane.

In this framework, the ionic liquids (ILs) show unique properties, such as a negligible vapor pressure, high thermal stability, and a wide electrochemical window. These molten electrolytes can be used as catalyst, cocatalyst, and solvents in different types of chemical conversions (Izák et al., 2018). However, the use of ILs in chemical processes at an industrial scale is associated with some disadvantages, because these compounds show high viscosity, high costs related to their synthesis, and high energy requirements for their recycling (Wang et al., 2016). All these drawbacks may be overcome by using ILs stabilized in a membrane structure. Thus, the use of ionic liquid membranes (ILMs) involves the stabilization of a limited amount of ILs in a porous support membrane, maintaining the original thermodynamic and transport properties of the ILs, or considers its quasi-solidification to endow material with good mechanical strength (Wang et al., 2016). The ILMs reported in literature have been prepared using different types of cations (imidazolium, pyridinium, pyrolidinium, piperidinum, ammonium, phosphonium, sulphonium, morpholinium, etc.) and anions [hexafluorophosphate, dicyanamide, chloride, bromide, trifluoromethylsulfonate, tetrafluoroborate, bis(trifluoromethylsulfonyl)imide, trifluoromethylacetate, etc.]. The following sections summarize the role of ILMs in different types of processes, as well as their specific applications in catalytic conversions.

11.2   MEMBRANES BASED ON IONIC LIQUIDS

There is a significant body of literature on the use of ionic liquid membranes for different Applications (Lozano et al., 2011). The huge variety of structures and cation-anion combinations of the ionic liquids allow a wide range of functionalization of these materials, which can be used to prepare membranes with specific features. Nowadays, it is possible to find mainly three different fields where the ILMs are used: separations, design of different electrochemical devices, and catalytic reactions. Wang et al. (2016) summarize the main applications of the ILMs in these fields. Figure 11.2 shows these specific applications.

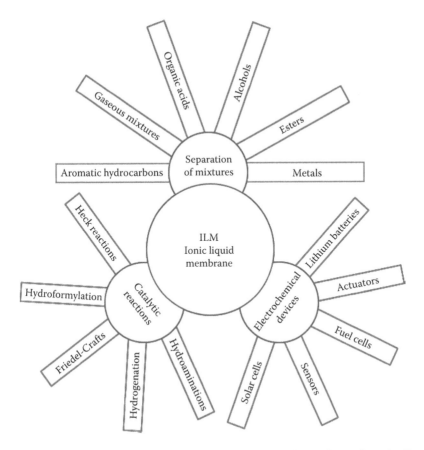

FIGURE 11.2 Properties and applications of ILMs. (Adapted from Wang, J. et al., *Green Energ. Environ.*, 1, 43–61, 2016. With permission.)

These applications involve different operations, such as processes of pervaporation (Rdzanek et al., 2018) and perstraction (Merlet et al., 2017), as well as analytical applications (Clark et al., 2016), separation methods of mixed gases (Bhattacharya and Mandal, 2017), electrochemical applications (Martins and Torresi, 2018), and separation of ions. In several applications described in the literature, the ILs are stabilized in a membrane support. However, the mechanical stability of the ILMs can be affected by the direct contact with circulating phases, which could drag the ILs retained in the support. Thus, the preparation of an ILM involves different stabilization strategies to retain an ionic liquid in a supported or self-supported structure. In this way, there are several preparation methodologies described in the literature to obtain different membrane structures (Voss et al., 2009; Plaza et al., 2013). Figure 11.3 shows an outline of the main configurations of IL-based membranes already reported.

In this framework, the following sections describe the use of ILs as catalysts and solvents combined with membrane processes, considering the preparation of catalytic membranes where the ILs can be part of their structure. Biocatalytic processes are separately described because of the specific interactions and synergy between

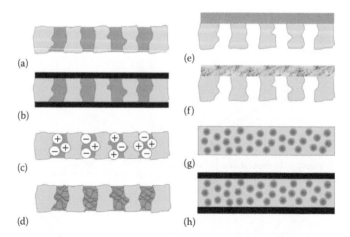

FIGURE 11.3 Different approaches for utilizing IL as membrane materials: (a) supported ionic liquid membrane (SILM); (b) silicon-coated ILM; (c) ion exchange membrane; (d) supported-gelled ionic liquid membrane; (e) covalent binding of IL; (f) inclusion of IL in polymer matrix; (g) quasi-solidified ionic liquid membrane (QSILM); and (h) silicon-coated QSILM. (Adapted from Heitmann, S. et al., *Sep. Purif. Technol.*, 97, 108–114, 2012. With permission; Wang, J. et al., *Green Energ. Environ.*, 1, 43–61, 2016. With permission.)

the properties of the ILs and the use of enzymes. These characteristics could represent several advantages to propose novel and greener operations. A last section describes some specific applications combining membrane processes, ionic liquids, and supercritical carbon dioxide in order to show the potential of biphasic IL/dense CO_2 systems in reaction and separation processes.

11.3 MEMBRANES AND IONIC LIQUIDS AS CATALYSTS AND SOLVENTS

The large amount of cation-anion combinations and the negligible vapor pressure of ILs offer an ensemble of advantageous characteristics for use as catalysts, which can be classified in three groups: homogeneous, heterogeneous, and biological. Thus, ionic liquids have opened a wide range of possibilities that conventional volatile organic solvents cannot reach because those features are capable of conferring selectivity in a specific product and inhibiting by-products, as well as stability and activity to catalysts in chemical reactions when they are playing a role as solvent or support.

Izák et al. (2018) propose that ILs can be classified in four groups: acidic ILs (aILs), polymeric ILs (pILs), supported ILs (sILs), and chiral ILs (cILs), depending on their functional groups in the cation and the anion. Table 11.1 shows some applications of these ILs. sILs are not considered in this table because they depend on the presence of a support.

The aILs are commonly used to catalyze reactions which require an acid environment, such as the esterification of aromatic acid reported by Zhou and collaborators, who considered the use of $[MimC_3SO_3H][HSO_4]$ or $[MimC_4SO_3H][HSO_4]$ (Zhou et al., 2013). Meanwhile, the pILs are constituted by repeating monomers, which are

TABLE 11.1

Examples of Acidic, Polymeric and Chiral Ionic Liquids

Ionic Liquid Group	Name	Abbreviation
Acidic ionic liquids	1-methylimidazolium tetrafluoroborate	[hmim][BF$_4$]
	1-(3-propylsulfonic)-3-methylimidazolium hydrogensulfate	[MimC$_3$SO$_3$H][HSO$_4$]
	1-butyl-3-methylimidazolium dihydrogen phosphate	[bmim][HSO$_4$]
Polymeric ionic liquids	Poly[1-(p-Vinylbenzyl)-3-butyl-imidazolium hexafluorophosphate]	P[VBBI][PF$_6$]
	Poly[1-(p-Vinylbenzyl)-3-butylimidazolium o-benzoicsulphimide]	P[VBBI][Sac]
	Poly(4-vinyl-benzyltriethylammonium hexafluorophosphate)	P[VBTEA][PF$_6$]
Chiral ionic liquids	1-butyl-3-methylimidazolium (T-4)- bis[(2S)-2-(hydroxy-κO]propanoato- κO]borate	[bmim][BLLB]
	1-butyl-3-methylimidazolium (T-4)- bis[(2R)-2-(hydroxy-κO]propanoato- κO]borate	[bmim][BRLB]
	1-ethyl-3-methylimidazolium (T-4)-bis[(αS)-α-(hydroxy-κO)benzeneacetato- κO)borate	[emim][BSMB]

concatenated to a polymeric backbone in order to form a macromolecule with some physical properties different to the monomer ones, expanding the scope of the ionic liquid monomer. Detailed reviews on this topic are presented by Yuan et al. (2013) and Qian et al. (2017).

On the other hand, sILs are related to a different approach where the ionic liquid is contained in a solid support in order to obtain a thin film that is spread out on the support surface. This becomes an advantage when the solid support has a large mass transfer surface which was conferred by the presence of pores, being one of the most important characteristics that catalytic membranes can show. The last group, the cILs, show chiral selectivity, being considered efficient enantioselective solvents used in a wide range of applications, which involve asymmetric synthesis, stereoselective polymerization, gas chromatography, shift reagents in NMR, spectroscopy, and liquid crystals (Baudequin et al., 2005; Ding and Armstrong, 2005; Yu et al., 2008; Podolean et al., 2013).

From the chemical properties described earlier, ionic liquids have been used in a broad range of catalytic reactions, showing an improvement in their yields, selectivity, inhibition of side reactions, and stability. Therefore, ILs have been used as solvents or catalysts in many chemical reactions. Table 11.2 summarizes some examples of these reactions and the used ILs.

Despite the advantages in the use of ILs as solvents and catalysts, there is still a point to overcome related to the separation and purification of the products from the ionic liquid phase. Some conventional separation processes, such as solvent extraction or thermal separations, show some drawbacks mainly represented by the large amount of solvent required and the energy consumption, respectively.

TABLE 11.2
Some Examples of Reactions Using ILs as Catalysts or Solvents

Reaction Name	Reaction	ILs	References
Epoxide reaction	$CO_2 +$ [epoxide structure] $\xrightarrow{\text{Cat.}}$ [cyclic carbonate structure]	[(CH$_2$COOH)DMDA]Br	(Chang et al., 2015)
Enantioselective Michael addition	[maleimide] + [aldehyde] $\xrightarrow[\text{room temperature, 20 h}]{\text{catalyst 20 mol\%}, \ [bmim]BF_4/H_2O \, (2:1 \text{ v/v})}$ [(S) product, R^1 R^2 $N-R^3$]	[bmim][BF$_4$]	(Kochetkov et al., 2017)
Fisher Esterification	[benzoic acid + acetic anhydride] $\rightarrow$ [product]	Cation: [MIMPS]$^+$, [PSPy]$^+$, [bmim]$^+$, [hmim]$^+$ Anion: [HSO$_4$]$^-$, [H$_2$SO4]$^-$	(Shi et al., 2010)
Henry Reaction	**1** [PhCHO + CH$_3$NO$_2$] $\xrightarrow[40^\circ C]{\text{Metalfluorides,} \ [bmim][BF_4],}$ **2** [product with OH and NO$_2$]	[bmim][BF$_4$]	(Shinde et al., 2014)

(Continued)

TABLE 11.2 (Continued)

Some Examples of Reactions Using ILs as Catalysts or Solvents

Reaction Name	Reaction	ILs	References
Aldol condensation	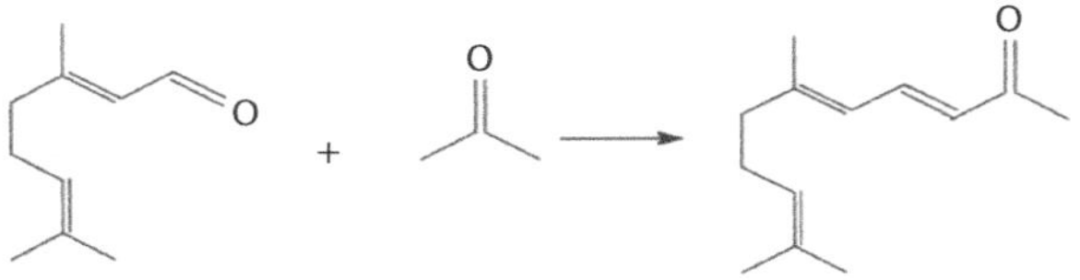	PILs: 2-HEAF, 2-HEAA, 2-HEAPr, 2-HEAB, 2-HEAiB, 2-HEAPe	(Cota et al., 2014)
Diels-Alder	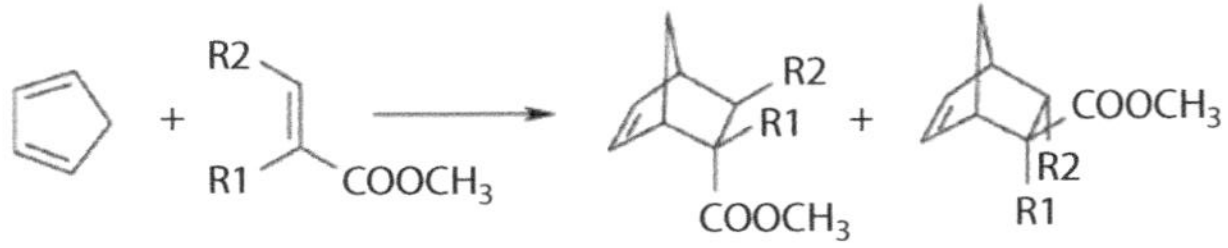	[bmim][BF$_4$], [bmim][PF$_6$], [bmim][Lactate], [emim][BF$_4$], [emim][TFA], [emim][Tf$_2$N], [omim][BF$_4$]	(Sarma and Kumar, 2008)
Oxidation	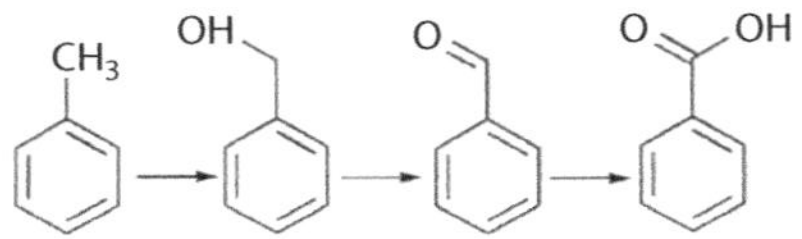	[C$_6$mim][Cl], [C$_4$mim][BF$_4$] [C$_4$mim][Br], [C$_4$mim][BF$_4$] [C$_4$dmim][BF$_4$]	(Seddon and Stark, 2002)
Hydrogenation	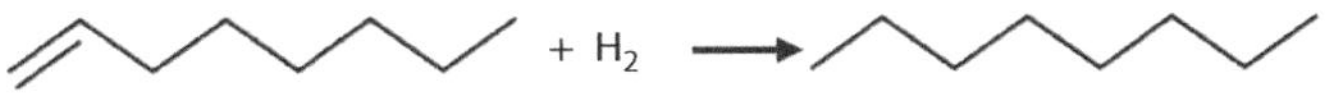	[hmim][Tf$_2$N]	(Ahosseini et al., 2009)

(Continued)

TABLE 11.2 (Continued)
Some Examples of Reactions Using ILs as Catalysts or Solvents

Reaction Name	Reaction	ILs	References
Organocatalysis	Basic chiral ionic liquid	Several chiral ILs	(Vasiloiu et al., 2013)
Suzuki-Miyaura	pd(0), aq Na_2CO_3 [bmim][BF_4]/DMF; TFA cleavage	[bmim][BF_4]	(Revell and Ganesan, 2002)
Friedel-Crafts	[bmim]Cl-Cl-AlCl$_3$	[bmim][Cl]	(Nara et al., 2001)
Metathesis	carbene catalyst ionic liquid	[bmim][BF_4], [bmim][PF_6], [nmim] [N($SO_2CF_3)_2$]	(Williams et al., 2006)

To overcome these problems, membrane processes could be considered a competitive alternative to obtain the reaction products from the IL phase, since these show a higher energy efficiency than conventional separation/purification processes. Furthermore, membrane processes involve a lower initial capital investment, meanwhile, their selectivity is higher depending on the chosen type of membrane. An example of this concept, ionic liquid as a solvent and a membrane as a separator, was proposed by Uragami et al. (2012), who propose the improvement in the conversion of butanol from acetic acid by Fischer esterification reaction, described in Equation 11.1, using a hybrid membrane of poly(vinyl alcohol) tetraethosilane and the ionic liquid 1-allyl-3-butylimidazolium bis(trifluoromethanesulfonyl)imide ([abim][Tf$_2$N]). The Brønsted-acidic IL plays the role of catalyst and separation agent in order to remove water, which represents a by-product of this reaction.

$$C_4H_9OH + CH_3COOH \leftrightarrow CH_3COOC_4H_9 + H_2O \tag{11.1}$$

In this study, Figure 11.4 shows an outline of the reactive system with the simultaneous separation of water as by-product through the membrane. Thus, water molecules were continually produced during the reaction and transferred through the hydrophobic IL phase to be preferentially extracted through the membrane, where the permeate side was under vacuum. Thus, the conversion was improved to 76.5%.

On the other hand, the electrochemical conversion/separation processes have also been studied considering the integration of a membrane based on ILs. Thus, Sakamoto et al. (2013) proposed the design of an ILM of [emim][Tf$_2$N], which is held between two electrodes in order to achieve the simultaneous removal and reduction of NOx from a gaseous stream of NOx mixture with excess oxygen. The removal of NOx, as well as the rejection of oxygen, seems promising due to the lower operating temperature and the high current efficiency of the process when compared with conventional separation methods.

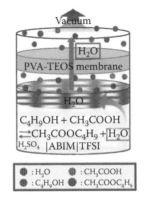

FIGURE 11.4 Outline of the Fischer esterification in a [abim][TFSI] phase coupled to a membrane system for the continuous separation of water. (From Knapp, R. et al., *J. Catal.*, 276, 280–291, 2010.)

11.4   CATALYTIC IONIC LIQUID MEMBRANES

One of the most efficient processes of configuration involves the simultaneous use of a membrane as catalyst and separator in order to increase the reaction yield. The hydroformylation reaction using a catalytic membrane, described in Figure 11.5, is a good example of this configuration process.

Catalytic membranes based on ILs can be prepared from multi-layer, asymmetric, or coating layer configurations, polymerized ILM or supported ionic liquid membrane (SILM). In this way, SILMs can be prepared by three different methods: impregnation of the IL using vacuum, direct immersion, or pressure procedures described in the literature and explained by Fortunato et al. (2005), Scovazzo et al. (2004), and Hernández-Fernández et al. (2009), respectively. In SILM, the main conceptual idea is always to take advantage of the large surface contact area available for mass transfer, which can reach values ranging from 1500 to 7000 $m^2 m^{-3}$ (Crespo and Noble, 2014). From this point of view, aILs, pILs, sILs, and cILs can all be considered in the preparation of catalytic membranes considering their tunable properties to propose task-specific ILMs. In this framework, the solubility of the ionic liquids in surrounding phases depends on their hydrophobic or hydrophilic character.

Ionic liquid catalytic membranes are ILMs where the IL plays the role of catalyst or the phase that contains it. Both cases involve the immobilization of the catalyst in order to avoid the recovery of the catalyst, which can be necessary in a homogeneous process. One of the first studies about a catalytic membrane based on ILs was reported by Carlin and Fuller (1997), who described the hydrogenation reaction from propene to propane (Equation 11.2) using 1-n-butyl-3-methylimidazolium hexafluorophosphate ([bmim][PF$_6$]) and poly(vinylidene fluoride)-hexafluoropropylene, which were polymerized to contain palladium on carbon (Pd/C) catalyst. Thus, the catalyst was homogeneously spread out in the whole membrane structure, reaching reaction conversions of 70%.

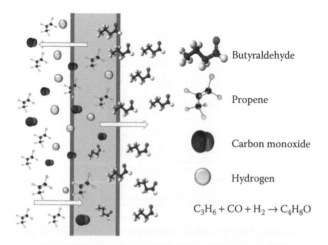

Butyraldehyde

Propene

Carbon monoxide

Hydrogen

$$C_3H_6 + CO + H_2 \rightarrow C_4H_8O$$

FIGURE 11.5   Hydroformylation reaction through a catalytic membrane.

$$C_3H_6 + H_2 \rightarrow C_4H_8 \tag{11.2}$$

The reduction of sugars from the polysaccharide to the monosaccharide, described in Equation 11.3, using a catalytic ionic liquid membrane is another topic recently studied by Lu and et al. (2018).

$$\tag{11.3}$$

These authors developed a polysulfone membrane (PSF) modified by an imidazolium-based IL. At first, a PSF-Im-SO$_3$H powder was prepared in a procedure based on four steps: (1) the preparation of PSF-Cl from PFS and zinc chloride (ZnCl$_2$); (2) the imidazolium IL was poured to obtain PSF-Im; (3) the mixture of PSF-Im with 1.4-butane sultone (C$_4$H$_8$O$_3$S) to obtain PSF-Im-SO$_3$H membrane; and (4) the membrane drying. This membrane shows high catalytic activity of the hydrolysis reaction.

Another important reaction was a water-gas shift reaction, described by Equation 11.4, which has also been studied using ionic liquid catalytic membranes (Nancarrow et al., 2014) by the preparation of a composite [bmim][TfO]/polyimide membrane containing RuCl$_3$. Thus, the [bmim][TfO]/polyimide membrane separates the product, and the Ru carbon complex plays the role of catalyst, which was formed by the contacting of RuCl$_3$ contained in the membranes with CO from the feed H$_2$O/CO stream.

$$CO + H_2O \leftrightarrow CO_2 + H_2 \tag{11.4}$$

Authors suggested that a polyimide membrane prepared with 2 wt% of RuCl$_3 \times H_2O$, 20 wt% of [bmim][TfO] offers the best commitment between reactivity and endurance when operated at 2 bar and 140°C.

11.4.1 Supported Ionic Liquid Membranes as Catalysts

There is a large body of literature on the use of supported ILs, membranes, and chemical reactions. Izák et al. (2018) summarized different approaches combining these aspects, describing the research carried out by Mehnert et al. (2002) on hydrogenation reactions, which are described in Equations 11.5–11.7, using a [Rh(NBD)(PPh$_3$)$_2$] [PF$_6$] complex in [bmim][PF$_6$]. The IL catalyst showed long-term stability, obtaining conversions ranging from 97% to 100% for hexane and a reaction yield increased from 68% to 99%. All these results were obtained using a low amount of IL phase.

$$\tag{11.5}$$

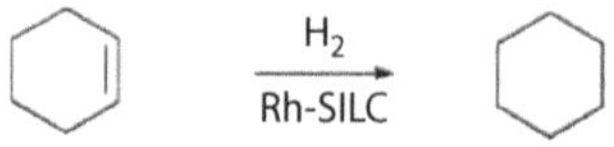

$$\text{(11.6)}$$

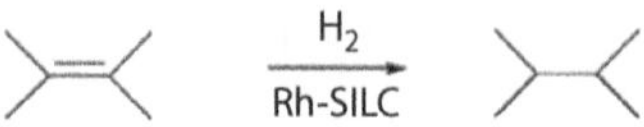

$$\text{(11.7)}$$

On the other hand, Podolean et al. (2013) proposed a chiral-supported ionic liquid phase using Ir, Ru, or Rh complexes and the ILs [emim][Tf$_2$N], [bmim][BF$_4$] and [bmim][PF$_6$] as support phase. Three different compounds, trimethylindolenina, 2-methylquinoline (quinaldine), and dimethylitaconate were hydrogenated by reactions described in Equations 11.8 through 11.10, respectively. The conversion and enantioselectivity mainly depended on the complex, the immobilization method, the selected ionic liquid, and experimental conditions.

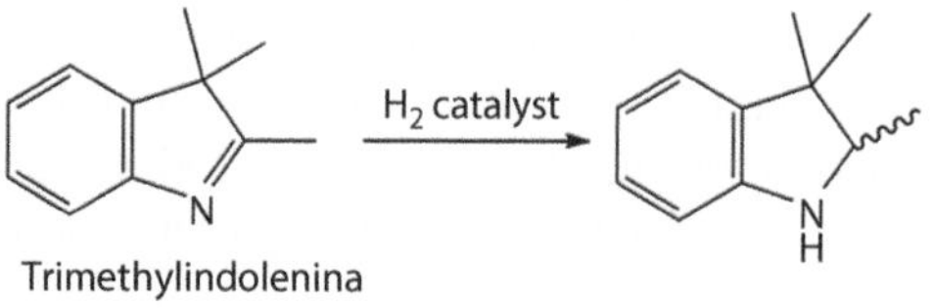

Trimethylindolenina

$$\text{(11.8)}$$

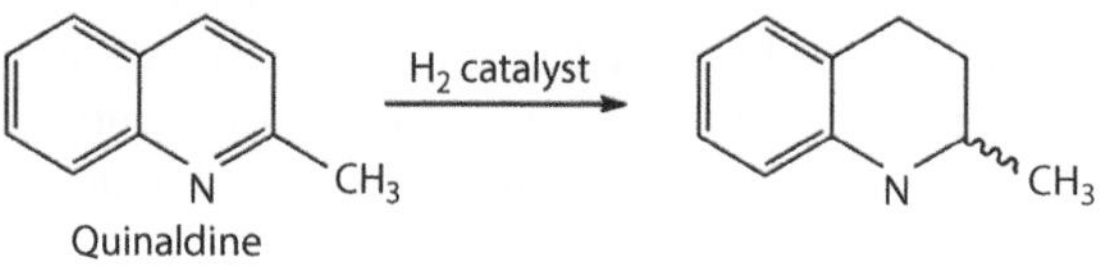

Quinaldine

$$\text{(11.9)}$$

Dimethylitaconate

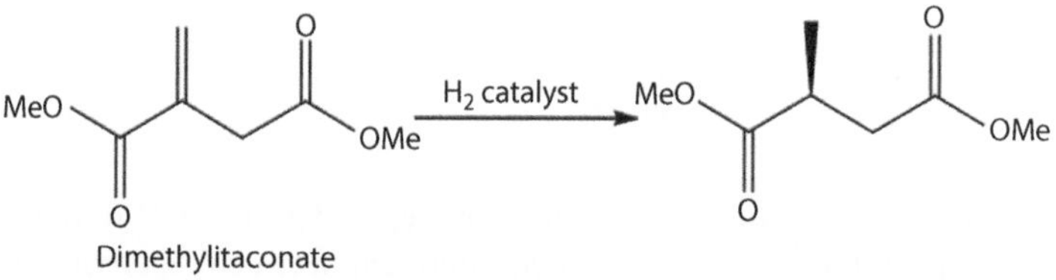

$$\text{(11.10)}$$

A recent contribution of hydrogenation reaction using sILs was reported by Brünig et al. (2018), who studied the catalytic hydrogenation and chemoselectivity of aromatic and aliphatic aldehydes to obtain alcohols. Fe(II) 2,6-bis(phosphonomethyl)pyridine pincer complex (Fe(II) PNP) dissolved in 1-butyl-2,3-dimethyl-imidazolium ([bm$_2$im] [Tf$_2$N]) was used as catalyst, stabilized in powdered silica, which was coated with a thin film of this catalyst. Furthermore, the same authors functionalized silica gel with 1,2-dimethyl-3-(3-trimethoxysilylpropyl)imidazolium-bis(trifluoro-methylsulfonyl) imide ([TMSpm$_2$im][Tf$_2$N]). These catalysts can be used in a wide variety of hydrogenation reactions, Equations 11.11 through 11.13 show only some of them.

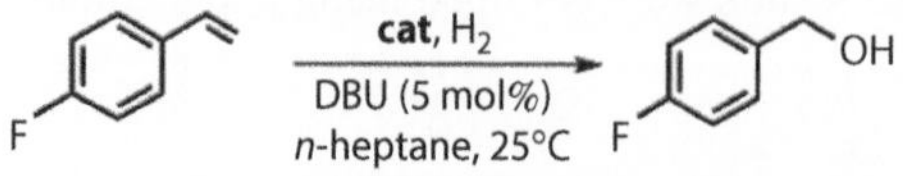

$$\text{(11.11)}$$

TABLE 11.3

Summary of the Supported Ionic Liquid Phases Used as Catalysts

Name	Reaction/IL/Catalyst	Comments	References
Hydro-formylation	$C_3H_6 + CO + H_2 \rightarrow C_4H_8O$ [bmim][PF$_6$]/ based on SILP Rhodium	- Long-term stability - No loss in selectivity for linear butanal	(Riisager et al., 2005)
	$C_3H_6 + CO + H_2 \rightarrow C_4H_8O$ [bmim][OctSO4]/ based on Rhodium	- An approach of factors influencing in activity	(Shylesh et al., 2012)
Reverse water gas shift	$CO_2 + H_2 \rightarrow CO + H_2O$ [bmim][Cl]/based on Rhodium	- Large surface area, good mass transport - 20 Cycles was achieved	(Yasuda et al., 2018)
Water gas shift	$CO + H_2O \rightarrow CO_2 + H_2$ DBiMIm/Cu	- A higher turn over frequency for the WGS at low temperatures	(Knapp et al., 2010)
Fischer-Tropsch	$nCO + 2nH_2 \leftrightarrow C_nH_{2n} + H_2O$ $nCO + (2n+1)H_2 \leftrightarrow C_nH_{2n+2} + H_2O$ /bmim[Tf$_2$N]/based on Cobalt	- Synthesis of more active and selective FT catalysts - A higher activity and selectivity in the FT catalysts	(Silva Dagoberto et al., 2008)

$$\text{1 (0.5 mol\%), H}_2\text{ (50 bar)} \quad \text{DBU (5 mol\%)} \quad n\text{-heptane, 25°C, 1 h}$$

(11.12)

$$R \diagdown O \xrightarrow[\text{DBU (5 mol\%)}]{\text{1 (0.5 mol\%), H}_2\text{ (50 bar)}} R \diagdown OH$$

$$n\text{-heptane, 25°C, 1 h}$$

(11.13)

Similar approaches have been reported in several studies, which are summarized in Table 11.3. These studies are relevant in terms of contributing in the proposal of catalytic system coupled to membrane separation processes.

Supported ionic liquid phases could play the simultaneous role of catalyst and separation phase. Figure 11.6a shows this option with double function of the IL layer. However, a top layer membrane could be placed at the permeate side of the reaction layer separating impurities or unreacted reagents from the products (Figure 11.6b). In a similar approach, the top layer membrane could be placed at the feed side in order to reject impurities before the catalytic step (Figure 11.6c) or on both feed and permeate sides (Figure 11.6d) to reject impurities before the catalytic step and retain the unreacted reagents in the catalytic phase, respectively.

According to the approach described earlier, Wilson et al. (2018) carried out a Suzuki-Miyaura carbon-carbon coupling reaction using a catalytic

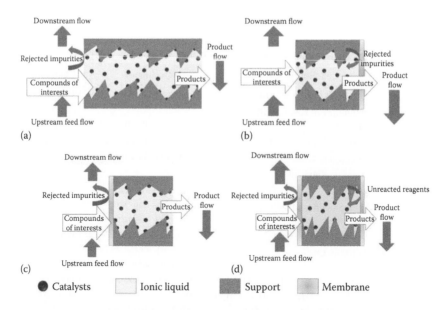

FIGURE 11.6 Configuration of supported ionic liquid catalytic membranes. (a) Supported IL fulfils double function, separation and catalytic phase; (b) Supported IL with a catalytic function and the top layer membrane placed in the permeate side as a selective barrier, which retains the unreacted reagents and rejects the impurities; (c) Supported IL with a catalytic function and the membrane as a top layer with the function of a selective barrier to reject the impurities before the catalytic step; and (d) Supported IL as catalyst and two membranes with different functions to reject impurities at the feed side, before the catalytic step, and to retain the unreacted reagents in the catalytic phase on the permeate side.

palladium-poly(ionic liquid) membrane. A Suzuki-Miyaura carbon-carbon coupling reaction can be described by the following equation:

$$\text{(11.14)}$$

The catalytic membrane and its mass transfer capacity are described in the outline presented in Figure 11.7.

In this case, the palladium-poly(ionic liquid) catalyst-coated membrane shows a product yield equal to 77% and a selectivity higher than 99%.

11.4.2 Nanoparticles-Supported Ionic Liquids and Their Applications as Catalytic Membranes

The combination of nanoparticles and ionic liquids as their support involves an interesting ensemble of advantages that considers aspects of homogeneous and

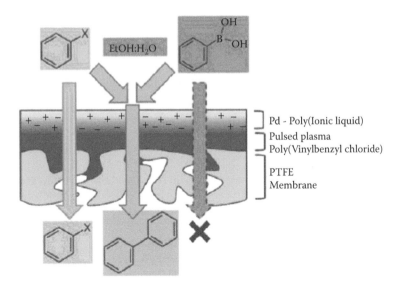

FIGURE 11.7 Structure of the membrane to carry out a Suzuki-Miyaura reaction. (From Wilson, M. et al., *Colloids Surf. A Physicochem. Eng. Asp.*, 545, 78–85, 2018.)

heterogeneous catalysis (Astruc and Aranzaes, 2005). Metal nanoparticles are stabilized in an ionic liquid phase because of a steric and electrostatic effect, which avoids their agglomeration. The interaction between the IL and different types of nanoparticles could enhance some physicochemical and transport properties, such as the diffusion coefficient of different species.

This combination of nanoparticles stabilized in an IL to be used as a catalyst involves a high number of active sites. Meanwhile, the separation of different species is facilitated, as well as the increased reusability of the catalyst, and the reaction rate and yield can be modified, even improving the selectivity of asymmetric reactions (He and Alexandridis, 2017). These positive effects were demonstrated by Chen et al. (2009), who combined [bmim][PF$_6$] with 0.08% w/w of copper nanoparticles noting an interesting improvement on the transport properties and the combination nanoparticle/IL involved an improved tunability as catalyst. Nevertheless, the use of nanoparticle/IL systems is also done within other fields, such as separation. Thus, the use of nanoparticles can improve the solubility of specific species in the membrane, which can be used in gas separation. Ji et al. (2016) and He and Alexandridis (2017) reported an improvement on the solubility of CO$_2$ from CO$_2$/N$_2$ mixtures given by the presence of AgO nanoparticles contained in a [omim][BF$_4$] phase, meanwhile, further research was reported on the combination of different ionic liquids with AgCl nanoparticles and their effect on the solubility of different gases (Erdni-Goryaev et al., 2015).

Membranes with an enhanced absorption capacity of CO$_2$ have been used in the selective removal of products from CO/O$_2$ streams (Perdikaki et al., 2016). These membranes with a zirconia top layer and a silylated ionic liquid, [spmim][Tf$_2$N], contained gold nanoparticles as catalyst. The experimental results obtained using these membranes show that reaction conversion depends on the size of the gold nanoparticles, as well as on the operation parameters.

On the other hand, hydrogenation catalyzed by Pd nanoparticles in a supported ionic liquid phase on carbon nanofibers has been applied to improve the selectivity in the conversion of acetylene to ethylene (Ruta et al., 2008), which is described in Equation 11.15:

$$C_2H_2 + H_2 \rightarrow C_2H_4 \tag{11.15}$$

The selectivity obtained with this catalyst increased from 60% to 85% using [bmim] [PF$_6$] or [bmimOH][Tf$_2$N].

Another study reported by Huang et al. (2004) reported the use of Pd nanoparticles contained in [bmim][PF$_6$]. This catalyst was immobilized onto molecular sieves using an IL layer to improve the reaction rate in the hydrogenation of olefins:

$$Olefin + H_2 \rightarrow Alkanes \tag{11.16}$$

This type of system can also be studied using nanoparticles supported in polymeric ionic liquids with the same purpose. However, the use of polymeric ionic liquid can involve a significant modification of the transport properties through the support phase. In this way, Pd nanoparticles have been supported in poly(ionic liquid) formed as a membrane by means of the photo-grafting process (Gu et al., 2015). With this catalytic membrane, the Suzuki-Miyaura reaction, described in Equation 11.17, was carried out obtaining total conversion in only 10s without formation of by-products.

$$\tag{11.17}$$

In a similar approach, nanoparticle/IL/polymer composite membranes were prepared using iridium nanoparticles, which were initially synthesized in [bmim][PF$_6$] and [Ir(cod)Cl]$_2$, and, subsequently, these nanoparticles were dispersed in [bmim][Tf$_2$N] and spread to form a cellulose acetate membrane (Faria et al., 2015). These membranes were used to catalyze the hydrogenation from 1-hexene to hexane:

$$C_6H_{12} + H_2 \rightarrow C_6H_{14} \tag{11.18}$$

The catalytic membrane showed improved catalytic activity and high stability.

On the other hand, rhodium and platinum nanoparticles were immobilized with a similar technique in an acetate cellulose membrane using [bmim][Tf$_2$N]. This membrane was used to catalyze the hydrogenation of cyclohexene to cyclohexane (Gelesky et al., 2009):

$$\tag{11.19}$$

In this case, the catalytic activity was significantly improved and the durability of the catalyst was enhanced.

All these examples using metal nanoparticles and ILs stabilized in membrane structure show only a fraction of the whole range of possibilities to propose novel catalytic systems, where the type of nanoparticle, its interaction with the IL, and the membrane support can be tuned for specific reactions improving reaction rates, conversions, and selectivity.

11.5 BIOCATALYSIS IN IONIC LIQUIDS AND MEMBRANE PROCESSES

An independent section of this chapter is dedicated to biocatalytic systems, because these cases can show a process configuration similar to the previously reported ones. However, their reaction mechanisms and the focus in green chemistry involve large potential for novel applications and growing interest considering their specificity (da Silva and de Castro, 2018). In biocatalytic processes, the presence of water has a direct effect on the reaction mechanisms, as well as it having an inhibitory action and difficulty for extracting reaction products from the aqueous system. From this condition, one of the greatest challenges is the proposal of an alternative liquid phase in order to design a new process in order to use the least water as possible (Lozano et al., 2007). Thus, the ILs can be considered as a real alternative to conventional organic solvents because their highly hydrophobic character avoids the denaturation of enzymes (De Diego et al., 2009) and improves enzymatic activity and selectivity. The use of ILs can also facilitate the formation of more stable biphasic systems with water, where the substrates and products can be transferred between phases to be incorporated or extracted from the reactive system, respectively. In this way, the immobilization of enzymes in ILs and support membranes seems to be an interesting alternative to reduce the operational costs and also to control contact surface area available for mass transfer and reaction (Mori et al., 2005).

11.5.1 MEMBRANE BIOREACTORS USING ILS

The combination of enzymatic reactions in an ionic liquid medium coupled to a membrane separation process involves several advantages, such as a positive effect of the process design on the chemical equilibrium. Thus, Bélafi-Bakó et al. (2002) reported the combination of the enzymatic esterification to produce 1-butyl-2-chloro-propanoate using *Candida rugosa* lipase in the ionic liquids [omim][PF$_6$] and [bmim][PF$_6$] as reaction media with a pervaporation step. The pervaporation process was implemented using a hydrophilic membrane able to remove water. The permanent removal of water allowed the displacement of the chemical equilibrium to the products, meanwhile, the extraction of specific products avoids the inhibition of the membrane activity. Furthermore, the use of the ILs as reaction media showed a significant improvement in the reaction performance, the enantioselectivity, and the enzymatic activity.

From these results, Fehér et al. (2009) incorporated an osmotic distillation process to obtain a higher extraction of by-products, such as water from the esterification of isoamyl alcohol. The raffinate of this system was contacted with a hydrophobic/organophilic pervaporation membrane to recover products, such as isoamylacetate obtained from isoamyl alcohol. The non-transferred species contained in an aqueous reaction medium were mixed with acetic acid and recirculated to a biocatalytic reactor containing *Candida antarctica* lipase in an acrylic resin with toluene immobilized in [bmim][PF$_6$], obtaining a 100% of selectivity.

In this context, the design of a continuous reactor for the esterification of ethanol and acetic acid catalyzed by Novozyme 435 coated with [bmim][PF$_6$] was also reported by Gubicza et al. (2008). This system was operated with two pervaporation membranes in series in order to break the azeotrope of the ternary system formed by the reaction products: ethyl acetate, water, and unreacted ethanol. Water and ethanol were removed from the stream in the first pervaporation step using a hydrophilic membrane, meanwhile the ester was removed in the second pervaporation step.

11.5.2 Biocatalytic Membranes Based on ILs

As mentioned in the previous sections, ionic liquids supported on porous or composed media have shown enormous potential in the development of selective membranes. These membranes could stabilize a certain group of enzymes and become biocatalytic. The use of lipases in ILs supported in porous membranes was reported earlier in Miyako et al. (2003) for the separation of (S)-ibuprofen from a feed phase containing (R)-ibuprofen, (S)-ibuprofen, ethanol, and *Candida rugose* lipase, producing an enantiomeric esterification at the interface of a supported [bmim][PF$_6$]/[omim][PF$_6$] polypropylene membrane. Thus, the ester obtained as product of S-ibuprofen was preferentially permeated through the membrane. This transferred compound was subsequently hydrolyzed in the aqueous receiving phase by the action of porcine pancreatic lipase.

From the successful enantiomeric separation achieved using ILMs, Hernández-Fernández et al. (2007) and de los Ríos et al. (2008) reported the enantiomeric separation of Rac-1-phenylethanol and Rac-2-pentanol from the transesterification with vinyl ester by the enzymatic action of the *Candida rugose* lipase B. The catalyzed reaction takes place in the feed phase, where the non-reactive (S)-enantiomers can permeate through a supported ionic liquid membrane prepared with a porous hydrophilic polyamide support.

On the other hand, Mori et al. (2005) proposed the production of butyl laurate from the acidolysis of butyl acetate and lauric acid in a hexane/water mixture (98:2 v/v) using *Candida antarctica* lipase, which was dissolved in hydrophobic imidazolium-based ILs with different anions, such as [PF$_6$]$^-$ and [Tf$_2$N]$^-$. These ILs were embedded in the pores of a tubular ceramic membrane. The permeate of this membrane system was rich in butyl laurate showing good stability when it was operated at a transmembrane pressure equal to 2 bar. Figure 11.8 shows an outline of the process at the proximities of the membrane. The best performance was obtained when [bmim][Tf$_2$N] was considered as enzyme support because of its higher hydrophobicity.

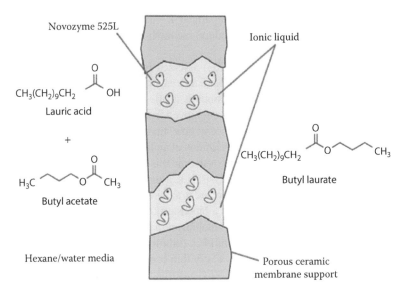

FIGURE 11.8   Outline of the butyl laurate production obtained from the acidolysis of butyl acetate and lauric acid through an enzymatic IL membrane.

11.6   CATALYTIC PROCESSES IN IONIC LIQUID-SUPERCRITICAL CO₂ SYSTEMS

A supercritical fluid is a compound in a thermodynamic state where its pressure and temperature are both higher than their critical values. Thus, the physicochemical properties are in an intermediate condition between liquid and gas phase. Carbon dioxide is the most used compound as supercritical fluid because it shows a high solvent power related to a high dielectric constant. Furthermore, carbon dioxide is cheap, inert, non-toxic, non-flammable, and its critical point involves a low temperature (304.25 K), a relatively low pressure (7.39 MPa), not difficult to reach, and a high density (10.6 mol L^{-1}). The solvent capacity of supercritical carbon dioxide (scCO$_2$) depends on its density. Thus, this capacity can be tuned in order to extract and fractionate a wide range of organic compounds. These properties suggest the use of scCO$_2$ as a green solvent, which can be combined with other clean technologies, such as ionic liquids and membrane processes (Cabezas et al., 2015). In this way, scCO$_2$ can be used as a transport medium for substrates in a supported ionic liquid phase (SILP) because substrates dissolved in dense CO$_2$ can be easily transferred to an IL phase containing the catalyst. Dense carbon dioxide can be easily dissolved in a wide range of ILs, but these molten salts are not dissolved in the CO$_2$ phase. Thus, an interesting set of process configurations can be proposed because two different solvents are combined in the same process without cross contamination and some properties of the ILs can be modified (e.g., decrease of viscosity) by the presence of scCO$_2$ (Solinas et al., 2004).

Several authors report the improvement of catalytic reactions using both ILs and scCO$_2$. Brown et al. (2001) studied the hydrogenation of catalyzed tiglic acid using an IL medium to produce 2-methylbutanoic acid with high conversion and enantioselectivity. The product was separated from the IL phase by using $scCO_2$, obtaining a clean and catalyst-free product, meanwhile, the IL/catalyst solution was reused without losing its activity.

On the other hand, Webb et al. (2003) assessed hydroformylation from 1-octane using a IL/$scCO_2$ continuous system, where the reagent was solubilized in $scCO_2$ to be circulated through the reactor, which contains an IL layer that supports the catalyst. The products and unreacted reactants are removed from the reaction in the CO_2 phase.

One of the main drawbacks of the processes mentioned earlier was the high volume of ILs required in the reactors. The immobilization of an IL used as catalyst or as a catalyst-container phase could be the solution to overcome this drawback. An early contribution describing a process that integrates catalysis, ILs, a supercritical fluid, and a membrane process in order to synthesize butyl propionate in a membrane reactor where the *Candida antarctica* lipase B was immobilized in a biphasic IL/$scCO_2$ system (Hernández et al., 2006). In a first step, the enzyme was immobilized in a α-Al_2O_3 membrane, which was covered with three different ILs: [bmim][PF_6], [bdimim][PF_6], and [omim][PF_6], which were used in small amounts to achieve a very thin coating film. The synthesis was carried out in a reactor with recirculation at 50°C and filled with $scCO_2$ at 80 bar, where substrates were pumped and solubilized in the CO_2 phase circulated through the membrane reactor. Samples were obtained by depressurization quantifying products and unreacted reagents from the experimental setup described in Figure 11.9. This system obtained a selectivity higher than 99.5%, a value that is higher than the selectivity obtained with the same process configuration, but using a membrane without IL (95%). In this membrane reactor system, the transport of substrates involves three consecutive steps: (1) diffusion from the bulk of the $scCO_2$ phase to the $scCO_2$/IL interface; (2) the thermodynamic equilibrium between both phases; and (3) diffusion through the IL phase to the immobilized enzyme. A loss of enzymatic activity has been reported in this biphasic $scCO_2$/IL system

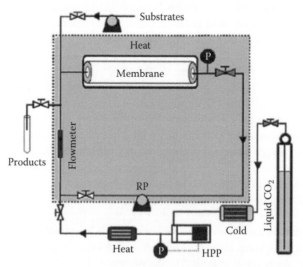

FIGURE 11.9 Experimental setup of the recirculating enzymatic reactor with $scCO_2$. (Adapted from Lozano, P. et al., *J Supercrit. Fluids*, 29, 121–128, 2004. With permission; Hernández, F.J. et al., *Appl. Catal. B Environ.*, 67, 121–126, 2006. With permission.)

and explained by Lozano et al. (2004), who described a limitation of the mass transfer rate through the IL layer that stabilized the enzyme. This limitation has to be taken into account in future designs of biphasic scCO$_2$/IL systems.

On the other hand, Kohlmann et al. (2011) developed a continuous flow system for the catalytic synthesis of (R)-2-octanol in an enzymatic membrane reactor (Jülich Chiral Solutions, Germany) using the ionic liquid AMMOENG TM 101 as cosolvent in order to increase the concentration of the substrate. Thus, this system showed great stability using *Lactobacillus brevis* alcohol dehydrogenase to reduce 2-octanone, obtaining selectivities higher than 99%. Furthermore, this continuous system generated 70% less waste to produce the same amount of product as a conventional one. The separation of products was achieved using a solid extraction column, based on different materials, which was coupled to the membrane reactor. After the extraction of the products, the column was regenerated with scCO$_2$ to obtain solvent-free products. The column used in this study was regenerated more than 30 times without a significant decrease of extraction capacity.

Finally, research that combines the use of an enzymatic reactor, a membrane, an ionic liquid, and scCO$_2$ was reported by Lozano et al. (2010) for the resolution of ketoprofen by means of esterification with different 1-alkanols. Thus, *Candida antarctica* lipase B (Novozym 435®) was immobilized in a high-pressure enzymatic reactor pressurized with scCO$_2$. The substrates and the immobilized enzyme were contacted in a tank containing an IL as reaction medium and recirculated through a membrane module, which contains a porous nylon membrane where the product (R)-ketoprofen was extracted in the dense gas phase and subsequently collected from the experimental setup described in Figure 11.10. Further research is necessary to improve

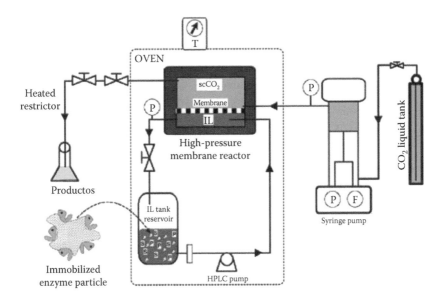

FIGURE 11.10 Experimental setup of the continuous enzymatic membrane reactor working with IL and scCO$_2$. (From Lozano, P. et al., Enzymatic membrane reactor for resolution of ketoprofen in ionic liquids and supercritical carbon dioxide, In *Ionic Liquid Applications: Pharmaceuticals, Therapeutics, and Biotechnology*, American Chemical Society, pp. 25–34, 2010.)

the enantioselectivity of this process. However, this type of process represents a step forward in the design of green operations, which can be used in the pharmaceutical industry for the resolution of different compounds with large processing capacity.

REFERENCES

Ahosseini, A., Wei, R., and Scurto, A. M. (2009) Hydrogenation in biphasic ionic liquid—Carbon dioxide systems. In *Gas-Expanded Liquids and Near-Critical Media*, Hutchenson, K. W. and Scurto, A. M. (Eds.), American Chemical Society, Washington, DC, pp. 218–234.

Astruc, D., Feng, L., and Aranzaes, J. R. (2005) Nanoparticles as recyclable catalysts: The frontier between homogeneous and heterogeneous catalysis, *Angewandte Chemie International Edition*, 44 (48), 7852–7872. doi:10.1002/anie.200500766.

Baudequin, C., Brégeon, D., Levillain, J., Guillen, F., Plaquevent, J.-C., and Gaumont, A.-C. (2005) Chiral ionic liquids, a renewal for the chemistry of chiral solvents? Design, synthesis and applications for chiral recognition and asymmetric synthesis, *Tetrahedron: Asymmetry*, 16 (24), 392–3945. doi:10.1016/j.tetasy.2005.10.026.

Bélafi-Bakó, K., Dörmő, N., Ulbert, O., and Gubicza, L. (2002) Application of pervaporation for removal of water produced during enzymatic esterification in ionic liquids, *Desalination*, 149 (1), 267–268. doi:10.1016/S0011-9164(02)00781-6.

Bhattacharya, M. and Mandal, M. K. (2017) Synthesis and characterization of ionic liquid based mixed matrix membrane for acid gas separation, *Journal of Cleaner Production*, 156, 174–183. doi:10.1016/j.jclepro.2017.04.034.

Brown, R. A., Pollet, P., McKoon, E., Eckert, C. A., Liotta, C. L., and Jessop, P. G. (2001) Asymmetric hydrogenation and catalyst recycling using ionic liquid and supercritical carbon dioxide, *Journal of the American Chemical Society*, 123 (6), 1254–1255. doi:10.1021/ja005718t.

Brünig, J., Csendes, Z., Weber, S., Gorgas, N., Bittner, R. W., Limbeck, A., Bica, K., Hoffmann, H., and Kirchner, K. (2018) Chemoselective supported ionic-liquid-phase (SILP) aldehyde hydrogenation catalyzed by an fe(ii) pnp pincer complex, *ACS Catalysis*, 8 (2), 1048–1051. doi:10.1021/acscatal.7b04149.

Cabezas, R., Plaza, A., Merlet, G., and Romero, J. (2015) Effect of fluid dynamic conditions on the recovery of ABE fermentation products by membrane-based dense gas extraction, *Chemical Engineering and Processing: Process Intensification*, 95, 80–89. doi:10.1016/j.cep.2015.04.003.

Carlin, R. and Fuller, J. (1997) Ionic liquid-polymer gel catalytic membrane, *Chemical Communications*, (15), 1345–1346. doi:10.1039/A702195J.

Chang, T., Gao, X., Bian, L., Fu, X., Yuan, M., and Jing, H. (2015) Coupling of epoxides and carbon dioxide catalyzed by Brönsted acid ionic liquids, *Chinese Journal of Catalysis*, 36 (3), 408–413. doi:10.1016/S1872-2067(14)60227-8.

Chen, F.-L., Sun, I. W., Wang, H. P., and Huang, C. H. (2009) Nanosize copper dispersed ionic liquids as an electrolyte of new dye-sensitized solar cells, *Journal of Nanomaterials*, 2009:4. doi:10.1155/2009/472950.

Clark, K. D., Nacham, O., Purslow, J. A., Pierson, S. A., and Anderson, J. L. (2016) Magnetic ionic liquids in analytical chemistry: A review, *Analytica Chimica Acta*, 934, 9–21. doi:10.1016/j.aca.2016.06.011.

Cota, I., Medina, F., Gonzalez-Olmos, R., and Iglesias, M. (2014) Alanine-supported protic ionic liquids as efficient catalysts for aldol condensation reactions, *Comptes Rendus Chimie*, 17 (1), 18–22. doi:10.1016/j.crci.2013.06.004.

Crespo, J. G. and Noble, R. D. (2014) Ionic liquid membrane technology. In *Ionic Liquids Further UnCOILed*, Plechkova, N.V. and Seddon, K.R. (Eds.), Wiley, Hoboken, NJ, pp. 87–116.

da Silva, V. G. and de Castro, R. J. S. (2018) Biocatalytic action of proteases in ionic liquids: Improvements on their enzymatic activity, thermal stability and kinetic parameters, *International Journal of Biological Macromolecules*, 114, 124–129. doi:10.1016/j.ijbiomac.2018.03.084.

De Diego, T., Lozano, P., Abad, M. A., Steffensky, K., Vaultier, M., and Iborra, J. L. (2009) On the nature of ionic liquids and their effects on lipases that catalyze ester synthesis, *Journal of Biotechnology*, 140 (3–4), 234–241. doi:10.1016/j.jbiotec.2009.01.012.

de los Ríos, A. P., Hernández-Fernández, F. J., Tomás-Alonso, F., Rubio, M., Gómez, D., and Víllora, G. (2008) On the importance of the nature of the ionic liquids in the selective simultaneous separation of the substrates and products of a transesterification reaction through supported ionic liquid membranes, *Journal of Membrane Science*, 307 (2), 233–238. doi:10.1016/j.memsci.2007.09.020.

Ding, J. and Armstrong, D. (2005) Chiral ionic liquids: Synthesis and applications. *Chirality*, 17, 281–292.

Erdni-Goryaev, E. M., Alentiev, A. Y., Bondarenko, G. N., Yaroslavtsev, A. B., Safronova, E. Y., and Yampolskii, Y. P. (2015) Facilitated transport of gases in polymer hybrid materials containing ionic liquid additives, *Petroleum Chemistry*, 55 (9), 693–702. doi:10.1134/S0965544115090029.

Faria, V. W., Brunelli, M. F., and Scheeren, C. W. (2015) Iridium nanoparticles supported in polymeric membranes: A new material for hydrogenation reactions, *RSC Advances* 5 (103), 84920–84926. doi:10.1039/C5RA16426E.

Fehér, E., Major, B. K., and Gubicza, L. (2009) Semi-continuous enzymatic production and membrane assisted separation of isoamyl acetate in alcohol-ionic liquid biphasic system, *Desalination*, 241 (1–3), 8–13. doi:10.1016/j.desal.2007.11.080.

Fortunato, R., González-Muñoz, M. J., Kubasiewicz, M., Luque, S., Alvarez, J. R., Afonso, C. A. M., Coelhoso, I. M., and Crespo, J. G. (2005) Liquid membranes using ionic liquids: The influence of water on solute transport, *Journal of Membrane Science*, 249 (1), 153–162. doi:10.1016/j.memsci.2004.10.007.

Gelesky, M. A., Scheeren, C. W., Foppa, L., Pavan, F. A., Dias, S. L. P., and Dupont, J. (2009) Metal nanoparticle/ionic liquid/cellulose: New catalytically active membrane materials for hydrogenation reactions, *Biomacromolecules*, 10 (7), 1888–1893. doi:10.1021/bm9003089.

Gu, Y., Favier, I., Pradel, C., Gin, D. L., Lahitte, J.-F., Noble, R. D., Gómez, M., and Remigy, J.-C. (2015) High catalytic efficiency of palladium nanoparticles immobilized in a polymer membrane containing poly(ionic liquid) in Suzuki–Miyaura cross-coupling reaction, *Journal of Membrane Science*, 492, 331–339. doi:10.1016/j.memsci.2015.05.051.

Gubicza, L., Belafi-Bako, K., Feher, E., and Frater, T. (2008) Waste-free process for continuous flow enzymatic esterification using a double pervaporation system, *Green Chemistry*, 10 (12), 1284–1287. doi:10.1039/B810009H.

He, Z. and Alexandridis, P. (2017) Ionic liquid and nanoparticle hybrid systems: Emerging applications, *Advances in Colloid and Interface Science*, 244, 54–70. doi:10.1016/j.cis.2016.08.004.

Heitmann, S., Krings, J., Kreis, P., Lennert, A., Pitner, W. R., Górak, A., and Schulte, M. M. (2012) Recovery of n-butanol using ionic liquid-based pervaporation membranes, *Separation and Purification Technology*, 97, 108–114. doi:10.1016/j.seppur.2011.12.033.

Hernández, F. J., de los Ríos, A. P., Gómez, D., Rubio, M., and Víllora, G. (2006) A new recirculating enzymatic membrane reactor for ester synthesis in ionic liquid/supercritical carbon dioxide biphasic systems, *Applied Catalysis B: Environmental*, 67 (1), 121–126. doi:10.1016/j.apcatb.2006.04.009.

Hernández-Fernández, F. J., de los Ríos, A. P., Tomás-Alonso, F., Gómez, D., Rubio, M., and Víllora, G. (2007) Integrated reaction/separation processes for the kinetic resolution of rac-1-phenylethanol using supported liquid membranes based on ionic liquids, *Chemical Engineering and Processing: Process Intensification*, 46 (9), 818–824. doi:10.1016/j.cep.2007.05.014.

Hernández-Fernández, F. J., de los Ríos, A. P., Tomás-Alonso, F., Palacios, J. M., and Víllora, G. (2009) Preparation of supported ionic liquid membranes: Influence of the ionic liquid immobilization method on their operational stability, *Journal of Membrane Science,* 341 (1), 172–177. doi:10.1016/j.memsci.2009.06.003.

Huang, J., Jiang, T., Gao, H., Han, B., Liu, Z., Wu, W., Chang, Y., and Zhao, G. (2004) Pd nanoparticles immobilized on molecular sieves by ionic liquids: Heterogeneous catalysts for solvent-free hydrogenation, *Angewandte Chemie International Edition,* 43 (11), 1397–1399. doi:10.1002/anie.200352682.

Izák, P., Bobbink, F. D., Hulla, M., Klepic, M., Friess, K., Hovorka, S., and Dyson, P. J. (2018) Catalytic ionic-liquid membranes: The convergence of ionic-liquid catalysis and ionic-liquid membrane separation technologies, *ChemPlusChem,* 83 (1), 7–18. doi:10.1002/cplu.201700293.

Ji, D. and Kang, S. W. (2016) 1-Methyl-3-octylimidazolium tetrafluoroborate/AgO nanoparticles composite membranes for facilitated gas transport, *Korean Journal of Chemical Engineering,* 33 (2), 666–668. doi:10.1007/s11814-015-0169-9.

Knapp, R., Wyrzgol, S. A., Jentys, A., and Lercher, J. A. (2010) Water–gas shift catalysts based on ionic liquid mediated supported Cu nanoparticles, *Journal of Catalysis,* 276 (2), 280–291. doi:10.1016/j.jcat.2010.09.019.

Kochetkov, S. V., Kucherenko, A. S., and Zlotin, S. G. (2017) Asymmetric Michael addition of aldehydes to maleimides in primary amine-based aqueous ionic liquid-supported recyclable catalytic system, *Mendeleev Communications,* 27(5), 473–475. doi:10.1016/j.mencom.2017.09.014.

Kohlmann, C., Leuchs, S., Greiner, L., and Leitner, W. (2011) Continuous biocatalytic synthesis of (R)-2-octanol with integrated product separation, *Green Chemistry,* 13 (6), 1430–1436. doi:10.1039/C0GC00790K.

Lozano, L. J., Godínez, C., de los Ríos, A. P., Hernández-Fernández, F. J., Sánchez-Segado, S., and Alguacil, F. J. (2011) Recent advances in supported ionic liquid membrane technology, *Journal of Membrane Science,* 376 (1), 1–14. doi:10.1016/j.memsci.2011.03.036.

Lozano, P., De Diego, T., Manjón, A., Abad, M. A., Vaultier, M., and Iborra, J. L. (2010) Enzymatic membrane reactor for resolution of ketoprofen in ionic liquids and supercritical carbon dioxide. In *Ionic Liquid Applications: Pharmaceuticals, Therapeutics, and Biotechnology,* Malhotra, S. V. (Ed.), American Chemical Society, Washington, DC, pp. 25–34.

Lozano, P., De Diego, T., Sauer, T., Vaultier, M., Gmouh, S., and Iborra, J. L. (2007) On the importance of the supporting material for activity of immobilized Candida antarctica lipase B in ionic liquid/hexane and ionic liquid/supercritical carbon dioxide biphasic media, *The Journal of Supercritical Fluids,* 40 (1), 93–100. doi: 10.1016/j.supflu.2006.03.025.

Lozano, P., Víllora, G., Gómez, D., Gayo, A. B., Sánchez-Conesa, J. A., Rubio, M., and Iborra, J. L. (2004) Membrane reactor with immobilized Candida antarctica lipase B for ester synthesis in supercritical carbon dioxide, *The Journal of Supercritical Fluids,* 29 (1), 121–128. doi:10.1016/S0896-8446(03)00050-0.

Lu, P., Cao, Y., and Wang, X. (2018) Kinetic model of biomass hydrolysis by a polysulfone membrane with chemically linked acidic ionic liquids via catalytic reactor, *RSC Advances,* 8 (21), 11714–11724. doi: 10.1039/C8RA00658J.

Martins, V. L. and Torresi, R. M. (2018) Ionic liquids in electrochemical energy storage, *Current Opinion in Electrochemistry.* doi:10.1016/j.coelec.2018.03.005.

Mehnert, C. P., Mozeleski, E. J., and Cook, R. A. (2002) Supported ionic liquid catalysis investigated for hydrogenation reactions, *Chemical Communications,* (24), 3010–3011. doi:10.1039/B210214E.

Merlet, G., Uribe, F., Aravena, C., Rodríguez, M., Cabezas, R., Quijada-Maldonado, E., and Romero, J. (2017) Separation of fermentation products from ABE mixtures by perstraction using hydrophobic ionic liquids as extractants, *Journal of Membrane Science,* 537, 337–343. doi:10.1016/j.memsci.2017.05.045.

Miyako, E., Maruyama, T., Kamiya, N., and Goto, M. (2003) Transport of organic acids through a supported liquid membrane driven by lipase-catalyzed reactions, *Journal of Bioscience and Bioengineering*, 96 (4), 370–374. doi:10.1016/S1389-1723(03)90139-3.

Mori, M., Gomez Garcia, R., Belleville, M. P., Paolucci-Jeanjean, D., Sanchez, J., Lozano, P., Vaultier, M., and Rios, G. (2005) A new way to conduct enzymatic synthesis in an active membrane using ionic liquids as catalyst support, *Catalysis Today*, 104 (2–4), 313–317. doi:10.1016/j.cattod.2005.03.039.

Nancarrow, P., Liang, L., and Gan, Q. (2014) Composite ionic liquid–polymer–catalyst membranes for reactive separation of hydrogen from carbon monoxide, *Journal of Membrane Science*, 472, 222–231. doi:10.1016/j.memsci.2014.08.044.

Nara, S. J., Harjani, J. R., and Salunkhe, M. M. (2001) Friedel–Crafts Sulfonylation in 1-Butyl-3-methylimidazolium chloroaluminate ionic liquids, *The Journal of Organic Chemistry*, 66 (25), 8616–8620. doi:10.1021/jo016126b.

Perdikaki, A. V., Labropoulos, A. I., Siranidi, E., Karatasios, I., Kanellopoulos, N., Boukos, N., Falaras, P., Karanikolos, G. N., and Romanos, G. E. (2016) Efficient CO oxidation in an ionic liquid-modified, Au nanoparticle-loaded membrane contactor, *Chemical Engineering Journal*, 305, 79–91. doi:10.1016/j.cej.2015.11.111.

Plaza, A., Merlet, G., Hasanoglu, A., Isaacs, M., Sanchez, J., and Romero, J. (2013) Separation of butanol from ABE mixtures by sweep gas pervaporation using a supported gelled ionic liquid membrane: Analysis of transport phenomena and selectivity, *Journal of Membrane Science*, 444, 201–212. doi:10.1016/j.memsci.2013.04.034.

Podolean, I., Hardacre, C., Goodrich, P., Brun, N., Backov, R., Coman, S. M., and Parvulescu, V. I. (2013) Chiral supported ionic liquid phase (CSILP) catalysts for greener asymmetric hydrogenation processes, *Catalysis Today*, 200, 63–73. doi:10.1016/j.cattod.2012.06.020.

Qian, W., Texter, J., and Yan, F. (2017) Frontiers in poly(ionic liquid)s: Syntheses and applications, *Chemical Society Reviews*, 46 (4), 1124–1159. doi:10.1039/C6CS00620E.

Rdzanek, P., Marszałek, J., and Kamiński, W. (2018) Biobutanol concentration by pervaporation using supported ionic liquid membranes, *Separation and Purification Technology*, 196, 124–131. doi:10.1016/j.seppur.2017.10.010.

Revell, J. D. and Ganesan, A. (2002) Ionic liquid acceleration of solid-phase suzuki–miyaura cross-coupling reactions, *Organic Letters* 4 (18), 3071–3073. doi:10.1021/ol0263292.

Riisager, A., Fehrmann, R., Haumann, M., Gorle, B. S. K., and Wasserscheid, P. (2005) Stability and kinetic studies of supported ionic liquid phase catalysts for hydroformylation of propene, *Industrial & Engineering Chemistry Research*, 44 (26), 9853–9859. doi:10.1021/ie050629g.

Ruta, M., Laurenczy, G., Dyson, P. J., and Kiwi-Minsker, L. (2008) Pd nanoparticles in a supported ionic liquid phase: Highly stable catalysts for selective acetylene hydrogenation under continuous-flow conditions, *The Journal of Physical Chemistry C*, 112 (46), 17814–17819. doi:10.1021/jp806603f.

Sakamoto, Y., Kumagai, H., and Matsunaga, S. (2013) An electrochemical ionic liquid membrane reactor for NO x selective separation under excess oxygen conditions, *Journal of Applied Electrochemistry*, 43 (9), 967–973. doi:10.1007/s10800-013-0584-8.

Sarma, D. and Kumar, A. (2008) Rare earth metal triflates promoted Diels–Alder reactions in ionic liquids, *Applied Catalysis A: General*, 335 (1), 1–6. doi:10.1016/j.apcata.2007.10.026.

Scovazzo, P., Kieft, J., Finan, D. A., Koval, C., DuBois, D., and Noble, R. (2004) Gas separations using non-hexafluorophosphate [PF6]⁻anion supported ionic liquid membranes, *Journal of Membrane Science*, 238 (1), 57–63. doi:10.1016/j.memsci.2004.02.033.

Seddon, K. R. and Stark, A. (2002) Selective catalytic oxidation of benzyl alcohol and alkylbenzenes in ionic liquids, *Green Chemistry* 4 (2), 119–123. doi:10.1039/B111160B.

Shi, H., Zhu, W., Li, H., Liu, H., Zhang, M., Yan, Y., and Wang, Z. (2010) Microwave-accelerated esterification of salicylic acid using Brönsted acidic ionic liquids as catalysts, *Catalysis Communications*, 11 (7), 588–591. doi:10.1016/j.catcom.2009.12.025.

Shinde, P. S., Shinde, S. S., Dake, S. A., Sonekar, V. S., Deshmukh, S. U., Thorat, V. V., Andurkar, N. M., and Pawar, R. P. (2014) CsF/[bmim][BF4]: An efficient and reusable system for Henry reaction, *Arabian Journal of Chemistry*, 7 (6), 1013–1016. doi:10.1016/j.arabjc.2010.12.028.

Shylesh, S., Hanna, D., Werner, S., and Bell, A. T. (2012) Factors influencing the activity, selectivity, and stability of RH-based supported ionic liquid phase (silp) catalysts for hydroformylation of propene, *ACS Catalysis*, 2 (4), 487–493. doi: 10.1021/cs2004888.

Silva, D. O., Scholten, J. D., Gelesky, M. A., Teixeira, S. R., Dos Santos, A. C. B., Souza-Aguiar, E. F., and Dupont, J. (2008) Catalytic gas-to-liquid processing using cobalt nanoparticles dispersed in imidazolium ionic liquids, *ChemSusChem*, 1 (4), 29–294. doi:10.1002/cssc.200800022.

Solinas, M., Pfaltz, A., Cozzi, P. G., and Leitner, W. (2004) Enantioselective hydrogenation of imines in ionic liquid/carbon dioxide media, *Journal of the American Chemical Society*, 126 (49), 16142–16147. doi:10.1021/ja046129g.

Uragami, T., Kishimoto, J., and Miyata, T. (2012) Membrane reactor for acceleration of esterification using a special ionic liquid with reaction and separation and microwave heating, *Catalysis Today*, 193 (1), 57–63. doi:10.1016/j.cattod.2012.03.020.

Vasiloiu, M., Rainer, D., Gaertner, P., Reichel, C., Schröder, C., and Bica, K. (2013) Basic chiral ionic liquids: A novel strategy for acid-free organocatalysis, *Catalysis Today* 200, 80–86. doi:10.1016/j.cattod.2012.07.002.

Voss, B. A., Bara, J. E., Gin, D. L., and Noble, R. D. (2009) Physically gelled ionic liquids: solid membrane materials with liquidlike CO_2 gas transport, *Chemistry of Materials*, 21 (14), 3027–3029. doi:10.1021/cm900726p.

Wang, J., Luo, J., Feng, S., Li, H., Wan, Y., and Zhang, X. (2016) Recent development of ionic liquid membranes, *Green Energy & Environment* 1 (1), 43–61. doi:10.1016/j.gee.2016.05.002.

Webb, P. B., Sellin, M. F., Kunene, T. E., Williamson, S., Slawin, A. M. Z., and Cole-Hamilton, D. J. (2003) Continuous flow hydroformylation of alkenes in supercritical fluid–ionic liquid biphasic systems, *Journal of the American Chemical Society*, 125 (50), 15577–15588. doi:10.1021/ja035967s.

Williams, D., Bradley G., Ajam, M., and Ranwell, A. (2006) Highly selective metathesis of 1-octene in ionic liquids, *Organometallics*, 25 (12), 3088–3090. doi:10.1021/om051089q.

Wilson, M., Kore, R., Ritchie, A. W., Fraser, R. C., Beaumont, S. K., Srivastava, R., and Badyal, J. P. S. (2018) Palladium–poly(ionic liquid) membranes for permselective sonochemical flow catalysis, *Colloids and Surfaces A: Physicochemical and Engineering Aspects*, 545, 78–85. doi:10.1016/j.colsurfa.2018.02.044.

Yasuda, T., Uchiage, E., Fujitani, T., Tominaga, K-I., and Nishida, M. (2018) Reverse water gas shift reaction using supported ionic liquid phase catalysts, *Applied Catalysis B: Environmental*, 232, 299–305. doi:10.1016/j.apcatb.2018.03.057.

Yu, S., Lindeman, S., and Tran, C. D. (2008) Chiral ionic liquids: Synthesis, properties, and enantiomeric recognition, *The Journal of Organic Chemistry*, 73 (7), 2576–2591. doi:10.1021/jo702368t.

Yuan, J., Mecerreyes, D., and Antonietti, M. (2013) Poly(ionic liquid)s: An update. *Progress in Polymer Science*, 38 (7), 1009–1036. doi:10.1016/j.progpolymsci.2013.04.002.

Zhou, F. Y., Xin, B. W., and Hao, J. C. (2013) HSO_3-functionalized Brönsted acidic ionic liquids promote esterification of aromatic acid, *Chinese Science Bulletin*, 58 (26), 3202–3207. doi:10.1007/s11434-013-5888-x.

Index

Note: Page numbers in italic and bold refer to figures and tables, respectively